Lecture Notes in Mechanical Engineering

Lecture Notes in Mechanical Engineering (LNME) publishes the latest developments in Mechanical Engineering—quickly, informally and with high quality. Original research or contributions reported in proceedings and post-proceedings represents the core of LNME. Volumes published in LNME embrace all aspects, subfields and new challenges of mechanical engineering.

To submit a proposal or request further information, please contact the Springer Editor of your location:

Europe, USA, Africa: Leontina Di Cecco at Leontina.dicecco@springer.com
China: Ella Zhang at ella.zhang@cn.springernature.com
India, Rest of Asia, Australia, New Zealand: Swati Meherishi at swati.meherishi@springer.com
Topics in the series include:

- Engineering Design
- Machinery and Machine Elements
- Mechanical Structures and Stress Analysis
- Automotive Engineering
- Engine Technology
- Aerospace Technology and Astronautics
- Nanotechnology and Microengineering
- Control, Robotics, Mechatronics
- MEMS
- Theoretical and Applied Mechanics
- Dynamical Systems, Control
- Fluid Mechanics
- Engineering Thermodynamics, Heat and Mass Transfer
- Manufacturing Engineering and Smart Manufacturing
- Precision Engineering, Instrumentation, Measurement
- Materials Engineering
- Tribology and Surface Technology

Indexed by SCOPUS, EI Compendex, and INSPEC.

All books published in the series are evaluated by Web of Science for the Conference Proceedings Citation Index (CPCI).

To submit a proposal for a monograph, please check our Springer Tracts in Mechanical Engineering at https://link.springer.com/bookseries/11693.

Pramod Bhingole · Kamlesh Joshi ·
Surya Deo Yadav · Ankit Sharma
Editors

Advances in Materials Engineering

Select Proceedings of ICFAMMT 2024

Springer

Editors
Pramod Bhingole
Department of Mechanical and Aerospace Engineering
Institute of Infrastructure Technology Research and Management
Ahmedabad, Gujarat, India

Kamlesh Joshi
Department of Mechanical and Aerospace Engineering
Institute of Infrastructure Technology Research and Management
Ahmedabad, Gujarat, India

Surya Deo Yadav
Department of Metallurgical Engineering
Indian Institute of Technology (BHU)
Varanasi, India

Ankit Sharma
Department of Mechatronics Engineering and Technology, Space Applications Centre
Indian Space Research Organisation (ISRO)
Ahmedabad, Gujarat, India

ISSN 2195-4356 ISSN 2195-4364 (electronic)
Lecture Notes in Mechanical Engineering
ISBN 978-981-97-7113-4 ISBN 978-981-97-7114-1 (eBook)
https://doi.org/10.1007/978-981-97-7114-1

This Springer imprint is published by the registered company Springer Nature Singapore Pte Ltd.
The registered company address is: 152 Beach Road, #21-01/04 Gateway East, Singapore 189721, Singapore

Preface

The 2nd International Conference on Futuristic Advancements in Materials, Manufacturing and Thermal Sciences (ICFAMMT 2024) was held at the Institute of Infrastructure Technology Research and Management, Ahmedabad, India during January 19–21, 2024. The conference served as an exceptional forum for academicians, researchers, industry professionals, and young scientists to delve into novel challenges and explore futuristic advancements in materials, manufacturing, and thermal sciences. With aspirations to rank among the foremost international conferences globally, ICFAMMT 2024 promised an invigorating setting for participants to showcase and deliberate on the latest engineering technologies and their practical applications.

The first edition of the book titled *Advances in Material Science and Metallurgy* was published using the selected manuscripts presented in ICFAMMT-2022. In continuation of the first edition, the present book contains carefully curated peer-reviewed proceedings from ICFAMMT 2024, offering insights into cutting-edge research within material science and metallurgy. The book is divided into four parts: *Material Characterization*, *Material Processing*, *Material Properties*, and *Material Selection and Design*. These parts cover a wide range of topics, from the study of natural fibers and alloys to additive manufacturing techniques and phase diagram calculations. They also delve into the properties of various materials, including the use of CFRP for CubeSat structures and the superplasticity of metals.

The final parts focus on material selection and design principles, including the use of composite materials for wind turbine blades and the design evaluation of dental implants. The book also presents a multi-criteria decision-making approach for material selection in construction and industrial robot selection for arc welding processes. This book is a valuable resource for those interested in material science and metallurgy.

Ahmedabad, India	Pramod Bhingole
Ahmedabad, India	Kamlesh Joshi
Varanasi, India	Surya Deo Yadav
Ahmedabad, India	Ankit Sharma

Contents

Material Characterization

Characterization Study of Natural Fibers and Glass Fibers an Experimental Approach 3
S. Sulthan, K. Balaji, Vardhman S. Chavaj, and Gaikwad Pranoti Vijaykumar

Influence of Tribological Parameters on As-cast, T4-Treated, and Aged AZ61 Magnesium Alloy 15
Manan Shah, Dhaval Makwana, and Bhingole Pramod

A Review on Pineapple and Banana Natural Fiber Composite 25
Abin Paul Boban, Nitish Kumar, and Durgesh Kumar Mishra

A Study on Surface Treatment Methods of Metal Surface for Metal Fiber Laminates 39
Renu Sahu, Gaurav K. Gupta, Pradeep Kumar Ghosh, and Mohit Jaiswal

Comparative Study of Fe_3O_4 and Calcium-Doped Fe_3O_4 53
S. D. Raut and S. G. Dahotre

Sound Absorption Performance of 3D-Printed PLA/Wood Composites 63
Hiral H. Parikh, Hiren Rana, and Axit Raj

A Review of Biomedical Applications of Zirconia-Based Nanomaterials 71
Mukund Kumar, Shibashis Halder, and Arvind Kumar

Influence of Sustainable Lubrication on the Tribological Behaviour of Phosphorus Bronze 83
Aman Kumar, Palash Biswas, Anal Ranjan Sengupta, and Thia Paul

Material Processing

Calculation of Ni–Cu, Ni–Ag, and Cu–Ag Phase Diagram Using CALPHAD Method ... 103
Uttam Kumar Tarai, Swagat Dwibedi, Subhranshu Kumar Singh, and Suyog Jhavar

A Comprehensive Review on Wire Arc Additive Manufacturing of Inconel Superalloys ... 113
Rupendra Singh Tanwar, Pramod Bhingole, and Suyog Jhavar

Additive Manufacturing of Al_2O_3 Ceramic Components: Rheological and Microstructural Characterization ... 131
Rupendra Singh Tanwar, Swaraj Sarkar, M. Vignesh, and Suyog Jhavar

Comprehensive Review of Foaming Agents in Melt Processing for Lightweight Metallic Foam Production ... 145
Manan Shah, Dhaval Makwana, and Bhingole Pramod

Interfacial Characteristics of Ti64–In82 Bimetallic Structure Fabricated Using Wire-Arc Additive Manufacturing ... 155
Rupendra Singh Tanwar, Swaraj Sarkar, and Suyog Jhavar

Experiment on Modified Bituminous Road with Additive Waste Material ... 169
Anant Kumar, Aryansh Chopkar, Khushi Vyas, Kuldeep Suthar, and Raunak Prakash

Material Properties

Investigation of CFRP as a Material for Fabrication of 1U CubeSat Structure for Reduction of Weight with Improved Strength ... 185
Anand Belsare, Shruti Kulkarni, Aakanksha Kajale, Deepak Hujare, and Pravin Hujare

Physical Phenomena Behind the Enhancement of Joint Properties in Solution-Treated Nickel-Based Superalloy ... 201
Swagat Dwibedi, Nilakantha Sahu, and Suyog Jhavar

Influence of Weld Time on Joint Behavior of IN625-SS316L Weldments ... 213
Vivekananda Haldar, Swagat Dwibedi, and Suyog Jhavar

Study on Dissimilar SS316L-Incoloy 800 Joints with Inconel 82 as Filler Metal ... 225
Nilakantha Sahu, Swagat Dwibedi, Sushant Kumar Badjena, and Suyog Jhavar

Influence of Post Bond Heat Treatment on IN718 Joints: An Experimental Study 233
Uttam Kumar Tarai, Dipankar Saha, Swagat Dwibedi, and Suyog Jhavar

Engineering Applications of Superplasticity of Metals: Review 245
Amruta Mahajan and Vishvesh Badheka

Mechanical Properties of Concrete by Partial Replacement of Cement with Dolomite 259
Toshan Singh Rathour, Fiza Khan, Khyati Kodopi, Sanjana Sahu, and Somprabha Sonkale

Cellular Mg-Based Biodegradable Bone Implant: Recent Development to Future Prospects 269
Dhaval Makwana, Bhingole Pramod, and Manan Shah

Behaviour Analysis of Functionally Graded Sandwich Plate for Free Vibration by Finite Element Method Solution 275
Charmi Vidja and Supen Kumar Sah

Material Selection and Design

Composite Material for Design and Manufacturing of the Wind Turbine Blades—A Review 293
Pranoti Shinde, Prashant Sharma, Arun Kumar Sharma, Ashok Soni, and Arpit Shukla

Design Evaluation of Splinted and Non-splinted Crown of Dental Implant to Improve Stress Distribution at Bone-Implant Interface Using FEA 309
Aditya Gujare, Bhargav Hindurao, Pankaj Dhatrak, and Sandipan Roy

Performance Evaluation of Powder-Lubricated Journal Bearing Using FDM and CFD Technique 319
Jijo Jose and Niranjana Behera

Designing Dental Implants: The Use of Finite Element Analysis, a Comprehensive Review 339
G. H. Waghmare, Kanak G. Waghmare, Sachin T. Bagde, and Mayuri N. Deshmukh

Assessment of Modified Concrete Block with Constructional Waste, Industrial Waste and Natural Fiber 363
Anant Kumar, Antariksh Chandrakar, Aastha Chandrawanshi, Saumya Verma, and Arvind Deb Mishra

An AHP and TOPSIS MCDM-Based Approach for the Selection of Material for Construction Under Fuzzy Environment 377
Durgeshwar Pratap Singh, Shwetank Avikal, Harvindra Singh, and Shivani Monga

A MCDM-Based Approach for the Selection of Industrial Robots for Arc Welding Process ... 389
Durgeshwar Pratap Singh, Shwetank Avikal, Harvindra Singh, Shivani Monga, and Amit Sharma

Statistical Analysis and Data-Driven Modeling of Hole Circularity in EDD of Aerospace Material ... 401
Kedari Lal Dhaker, Amit Sharma, Tasnim Arif, Gaurav Shukla, Akshat Shukla, and Shwetank Avikal

Numerical Investigation of Thermal Hydraulic Performance of Solar Air Heater with M-Shaped Roughness ... 417
Harvindra Singh, Durgeshwar Pratap Singh, Narendra Gariya, Chandra Kishore, Shwetank Avikal, and Shivani Monga

About the Editors

Dr. Pramod Bhingole is Renowned Academic and Researcher and currently serving as Associate Professor at the Institute of Infrastructure Technology Research and Management (IITRAM), Ahmedabad. He has earned his Ph.D. in Metallurgical and Materials Engineering from the Indian Institute of Technology (IIT) Roorkee. His research interests are broad and interdisciplinary, spanning across material processing, composite materials, nanomaterials, and applications of materials. He has led several research projects, one of the notable ones being on the fabrication, surface modification, and evaluation of bioceramic hydroxyapatite coated with HVOF on Mg/Al alloy foam sandwich composite structure. He also holds patents in his field. In addition to his research, Dr. Bhingole is Editor of the book *Advances in Material Science and Metallurgy: Select Proceedings of ICFAMMT 2022*. He is Lifetime Member of several professional bodies, demonstrating his commitment to his field. His work continues to inspire students and researchers alike, and his contributions to Metallurgy and Metallurgical Engineering have been impactful, making him a respected figure in his field.

Dr. Kamlesh Joshi is currently working as Assistant Professor at the Institute of Infrastructure, Technology, Research and Management (IITRAM), India. He has an experience of around seven years in teaching and research. He obtained his doctoral degree from Indian Institute of Technology Bombay (IITB) in Mechanical Engineering and his M.Tech. from Indian Institute of Technology Roorkee in Metallurgical Engineering. His research interest lies in machining of difficult to cut materials for photovoltaic, aerospace, and medical applications. He has authored more than 15 research articles and organised two international conferences. Currently, he is handling three research projects sponsored from Department of Science and Technology (DST) India.

Dr. Surya Deo Yadav is working as Assistant Professor at Department of Metallurgical Engineering, Indian Institute of Technology (BHU), Varanasi. He received Ph.D. degree from Institute of Material Science and Welding, TU Graz, Austria

(2016), M.Tech. from MMED IIT Roorkee (2012), and B.Tech. in Mechanical Engineering from IPEC (2008). After Ph.D., he continued his research work as Postdoctoral Scientific Researcher at TU Graz and afterwards joined IGCAR as Visiting Scientist for two years. Dr. Yadav's research interests are in the area of microstructural based modelling of hot deformation and creep, precipitation kinetics, creep cavitation, design of tempered martensitic steels, thermomechanical processing using Gleeble-3800, and characterisation using XRD, EBSD, and TEM. Dr. Yadav has received few awards; those are best paper award in ICONS (2018) by SFA, InSIS and IGF, THERMEC (2016) Student Grant and Inspire Faculty Award (2018–2019).

Ankit Sharma has done B.Tech. in Metallurgy and Materials Engineering, National Institute of Advanced Manufacturing Technology, Ranchi (Erstwhile NIFFT Ranchi). He is Pursuing M.Tech. in Structures and Design (Aerospace engineering) Indian Institute of Space Science and Technology, Trivandrum. He is currently working as Scientist/Engineer-SE at Mechatronics Engineering and Technology Section of Space Applications Centre, ISRO, Ahmedabad. His main area of work is design, development, and realisation of mechanisms for different satellite payloads. He also works on modelling and structural analysis of mechanical subsystems. He has worked at Vikram Sarabhai Space Centre, ISRO, Thiruvananthapuram, from year 2010 to 2017 where he worked mainly on design, development, and realisation of Cast Components for Cryogenic and Semi-cryogenic Engines of ISRO Satellite Launch vehicles. He worked on modelling and simulation of various space grade cast components, characterisation, and optimisation of space grade cast alloys. He has also worked in the area of Rapid Prototyping and 3D Printing. He has published papers in referred journals and conference proceedings. He is presently Joint Secretary of Space Society of Mechanical Engineers (SSME). He is presently Active Member of Vigyan Bharti and Mission Kartavya Foundation which works for education of underprivileged children.

Material Characterization

Characterization Study of Natural Fibers and Glass Fibers an Experimental Approach

S. Sulthan, K. Balaji, Vardhman S. Chavaj, and Gaikwad Pranoti Vijaykumar

Abstract This paper's goal is to investigate the behavior of different natural fiber in order to replace glass fiber. The experimental test is done with a universal testing device and the Charpy test to determine the different fibers' strengths and stresses. The various test results proved that sisal fiber provides better results than banyan and century fiber. It reveals that the natural fiber strength is 60% lower than glass fiber. The present study is to identify the suitable material used for the replacement of glass fiber. The sisal fiber can be used for interior components of aviation applications in place of glass fiber wherever less strength is required.

Keywords Glass fiber · Natural fiber · Sisal fiber · Stress · Strain

1 Introduction

Composite materials are used for many applications due to its weight and strength-to-weight ratio. The Glass fiber is used for many aviation and general applications. It has good physical and chemical properties to carry the strength. At the same time, glass fiber is having higher cost and little bit dangerous to humans while handling. In order to solve the above-mentioned problem many researchers are focusing on identifying the alternate solution. The solution for the alternative reinforcement is Natural fibers. It is available more and less cost compared to the glass fiber. For finding new materials the mechanical properties are important, for that, we investigate the mechanical properties of ultimate stress and impact strength of the new finding materials. Ho et al.

S. Sulthan (✉)
Department of Aerospace Engineering, Sanjay Ghodawat University, Kolhapur 416118, India
e-mail: sulthanaero@gmail.com

K. Balaji
Department of Aeronautical Engineering, Parul Institute of Engineering and Technology, Parul University, Vadodara, Gujarat 391760, India

V. S. Chavaj · G. P. Vijaykumar
Department of Aerospace Engineering, Sanjay Ghodawat University, Kolhapur, India

P. Bhingole et al. (eds.), *Advances in Materials Engineering*, Lecture Notes in Mechanical Engineering, https://doi.org/10.1007/978-981-97-7114-1_1

[1] Several production procedures and their suitability for natural fiber composites were recommended based on the composition, mechanics, and temperature of the fibers and matrices. Tamanna et al. [2] proposed using fiber from the Corypha taliera fruit as a replacement for dangerous synthetic fiber. Young's modulus was reported to be 572.21 MPa for GL 30 mm and a GL 20 has a maximum tensile strength of 53.55 MPa. Additionally, it was discovered that the CTF fiber can withstand heat up to 2300 °C. Ayalew and Wodag [3] carried out a thorough experimental investigation on sisal fiber extraction and composite development. 6% NaOH concentration for fiber with 30:70 polyester to nylon ratio and a 48-h soaking period, optimal tensile and bending strengths of 44.0 MPa and 50.8 MPa, respectively, were found. It demonstrated improved tensile, bending, and water absorption performance than untreated fibers. de Andrade et al. [4] conducted a study on sisal fiber, a high-performance natural fiber, to determine its monotonic tensile properties. When the gauge length was raised from 10 to 40 mm, the risk of failure was lower from around 5.2% to 2.6%, and Young's modulus of sisal was found to be 18 GPa. To evaluate the degree of fiber strength variability, Weibull statistics were applied at different gauge lengths. It was found that the modulus of weibull gauge length went from 10 to 40 mm and reduced from 4.6 to 3. Naraganti et al. [5] did a comparison of the sisal fibers' performance with steel and polypropylene (PP) fibers. With fiber dosages of 0%, 0.50%, 1.00%, 1.25%, and 1.50%, both the performance and impact resistance of structural concrete were taken into consideration. The impact resistance of FRC was increased by adding more fiber and Sisal FRC (SiFRC) and HyFRC containing S-PP at a fiber dosage of 1.5% both outperformed and showed superior performance. Palanisamy [6] examined the impacts of sisal fiber reinforcement in random-oriented epoxy composites. Compression strength, water absorption, and tensile strength capability of the fiber surfaces after chemical treatment with an alkali 5% NaOH solution were evaluated and compared to those of pure epoxy. Tensile strength was seen to increase from 95.33 MPa (sisal fiber reinforced epoxy) to 81.23 MPa (pure epoxy). 52.12 MPa for pure epoxy and 76 MPa for epoxy reinforced with sisal fiber were measured in the compression test. Sisal fiber-reinforced epoxy has a 1.5% higher water absorption capacity than pure epoxy. Sisal fiber-reinforced epoxy is also proven to perform thermally 1.12% better than pure epoxy. Ahmad et al. [7] studied the properties such as mechanical strength and durability have all been examined and the reinforced sisal fiber showed increased strength along with extended durability but decreased flow ability. James et al. [8] conducted research on the morphological and sisal's mechanical characteristics and bagasse-reinforced acrylic composites. Researchers examined the impact, compression, water absorption, flexural, tensile, flexural, impact, and biodegradation capabilities of composites containing a core layer made of sisal fibers, three layers of sisal fiber, and and they discovered that they had good properties. de Andrade Silva et al. [9] carried out an experimental investigation into the mechanical and physical properties of a modified cement matrix reinforced with long, straight sisal fibers for structural and semi-structural uses. Two matrices were taken into consideration, one consisting without additives, made of Portland cement and designated M0 (composites of PC), and the other with calcium hydroxide and designated M1 (CH-free composite). The behavior of strain hardening in tension

and bending at 13.95 MPa for CH-free composites and 9.24 MPa for PC composites, the material manifested many cracks. Due to the ultimate bending strain, the sheets' corrugation, increased by 260%. An outstanding ultimate bending strength of CH-free composite, which is 3.8 times more than that of PC-based composite, and a toughness. Almusawi et al. [10] conducted an experimental investigation of sisal fiber's effects and temperature on Plaster of Paris's mechanical and physical characteristics. For various sisal percentages of 0, 2, 4, 6, 8, and 10 weight percent of POP was considered for a range of temperatures from 25, 100, 200, 300, 400, and 500 °C. The sisal fiber having 10% showed good mechanical properties and also noted that the temperature resistance improved with an increase in fiber content. At higher temperatures above 3000 °C it was found that the mechanical properties are getting adversely affected. Agilan et al. [11] research was done on the sisal fiber reinforced composites' mechanical characteristics. Here, epoxy and sisal fiber were combined to create a mat, and a hand layup approach was chosen. The results of the tensile, hardness, and impact tests were encouraging. Balaji et al. [12] lighted weight materials were used for manufacturing multipurpose drone. Liang et al. [13] compared the mechanical properties of continuous long fiber (SF)/Poly lactic acid (PLA) made using Long sisal fiber reinforced PLA composites with a single direction of reinforcement (LSFCs), an extrusion-rolling mixing procedure made using compression molding SSFCs, which were produced in an experiment using open milling and molding, and random short sisal fiber-reinforced PLA composites. The experimental results showed that the presence of long sisal fiber, along with a small tensile elongation and fiber extension direction, can increase the material's tensile strength for LSFC. The influence of long fiber on mechanical properties has been explored. Utilizing continuous long fibers increased the load-bearing capacity of LSFC as well, but it was shown that SSFC was negatively impacted by the mechanical properties of fiber reinforcement because of the poor compatibility between fiber and matrix. Padanattil et al. [14] identified a unique method for retrofitting concrete structures using a hybrid sisal-fiber reinforced polymer that has been researched and proposed. When natural and hybrid FRP's axial load-carrying capability, ductility, energy absorption rate, and durability were examined, hybrid FRP showed an increase in axial load capacity. Due to an increase in axial strain of 484% in HSGRF, the ductility has also increased. Ibrahim et al. [15] presented a study on a mixed blend of recycled polypropylene (rPP) and sisal fiber at four different fiber loadings, including Nanoclay was applied at 1, 2, and 5%, while other amounts were 10, 20, 30, and 40%. The grafted polypropylene with maleic anhydride (MAPP) is utilized as a reconcilier. The mechanical and thermal stability of the SF/rPP nano-composites were improved, and water absorption was decreased, through fiber treatment involving the inclusion of MAPP and nanoclay. Compared to SF/rPP composites that haven't been treated, it was found that the tensile strength and tensile modulus had improved by 32.80, 37.62, and 5.48%, respectively. Additionally, the water resistance improved significantly. Kovačevic et al. [16] conducted an experimental investigation to strengthen modified Spartium junceum L fiber by treating it with nanoclay (MMT) and citric acid (CA), hence improving the mechanical properties of Poly Acid (PLA) biocomposite. When compared to pure PLA, The findings indicated that the modified composite's tensile strength and tensile modulus were both improved by 164.0% and 85.7%, respectively. Sulthan et al. [17] carbon fiber has given good strength and light weight in all the components of the aircraft compared with aluminium alloys. Olszewski et al. [18] analyzed a novel technique for making polyurethane materials that has been proposed, and the effects of real sisal fiber and synthetic glass fiber are investigated. The composites performed better when the reaction temperature was 150 °F and the reaction time was 6 h. The composites are created via 15 min of hot pressing at 100 °C and 10 MPa pressure. Flexural strength, impact strength, hardness, and glass transition temperature of hard segments are seen to increase with an increase in the number of fibers. Lima et al. [19] cellulose acetate, hydrophobic starch, and cassava starch biopolymers were used to coat sisal fibers in an experimental investigation to see if the fiber's propensity to absorb water might be decreased by utilizing biodegradable resin. The fiber's mechanical properties were diminished as a result, but their hydrophilicity was increased thanks to the biopolymer coating made of cellulose acetate that had formed on their surface. Padanattil et al. [14] presented an innovative method using hybrid Sisal-glass fiber reinforced polymer (HSGFRP) for retrofitting concrete structures. When findings from CFRP and natural FRP are compared, an axial strain improvement of 484% is seen. Additionally, by conducting durability performance evaluations, it was discovered that confined FRP outperformed unconfined FRP when subjected to wet/dry conditions and temperature fluctuations. Gudayu et al. [20] examined the sisal fiber's thermal, mechanical, and moisture absorption properties using modification approaches. We take into account Untreated, acetylated, alkaline-treated, and combined acetylation/alkaline-treated modification techniques. Alkaline-treated/acetylated samples and acetylated samples moisture absorption decreased by 42% and 28%, respectively, and the thermal properties of additionally improved were the alkali-treated and alkali-treated/acetylated samples. The results indicated that the acetylated sample had improved tensile strength. Bakare et al. [21] Polyurethane resin with rubber seed oil created utilizing monoglyceride from rubber seed generated from the oil's glycerolysis as matrix and composite made of polyurethane based on rubber seed oil and reinforced with unidirectional sisal fibers are both taken into consideration in the [high performance of composite material] investigation. The polyurethane composites had ten times higher flexural and tensile strengths and roughly twice as high as unreinforced urethane using rubber seed oil as its basis. The polyurethane composites with rubber seed oil-based unidirectional fiber reinforcement demonstrated good mechanical and thermal properties.

The literature survey reported that in many places natural fibers are used for many applications. It is used to identify a research gap in the literature that indicates a lack of work on the fabrication and testing of various natural fibers for aviation applications. A new material was discovered that contains three natural fibers that can be used as a replacement for glass fiber.

2 Method

2.1 Selection of Material

Natural fibers are selected from easily available fibers of Banyan, Century, and sisal fiber. Those fibers are collected from the agricultural land and fibers are extracted by traditional procedure. To fabricate the specimen the fiber is mixed with resin and hardener. Natural fiber combinations are fabricated using Epoxy resin and HY 951 hardener to get the expected physical and chemical properties.

2.2 Fabrication Method

The natural fibers are fabricated into the required shape using the molding setup using the traditional hand layup method. The fibers are settled into the mold using the universal testing compression process through various manufacturing processes as shown in Fig. 1a–d. The fibers and resins are mixed with proper weight ratio and organized in 90° orientation and cure with room temperature for 48 h (Table 1).

2.3 Specimen Preparation

The fabricated natural fiber square sheets are cutter into the ASTM D638 dog bone shape to conduct the tensile test specimen as shown in the Fig. 2. The impact specimen is prepared in a regular rectangular shape depending on the testing machine specifications as shown in Fig. 2. The fabricated models are resized into the test specimen shape of tensile test specimen as shown in Fig. 3a and charpy test specimen as shown in Fig. 3b for further process.

3 Experimental Test

The experimental test was conducted by using the universal testing machine capacity of 80KN and for impact test was conducted with Charpy testing equipment which is available at Sanjay Ghodawat University, Kolhapur. The natural fibers are placed at specific locations and slow loads are applied to check the tensile strength and impact strength of different natural and glass fibers.

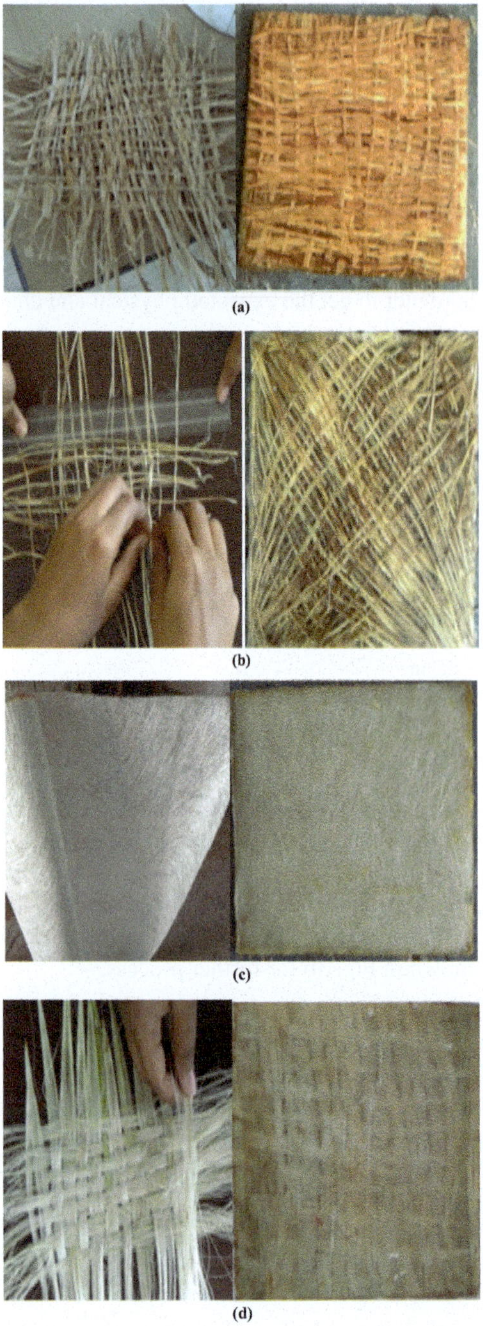

Fig. 1 **a** Banyan fiber with different orientation, **b** Preparation of century fiber, **c** Glass fiber mat preparation, **d** Preparation of Sisal fiber

Table 1 Specification of various fibers used for this study

Sr. No	Fiber name	Length (mm)	Width (mm)	Thickness (mm)
1	Banyan fiber	170	4	1
2	Century fiber	169	3.5	1
3	Glass fiber	170	4	1
4	Sisal fiber	170	4.2	1

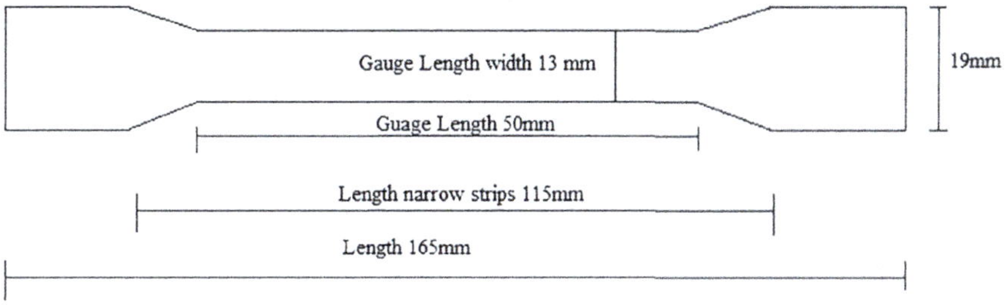

Fig.2 ASTM dog bone shape

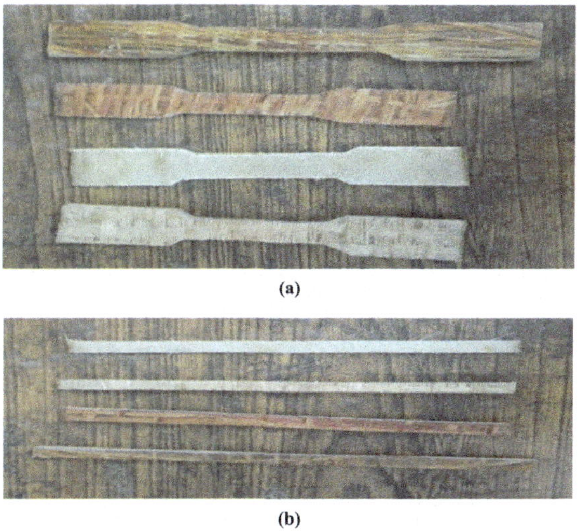

Fig. 3 **a** Universal testing machine specimen, **b** Charpy impact test specimen

4 Result and Discussion

The specimen is tested using destructive method and values are noted from the equipment until it gets damaged. The tensile test results are used to calculate the ultimate and breaking load as well as the yield stress and elongation as shown in Table 2. The impact strength is used to calculate the maximum capacity of the specimen to carry

Table 2 Specifications of experimental specimen used for this study

Sr. No	Types of fiber	Width (mm)	Length (mm)	Thickness (mm)	Area (mm^2)	Gauge length (mm)
1	Glass fiber	21	181	5	105	80
2	Banyan fiber	18	181	8	144	80
3	Century fiber	19	181	9	171	100
4	Sisal fiber	16	181	9	144	80

Table 3 Variation of stress levels of different specimen

Sr. No	Types of fiber	Ultimate load (KN)	Tensile strength (N/ mm^2)	Yield stress (N/mm^2)	Load at break (KN)	Elongation (%)
1	Glass fiber	24.2	230.48	85.24	9	3.75
2	Banyan fiber	12.25	85.07	55.90	0	1.25
3	Century fiber	11.5	67.25	53.80	0	1
4	Sisal fiber	14.55	101.04	62.15	0.1	5

Table 4 Variation of impact test results

Sr. No	Name of fiber	Load (J)	Area (mm^2)	Stress (J/mm^2)
1	Glass fiber	2	45.1	0.044
2	Banyan fiber	1	77.35	0.01
3	Century fiber	1	77.35	0.012
4	Sisal fiber	1	73.6	0.013

the sudden load. In that areas are calculated manually and stress values are identified as shown in Tables 3 and 4.

4.1 *Tensile Test*

Figure 4 shows the comparison of natural fibers tensile properties with glass fiber. The strength of materials is measured in terms of ultimate load, tensile load, and yield stress. It shows that the properties of sisal fiber are higher than century and banyan fiber at the same time less than the glass fiber. The reason behind this is that the sisal fiber has the tensile and yield strength naturally due to the biological behavior of sisal fiber. It concluded that sisal fiber is the best one among the three natural fibers.

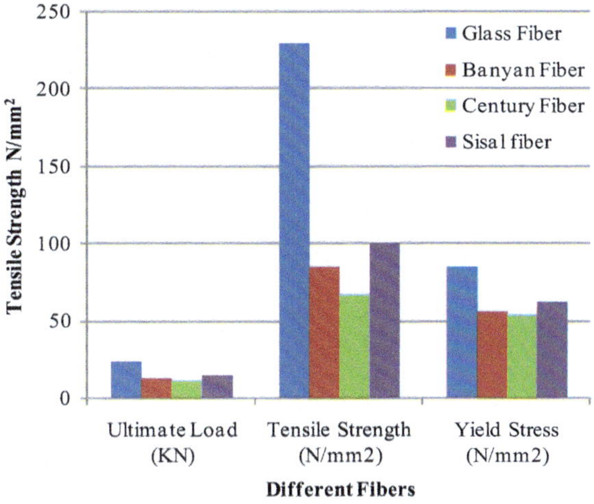

Fig. 4 Tensile strength of different fiber

4.2 *Impact Test*

The Impact strength of fibers is measured from the Charpy test. Figure 5 shows that the sisal fiber has the capacity to carry the greater impact load due to its high strength. At the same time, banyan fiber has very less strength. The overall results shows that the impact strength of sisal fiber is best and it is less than the glass fiber due to the internal structure arrangement by naturally. Impact strength of the glass fiber is higher than 40% of sisal fiber due to the internal molecules orientation arrangement over the surface.

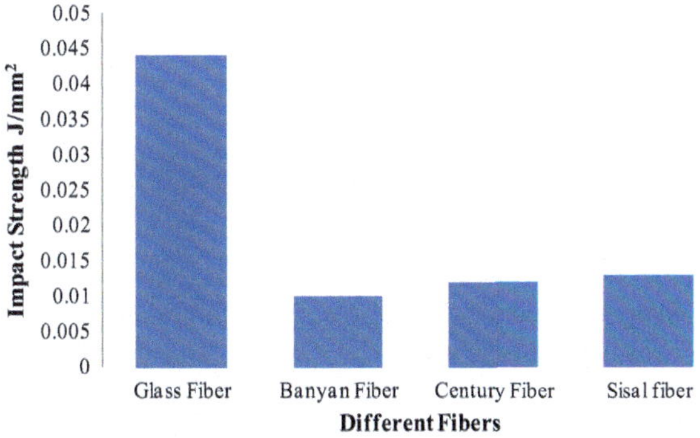

Fig. 5 Impact test results of different fiber

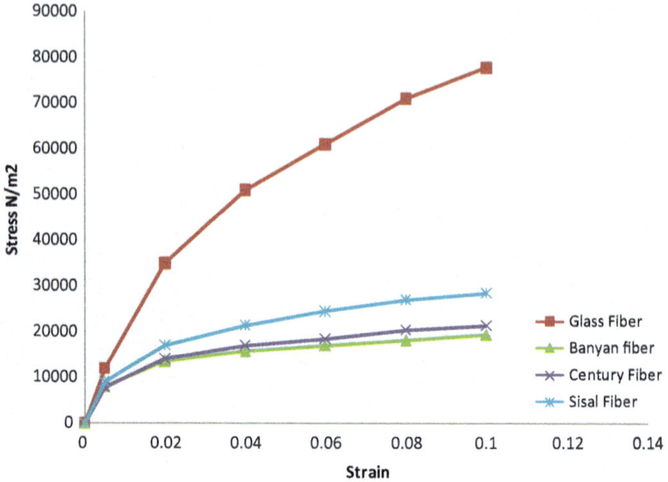

Fig. 6 Stress strain curve of different fiber

4.3 Validation

Figure 6 shows that the validation of test is done by using stress strain diagram. It reveals that the sisal fiber and glass fiber following the same pattern of stress strain profile. Century and banyan fibers carry less stress and strain during to that tensile strength are low. But in the sisal fiber provides good strength compared to banyan and century fiber and also which is less than glass fiber. The overall behavior of sisal fiber nearly matches that with glass fiber due to high tensile strength and stiffness, making them suitable for reinforcement applications, although their composition and processing methods differ.

5 Conclusion

The experimental study of the fiber is carried out by fabrication using the hand layup method and then specimen is reshaped into the ASTM standard size [4]. Those specimens were tested with different equipment and it brought the following conclusions.

- Sisal Natural fiber has good tensile and impact strength compared to the other natural fiber
- Tensile stress of sisal fiber increased by 80% compared to remaining natural fiber and decreases 40% compared to glass fiber
- The yield stress of sisal fiber is 62.15 which is 30% less than the glass fiber.
- Elongation properties of the sisal fiber is 80% compared to the glass fiber.

- Impact strength of the glass fiber is higher that is 40% higher than sisal fiber.
- The results would be changes 1% based on the quality of fibers selection and fibers orientation and environmental conditions.

The overall study concluded that sisal fiber is the suitable natural fiber in order to replace the glass fiber which is used at the aviation application. The limited usage of natural fiber is used at the minimum load carrying structural components.

References

1. Ho M, Wang H, Lee J-H, Ho C, Lau K, Leng J, Hui D (2012) Critical factors on manufacturing processes of natural fibre composites. Compos B Eng 43:3549–3562
2. Tamanna TA, Belal SA, Shibly MA, Khan AN (2021) Characterization of a new natural fiber extracted from Corypha Taliera Fruit. Sci Reports 11
3. Adugna Ayalew A, Fenta Wodag A (2022) Characterization of chemically treated sisal fiber/polyester composites. J Eng 1–11
4. de Silva F, Chawla N, Filho RD (2008) Tensile behavior of high performance natural (sisal) fibers. Compos Sci Technol 68:3438–3443
5. Naraganti SR, Pannem RM, Putta J (2019) Impact resistance of hybrid fibre reinforced concrete containing sisal fibres. Ain Shams Eng J 10:297–305
6. Manjula P, Venkatachalam CD, Anbupazhani MS, Chandran SS, Lakshmanan C, Naveen Kumar NK (2020) Fabrication and properties testing of sisal fiber reinforced epoxy composites with random orientation. In: AIP conference proceedings
7. Ahmad J, Majdi A, Deifalla AF, Ben Kahla N, El-Shorbagy MA (2022) Concrete reinforced with sisal fibers (SSF): overview of mechanical and physical properties. Crystals 12:952
8. James DJD, Manoharan S, Saikrishnan G, Arjun S (2019) Influence of Bagasse/sisal fibre stacking sequence on the mechanical characteristics of hybrid-epoxy composites. J Nat Fibers 17:1497–1507
9. de Silva F, Filho RD, de Filho J, de Fairbairn E (2010) Physical and mechanical properties of durable sisal fiber–cement composites. Constr Build Mater 24:777–785
10. ALmusawi AM, Hussein TS. Shallal MA (2018) Effect of temperature and sisal fiber content on the properties of plaster of Paris. Int J Eng Technol 7:205
11. Agilan H, Arun R, Kumar R, Alexraja G (2018) Automotive application and mechanical property characterisation of sisal fiber reinforced epoxy composite material. Int J Eng Res Technol (IJERT) 6:1–4
12. Balaji K, Babu V, Sulthan S (2022) Design and development of multipurpose agriculture drone using lightweight materials. SAE Int J Aerosp 16
13. Liang Z, Wu H, Liu R, Wu C (2021) Preparation of long sisal fiber-reinforced polylactic acid biocomposites with highly improved mechanical performance. Polymers 13:1124
14. Padanattil A, Karingamanna J, Mini KM (2017) Novel hybrid composites based on glass and sisal fiber for retrofitting of reinforced concrete structures. Constr Build Mater 133:146–153
15. Ibrahim ID, Jamiru T, Sadiku ER, Kupolati WK, Agwuncha SC (2016) Impact of surface modification and nanoparticle on sisal fiber reinforced polypropylene nanocomposites. J Nanotechnol 2016:1–9
16. Kovačević Z, Bischof S, Vujasinović E, Fan M (2019) The influence of pre-treatment of spartium junceum L. Fibres on the structure and mechanical properties of PLA Biocomposites. Arab J Chem 12:449–463
17. Sulthan S, Balaji K, Akhil CK, Babu V (2024) Composite fuselage proposal—part 1: static and modal finite element analysis. SAE Technical Paper Series

18. Olszewski A, Kosmela P, Mielewczyk-Gryń A, Piszczyk Ł (2020) Bio-based polyurethane composites and hybrid composites containing a new type of Bio-Polyol and addition of natural and synthetic fibers. Materials 13:2028
19. Lima PR, Santos HM, Camilloto GP, Cruz RS (2017) Effect of surface biopolymeric treatment on sisal fiber properties and fiber-cement bond. J Eng Fibers Fabr 12:155892501701200
20. Gudayu AD, Steuernagel L, Meiners D, Gideon R (2020) Effect of surface treatment on moisture absorption, thermal, and mechanical properties of sisal fiber. J Ind Text 51
21. Bakare IO, Okieimen FE, Pavithran C, Abdul Khalil HPS, Brahmakumar M (2010) Mechanical and thermal properties of sisal fiber-reinforced rubber seed oil-based polyurethane composites. Mater Des 31:4274–4280

Influence of Tribological Parameters on As-cast, T4-Treated, and Aged AZ61 Magnesium Alloy

Manan Shah, Dhaval Makwana, and Bhingole Pramod

Abstract The demand for AZ61 Magnesium alloy in lightweight structural applications is driven by its exceptional strength-to-weight ratio and enhanced stiffness. In this study, AZ61 Mg alloy was produced through the stir casting technique. Three variations of the alloy were analyzed: as-cast, solutionized at 420 °C, and subsequently aged at 200 °C. The wear behavior of these variations was assessed using a Pin-on-disk tribometer, focusing on changes in tribological parameters such as load and sliding velocity while maintaining a constant sliding distance of 1000 m, with wear rates measured in mm^3/min. The experimental investigation employed the Taguchi Approach to analyze wear behavior. Wear rates were compared across the AZ61 as-cast, T4-treated, and aged samples. Results indicated that wear rates increased with higher loads and sliding velocities. Microstructural examination via optical microscopy revealed significant findings: the solution heat-treated sample showed dissolution of the β phase, correlating with increased wear rates, while aging resulted in the diffusion of intermetallic phases, thereby enhancing wear resistance in the T6-treated sample.

Keywords AZ61 · Wear · Heat treatment · Microstructure

1 Introduction

A magnesium alloy designated as "AZ61" is composed primarily of magnesium (Mg), aluminum (Al-6 wt.%), and zinc (Zn-1wt. %). This alloy is widely used in several applications because of its favorable combination of properties, including lightweight, good strength-to-weight (S/W) ratio. [1]. AZ61 magnesium alloy is utilized in aerospace applications such as aircraft seats, cabin interiors, and structural

M. Shah · D. Makwana
Mechanical and Aero-Space Engineering, Institute of Infrastructure Technology Research and Management, Ahmedabad, Gujarat 380026, India

B. Pramod (✉)
Institute of Infrastructure Technology, Research and Management, Ahmedabad, India
e-mail: pramod_bhingole@yahoo.com

P. Bhingole et al. (eds.), *Advances in Materials Engineering*, Lecture Notes in Mechanical Engineering, https://doi.org/10.1007/978-981-97-7114-1_2

parts due to its lightweight nature and high S/W ratio [2]. In the automotive industry, parts like transmission cases, steering wheels, and engine blocks can be made from AZ61, contributing to overall vehicle weight reduction [3]. However, proper handling and processing techniques are required due to magnesium's flammability and reactivity with certain substances.

AZ61 magnesium alloy typically exhibits lower strength and ductility compared to some other magnesium alloys and engineering materials like aluminum alloys. Heat treatment processes involve heating the AZ61 alloy to a certain temperature to dissolve atoms into the magnesium matrix (solution treatment), followed by a controlled cooling and aging process to precipitate strengthening phases, which leads to upgradation of the mechanical properties [4, 5]. Heat treatment can be a valuable tool for addressing the limitations of AZ61 magnesium alloy by tailoring its microstructure and properties to meet specific application requirements. The microstructural evolution induced by heat treatment plays a critical role in determining the tribological properties of AZ61 magnesium alloy. Heat treatment processes such as T4 treatment and aging can result in changes in grain size, phase distribution, and dislocation density, which can influence friction, wear, and lubrication behavior [6, 7].

Optimizing heat treatment parameters is therefore of utmost importance as it can help to achieve the desired microstructural characteristics to enhance tribological performance [8, 9]. Based on the various literatures [10, 11], it is found that the T4 heat treatment temperature is 420 °C as the $Mg_{17}Al_{12}$ phase starts to dissolve after 413 °C and as β phase starts to precipitate out around 200 °C. Thus, in this research, the solution heat treatment conditions were fixed at 420 °C and that for aging at 200 °C.

The tribological performance of AZ61 magnesium alloy is influenced by a combination of parameters including load, sliding speed, and heat treatment conditions [12, 13]. Understanding the interactions between these parameters and their effects on wear degradation is essential for optimizing the performance and reliability of AZ61 alloy components in practical applications. Thus, a Taguchi approach is being utilized for designing the tribological experiments and determining the optimum conditions to yield the lowest wear rate [14–16].

2 Experimental Procedure

The pure magnesium ingots were melted in the crucible in the muffle furnace at the 750 °C temperature. Aluminium and Zinc granules were added as the alloying element in the proportions of 6 wt.% and 1 wt.% respectively in the balance Mg base alloy. This preparation of AZ61 alloy has been achieved using the stir casting technique. The schematic of the setup of Stir Casting is depicted in Fig. 1a. In the Stir casting setup, a graphite stirrer is driven by a DC motor and controlled using an RPM controller unit. K type thermocouple is being used to measure the temperature of the melt. The whole process of Stir casting was being performed in the controlled argon atmosphere [17, 18]. After the casting of AZ61, the crucible was quenched in the water at room temperature. The specimens of rectangular in shape (60 × 15 × 15 mm) were prepared from the as-cast alloy using a precision band saw machine. The rectangular specimens were then kept in an electric muffle furnace and were T4 or solution heat treated at 420 °C for 8 h. The specimens were air-cooled post-solution heat treatment. The samples then underwent an aging process at a temperature of 200 °C for a duration of 12 h, followed by air cooling. Both the heat treatments solution heat treatment and aging treatment of as-cast AZ61 were performed in a controlled argon atmosphere. The schematic of the heat treatment furnace is illustrated in the Fig. 1b.

The specimens of each as-cast, solution treated and T6 treated AZ61 alloy were then sectioned and were subjected to fine polishing for microstructural characterization using the different grit sizes paper progressively from grit sizes #100 to #2400. It was then followed by cloth and diamond polishing to achieve a mirror-like finish. The microstructural characterization was performed using Olympus GX51 optical microscope. Also, all the as-cast, T4, and aged specimens were subjected to measure the microhardness as per ASTM E384 with a load of 1 kg for 10 s dwell time using Mitutoyo Micro Vickers Hardness Machine at Institute of Plasma Research, Bhat, Gandhinagar. The rectangular specimens were machined to prepare a 6 mm diameter and 50 mm height pin for tribological testing. Three cylindrical pins of each as-cast, T4-treated, and aged AZ61 samples were tested on Ducom Pin on Disk tribometer in an ambient condition at room temperature as per ASTM G99 standard [19, 20]. The loads on the pin were varied as 15 N, 30 N, and 45 N with speed varying from 0.5, 1.25 to 2 m/s for 1000 m fixed sliding distance [17]. The wear rate was calculated in terms of mm^3/m.

The tribological tests were governed based on the L9 Taguchi Orthogonal Array (OA) of 3 levels and 3 parameters namely Treatment, Load, and Sliding velocity. The Taguchi OA for this experiment is shown in Table 1 [21]. The outcomes of

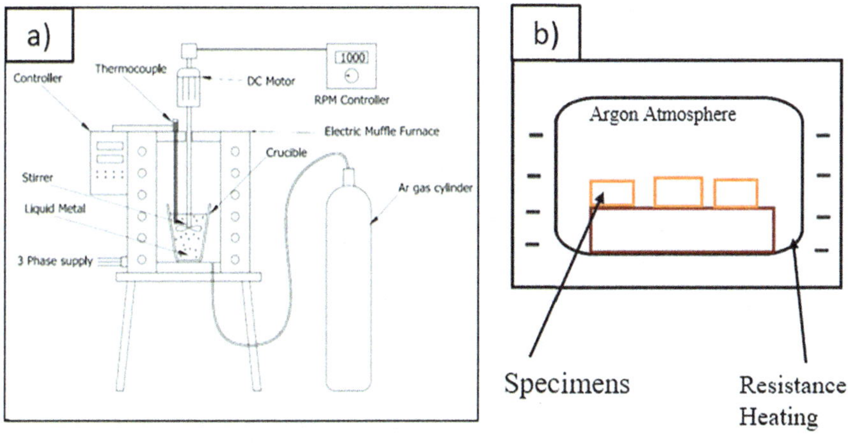

Fig. 1 Visual illustration of **a** Stir casting setup, **b** Heat treatment electric Muffle furnace

Table 1 Taguchi L9 orthogonal array for tribology experiments for AZ61 alloy

Sr. No	Treatment	Load (N)	Sliding velocity (m/s)
1	As-cast	15	0.5
2	As-cast	30	1.25
3	As-cast	45	2
4	T4	15	1.25
5	T4	30	2
6	T4	45	0.5
7	T6	15	2
8	T6	30	0.5
9	T6	45	1.25

the microstructural characterization, microhardness, and tribological behavior are discussed in the next section.

3 Results and Discussions

3.1 Impact of Heat Treatment on Microstructure of AZ61 Alloy

The finely polished samples were immersed in an acetic picral solution for a duration of 12–15 s for chemical etching and then were subjected to optical microscopy as shown in Fig. 2. In as-cast AZ61 alloy (Fig. 2), the dendritic structure is being formed due to the way its atoms arrange themselves during the cooling process during the solidification. Dendrites are tree-like structures that form as the alloy solidifies, with primary arms extending outward from nucleation sites. These dendrites grow as the alloy cools, and their shape and orientation are influenced by factors such as cooling rate, alloy composition, and processing conditions [5].

When the AZ61 alloy was being heat treated at 420 °C, the breakdown of β phase ($Mg_{17}Al_{12}$) occurred and following an extended treatment lasting 8 h, the β phase entirely dissolves at the grain boundaries, resulting in the formation of a homogeneous structure. (Fig. 2b). After homogenization the AZ61 alloy was subjected to aging at 200 °C for 12 h. In the aging treatment, the dissolved β phase starts precipitating near the grain boundaries [22, 23] and a discontinuous structure is formed near the grain boundaries (Fig. 2c).

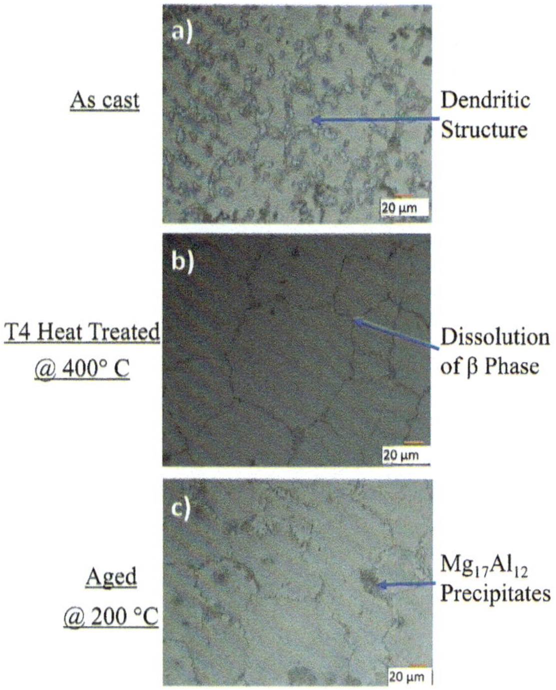

Fig. 2 Microstructural characterization obtained by optical microscopy of **a** as-cast, **b** T4 heat treated, **c** Aged sample

3.2 Influence of Heat Treatment on Micro Hardness of the AZ61 Mg Alloy

The as-cast, solution heat-treated and aged samples were subjected to measure micro hardness using Vickers hardness machine at 1 kg load for dwell time of 10 s. The results of micro hardness in HV are demonstrated in the graph in Fig. 3. The hardness value of the T4 heat-treated sample decreased to 64 HV as compared to as-cast AZ61 Mg alloy. After aging increment was found in 10% HV in relative to as-cast alloy and a 20% increment as compared to T4 heat-treated alloy.

As seen in the microstructure, due to the breakdown of intermetallic phases in the solution heat-treated specimen, the refinement of grain occurs homogenously and thus the dislocations can move freely resulting in higher ductility, resulting in the decrement of hardness. After aging, precipitation of phases occurs, which results in hindrance of dislocations and eventually increases the hardness of the AZ61 [4, 8].

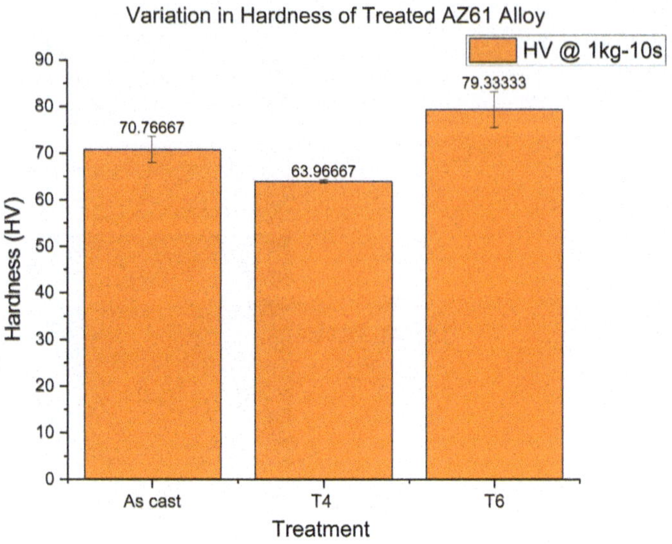

Fig. 3 Comparison of micro-hardness of as-cast, T4 heat treated and aged AZ61 alloy

3.3 Tribological Behavior of Treated AZ61 Mg Alloy Using Taguchi Approach

The tribological experiments were carried out according to the Taguchi-designed L9 orthogonal array. The wear rate (measured in mm^3/min) resulting from these experiments has been documented in Table 2. A main effects plot, generated using Minitab software, was utilized to examine the individual impact of each parameter on the wear rate (Fig. 4a).

Table 2 Results for tribology experiments for AZ61 alloy as per Taguchi L9 OA

Sr. No	Treatment	Load (N)	Sliding velocity (m/s)	Wear rate (mm^3/m)	Error (Wear rate) ($*10^{-5}$)	S/N ratio
1	As-cast	15	0.5	0.00891	3.443	41.0024459
2	As-cast	30	1.25	0.01163	6.2097	38.6884057
3	As-cast	45	2	0.01892	2.2258	34.4615774
4	T4	15	1.25	0.00821	3.6	41.7131369
5	T4	30	2	0.01148	2.69	38.8011622
6	T4	45	0.5	0.01395	1.265	37.1085158
7	T6	15	2	0.00832	5.469	41.5975335
8	T6	30	0.5	0.01122	2.5847	39.0001429
9	T6	45	1.25	0.01359	5.73	37.3356109

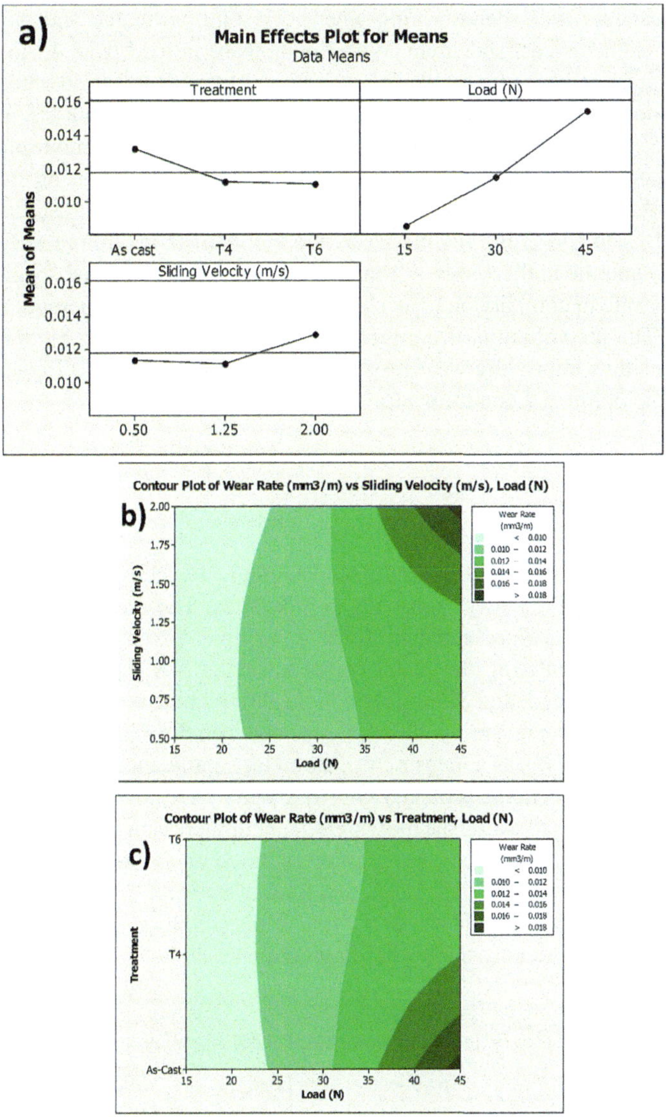

Fig. 4 A Main effects plot of variation in wear rate due to **a** Treatment, load, sliding velocity, contour plot of **b** Wear rate versus sliding velocity, load, **c** Wear rate versus treatment, Load plotted for tribological experimentation of AZ61 alloy

The wear rate starts deteriorating when it is solution-treated and then aged as compared to the as-cast specimen. With the increase in load from 15 to 45 N, the wear rate drastically increases [8, 24]. A different trend is followed when the wear rate is reduced initially with the increment in sliding velocity from 0.5 to 1.25 m/s and increased rapidly from 1.25 to 2 m/s sliding velocity. The contour plot is made to understand the combined effect of Tribological parameters using the Minitab Software. As per the Wear map of Load (N), wear rate (mm^3/m) and sliding velocity (m/s), and Load (N) (Fig. 4b) the wear rate recorded is maximum at the highest load (45N) and the highest speed (2 m/s) combined. Also, the aged specimen has a lesser wear rate in comparison to as-cast and T4 treated with an increase in the load (Fig. 4c). The precipitation of the β phase during aging creates a hindrance to the dislocations thus improving wear resistance. The goal of using the Taguchi method is to find the optimized conditions for the lower wear rate.

$$\eta = -10 \log\left[\frac{1}{n}\sum_{i=1}^{n} Y_i^2\right] \tag{1}$$

Thus, the wear rate data obtained from tribological experiments were converted using the "smaller the better" approach to S/N ratios. These ratios were calculated using Eq. (1) and are presented in Table 2. Optimal parameter conditions were determined by calculating the mean response of the S/N ratios for each parameter—Treatment, Load, and Sliding speed—resulting in the response table shown in Table 3. The delta, representing the difference between the maximum and minimum mean responses, serves as a measure of parameter significance. Accordingly, Load, exhibiting the highest delta, is ranked as 1, followed by Treatment and Sliding Speed. This indicates that load exerts the most significant influence on the wear rate of AZ61 alloy. The selection of optimized conditions is based on the highest mean response for each parameter—Treatment, Load, and Sliding Speed. Consequently, according to Table 3, it is inferred that T6 @ 15 N with a 2 m/s condition represents the optimal configuration for achieving the lowest wear rate.

Table 3 Response table for S/N ratios of wear rate of AZ61 Mg alloy

Outcome		Treatment	Load (N)	Sliding velocity (m/s)
(Wear rate)	Level 1	38.0508097	41.43771	39.03703488
	Level 2	39.207605	38.8299	39.24571781
	Level 3	39.3110957	36.3019	38.28675769
	Delta	1.26028607	5.135804	0.958960119
	Rank	2	**1**	3

4 Conclusions

The AZ61 Mg alloy is being fabricated using Stir casting technique followed by solution heat treatment and aging heat treatment. It is found that T4 Heat treatment lowers the hardness (9.6%) due to dissolution of $Mg_{17}Al_{12}$ phases as compared to as-cast condition. After aging there was increment of 10% in Hardness Value as compared to as-cast condition. Load is the most significant parameter affecting the wear rate of AZ61 alloy followed by the type of heat treatment and sliding velocity. Optimization of tribological conditions yields optimum value at T6 treated at 15 N load and 2m/s sliding velocity.

Acknowledgements This work was supported by the Department of Science and Technology; Science and Engineering Research Board, Government of India (Grant No. ECR/2016/001518). Authors acknowledge the Siemens Centre of Excellence; Centre of Advanced Defence Technology (CADT), IITRAM (Project No: IITRAM/COE/CADT/22-23/IRP05) for support.

References

1. Gupta N, Luong DD, Cho K (2012) Magnesium matrix composite Foams—Density, mechanical properties, and applications. https://doi.org/10.3390/met2030238
2. Hu H-J, Sun Z, Ou Z-W, Wang X-q (2016) Wear behaviors and wear mechanisms of wrought magnesium alloy AZ31 fabricated by extrusion-shear
3. Baghni IM, Wu Y-S, Jiuqing L, Cuiwei D, Zhang W. Mechanical properties and potential applications of magnesium alloys. Trans Nonferrous Metals Soc China (English Edition) 13:1253–1259
4. Senthil R, Gnanavelbabu A (2014) Effect of heat treatment on tensile strength and microstructure of AZ61A magnesium alloy. In: Applied mechanics and materials. Trans Tech Publications Ltd. pp 55–59. https://doi.org/10.4028/www.scientific.net/AMM.606.55
5. Rzychoń T, Kiełbus A (2007) Microstructure of WE43 casting magnesium alloy
6. Zhang Z, Zeng X, Ding W (2005) The influence of heat treatment on damping response of AZ91D magnesium alloy. Mater Sci Eng A 392:150–155. https://doi.org/10.1016/j.msea.2004.09.056
7. Vyshak TP, Sankar R, Babu AC, Harilal M, Devasia J (2020) A review on heating and rolling effect of AZ31 magnesium alloy. Int Res J Eng Technol
8. Fahad M, Bavanish B (2021) Tribological and ageing behavior of Az91D magnesium alloy fortified with nano lanthanum and nanoceria by stir casting for aviation application. Ind Lubr Tribol 73:635–641. https://doi.org/10.1108/ILT-12-2020-0475
9. Li ZM, Luo AA, Wang QG, Peng LM, Fu PH, Wu GH (2013) Effects of grain size and heat treatment on the tensile properties of Mg-3Nd-0.2Zn (wt%) magnesium alloys. Mater Sci Eng A 564:450–460. https://doi.org/10.1016/j.msea.2012.12.008
10. Shah M, Pandya A, Pramod B, Srivastava N (2023) Processing techniques of magnesium matrix composites and its heat treatment—A review. In: Advances in material science and metallurgy. Lecture Notes in Mechanical Engineering. Springer, pp 63–75. https://doi.org/10.1007/978-981-19-4918-0_6
11. Pandya A, Shah M, Pramod B, Srivastava N (2023) A review on heat treatment of magnesium alloys and its effect on various properties. In: Advances in material science and metallurgy. Lecture Notes in Mechanical Engineering. Springer, pp 77–87. https://doi.org/10.1007/978-981-19-4918-0_7

12. Su J, Teng J, Xu Z, Li Y (2022) Corrosion-wear behavior of a biocompatible magnesium matrix composite in simulated body fluid. Friction 10:31–43. https://doi.org/10.1007/s40544-020-0361-8
13. Xiu K, Wang HY, Sui HL, Wang Y, Xu CL, Wang JG, Jiang QC (2006) The sliding wear behavior of TiCp/AZ91 magnesium matrix composites. J Mater Sci 7052–7058. https://doi.org/10.1007/s10853-006-0946-2
14. Kaushik N, Singhal S (2018) Hybrid combination of Taguchi-GRA-PCA for optimization of wear behavior in AA6063/SiCp matrix composite. Prod Manuf Res 6:171–189. https://doi.org/10.1080/21693277.2018.1479666
15. Sathish T, Karthick S (2020) Wear behaviour analysis on aluminium alloy 7050 with reinforced SiC through taguchi approach. J Mater Res Technol 9:3481–3487. https://doi.org/10.1016/j.jmrt.2020.01.085
16. Girish BM, Satish BM, Sarapure S, Basawaraj (2016) Optimization of wear behavior of magnesium alloy AZ91 hybrid composites using taguchi experimental design. Metall Mater Trans A Phys Metall Mater Sci 47:3193–3200. https://doi.org/10.1007/s11661-016-3447-1
17. Makwana D, Pramod B (2023) Dry sliding wear and heat flux mapping of closed-cell Mg-2Zn-2Ca foam fabricated by stir casting route. Mater Lett 330:133379. https://doi.org/10.1016/j.matlet.2022.133379
18. Huang SJ, Abbas A, Ballóková B (2019) Effect of CNT on microstructure, dry sliding wear and compressive mechanical properties of AZ61 magnesium alloy. J Mater Res Technol 8:4273–4286. https://doi.org/10.1016/j.jmrt.2019.07.037
19. Gopi KR, Shivananda Nayaka H, Sahu S (2017) Wear properties of ECAP-processed AM80 magnesium alloy. J Mater Eng Perform 26:3399–3409. https://doi.org/10.1007/s11665-017-2764-x
20. Guicciardi S, Melandri C, Lucchini F, De Portu G (2002) On data dispersion in pin-on-disk wear tests
21. Rajesh AM, Kaleemulla MK, Doddamani S (2019) Effect of heat treatment on wear behavior of hybrid aluminum metal matrix composites. Tribol Ind 41:344–354. https://doi.org/10.24874/ti.2019.41.03.04
22. Mohd Daud MA, Talib Din A, Jusoff K, Selamat MZ (2013) Precipitation strengthening in extruded AZ61 magnesium alloys during heat treatment process. World Appl Sci J 21:41–46. https://doi.org/10.5829/idosi.wasj.2013.21.1005
23. Metall A, Gonzdez-Martinez R, Go ken J, Letzig D, Timmerberg J, SteinhofS K, Kainer KU (2007) Influence of heat treatment on damping behaviour of the magnesium wrought alloy AZ61
24. Turan ME, Zengin H, Sun Y (2020) Dry sliding wear behavior of (MWCNT + GNPs) reinforced AZ91 magnesium matrix hybrid composites. Met Mater Int 26:541–550. https://doi.org/10.1007/s12540-019-00338-8

A Review on Pineapple and Banana Natural Fiber Composite

Abin Paul Boban, Nitish Kumar, and Durgesh Kumar Mishra

Abstract Composite materials blend diverse materials to create a new one with unique properties. These materials are divided into two types: synthetic and natural. Synthetic composites, like glass fiber, are strong and durable but environmentally unfriendly due to their non-degradable nature and the high energy required for production. Conversely, natural composites are made from renewable plant fibers and are environmentally friendly, offering advantages such as degradability and lower production energy. The shift toward natural composites is driven by the global push for sustainability and reduced environmental impact. This study examines Pineapple Leaf Fiber (PALF) and banana fiber as sustainable alternatives to synthetic fibers. It explores their historical background, extraction methods, and applications, highlighting their environmental benefits, including sustainability and degradability. PALF is noted for its excellent mechanical strength and cost-effectiveness, making it ideal for the textile and paper industries, among others. Banana fibers, known for their flexibility and strength, are versatile in textile production, craftwork, and composite reinforcement. This research underlines the growing interest in these fibers for their eco-friendly properties, contributing to the development of sustainable materials science. It addresses the challenges in enhancing these composites' performance and suggests future research directions. Adopting natural fiber composites is a significant step toward sustainable material engineering, emphasizing the need to move from synthetic to natural materials in various sectors.

Keywords Composite materials · Synthetic composites · Natural composites · Pineapple fiber (PALF) · Banana fiber

---

A. P. Boban · D. K. Mishra
Department of Aeronautical Engineering, Hindustan Institute of Technology and Science, Chennai, India

N. Kumar (✉)
Department of Mechanical Engineering, Dayananda Sagar College of Engineering, Bangalore, Karnataka, India
e-mail: kumar.nitish492@gmail.com

P. Bhingole et al. (eds.), *Advances in Materials Engineering*, Lecture Notes in Mechanical Engineering, https://doi.org/10.1007/978-981-97-7114-1_3

1 Introduction

Composite refers to a group of materials which can be made of or extra separate materials that have diverse physical or chemical traits. The matrix and the reinforcement, constituents, are joined to supply a fabric with improved qualities that aren't typically present in any of the elements alone. The macroscopically blending of extra chemically unique additives results in composites, which might be valuable materials. Composites may be produced synthetically or naturally. Natural fibers were being used widely before the rise of synthetic fibers, after the rise of synthetic fibers natural fibers lost their importance in the midway of the twentieth century. This made the entire natural fiber market collapse. In the year 2009, it is being considered as the International Year of natural fiber for supporting the natural fiber market for benefiting the farmers and markets. A significant aspect of natural fibers is their low weight, essential for minimizing energy use [1]. The growing urgency for environmental sustainability in industrial practices has catalyzed a paradigm shift toward materials and processes that are eco-friendly, renewable, and carbon–neutral. One significant development in this realm is the utilization of natural fibers in composite materials, traditionally dominated by synthetic fibers like glass and carbon. This paper seeks to introduce and explore the various aspects of natural fibers as they are employed in composite materials, highlighting their sources, inherent properties, advantages, potential applications, and the challenges that lie ahead in their adoption and optimization. Composite materials, known for their high strength-to-weight ratio and superior performance in various conditions, have been extensively used in industries such as automotive, aerospace, construction, and sports equipment (Fig. 1).

Composites, which blend two or more chemically distinct components, offer enhanced properties not found in individual materials alone. This includes both naturally occurring and synthetically produced composites, such as Fiber-Reinforced Polymer (FRP) composites, where synthetic or natural fibers reinforce a polymer matrix. This combination allows for the distribution of load across the fibers, protecting them from damage and environmental factors, and can be engineered for specific features like high strength-to-weight ratios, stiffness, corrosion resistance, thermal stability, and electrical conductivity. Historical usage of composites, such as

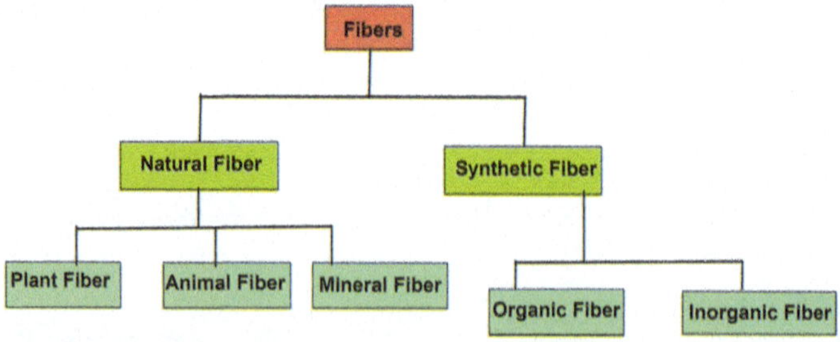

Fig. 1 Classification of fibers

mud bricks for construction, demonstrates early recognition of their benefits. Nowadays, the focus has shifted toward natural fibers like flax and ramie, particularly in industries like automotive, aerospace, and construction, due to their sustainable and eco-friendly nature. This shift is driven by the demand for cost-effective and environmentally conscious materials, with about 30 million tons of natural fibers produced annually for various applications, including clothing, packaging, and even as reinforcement in concrete. The preference for natural over synthetic fibers, such as glass or carbon, marks a significant trend in material science, emphasizing sustainability and eco-friendliness in engineering and design. Natural fibers have now achieved a competitive status alongside their synthetic counterparts [2]. In recent decades, there has been a resurgence of interest in the use of natural fibers due to their eco-friendly and sustainable characteristics. This review paper develops into the various aspects of natural fibers, shedding light on their sources, properties, applications, and the environmental benefits they offer. Natural fibers, derived from plant, animal, or mineral sources, have been fundamental to human activity for thousands of years, offering a range of uses from textiles and construction to a variety of modern industrial applications. Historically, these fibers were paramount before the advent of synthetic fibers, and they are now coming back to the forefront due to their biodegradability, renewable nature, and minimal environmental impact. The significance of natural fibers is multifaceted. They are not only crucial in supporting local economies and traditional crafts worldwide, but they also present an ecological alternative to synthetic fibers, whose production and disposal raise serious environmental concerns, including pollution and resource depletion. Natural fibers, in their various forms, contribute to the global push for sustainability, reducing the carbon footprint and waste, and presenting significant opportunities for recycling and reuse. Table 1 shows the annual production of the natural fiber and its sources.

As one moves from the S1 layer to the S2 layer, there is a consistent increase in cells Natural fibers possess a complex internal structure within their cell walls, comprising three primary structural components. The arrangement and angle of micro fibrils within these walls are crucial determinants of the fibers' attributes. The cell wall is primarily divided into a primary section (S1) and a secondary section (S2), each with distinct characteristics. The S1 layer develops during the plant's growth phase, while the S2 section consists of three distinct layers, each containing elongated micro fibril chains lose content, whereas the concentration of hemicelluloses remains

Table 1 Annual production natural fiber and sources [3–5]

Fiber source	World production (10^3 Tons)	Origin
Pineapple	Abundant	Leaf
Banana	200	Fruit
Kenaf	770	Stem
Sugarcane bagasse	75,000	Stem
Bamboo	10,000	Stem
Sisal	380	Stem

Table 2 Properties comparison of the natural fiber and glass fiber [8]

Properties	Natural fiber	Glass fiber
Density	Low	Double
Cost	Low	High
Renewability	Yes	No
Recyclability	Yes	No
Distribution	Wide	Wide
Energy consumption	Low	High
Disposal	Biodegradable	Biodegradable

relatively constant across the layers. However, the presence of lignin demonstrates an inverse relationship to that of cellulose. Hemicelluloses present a unique network-like molecular structure, establishing bonds with cellulose fibrils. This interaction between cellulose and hemicelluloses creates an integrated network, with lignin and pectin contributing adhesive properties. It's these adhesive characteristics that are pivotal in providing cellulosic fibers with their strength and stiffness.

The secondary layer, S2, is particularly influential in determining the fibers' physical and mechanical robustness. Generally, fibers exhibit enhanced strength properties when there's a high proportion of cellulose and a reduced micro fibrillar angle. Though glass fiber has very good density than the natural fiber with very high cost it is been replaced by using the natural fiber with very low cost compared to synthetic fibers. But the density is low (of 1.5 g/cm^3) compared to synthetic fiber [6]. The price of glass fibers typically ranges from approximately $1,200 to $1,800 per ton, with a density of roughly 2,500 kg/m^3. In contrast, natural fibers are more economical, with prices spanning $200 to $1,000 per ton, and have a lighter density that fluctuates between 1,200 and 1,800 kg/m^3 [7]. Table 2 shows the comparison of Natural Fiber and Glass Fiber.

Review is done on various papers and the comparison graph is generated according to the data that has been obtained from the research as delineated in Figs. 2 and 3.

2 Pineapple Plant

Pineapple is a family of Bormeliaceae [9]. The pineapple plant, scientifically known as Ananas comosus, is a tropical fruit-bearing species that belongs to the Bromeliaceae family. Originating from South America, likely between southern Brazil and Paraguay, the plant has gained immense popularity and is now cultivated in various warm climates around the globe. About 2,250,000 acres in India have been cultivated by pineapple plant, In Kilifi County, pineapple cultivation by small-scale farmers, especially in areas like Chamari and Adu-Mbuuni, under the Micro-Enterprise Support Programmed Trust, represents a vital agricultural activity on the Kenyan coast, with significant untapped potential and a strong market for its fresh and

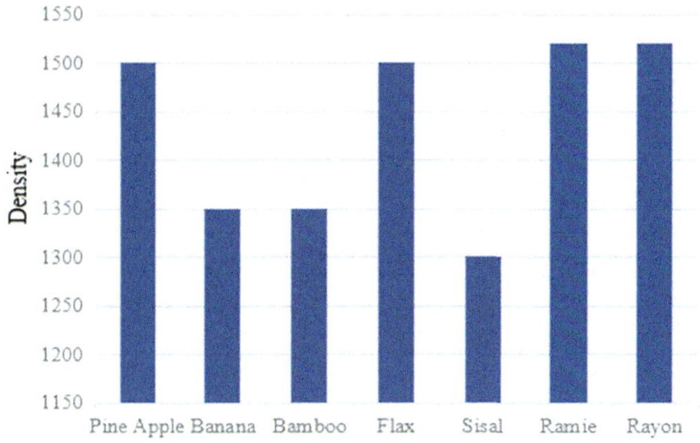

Fig. 2 Natural fiber versus density (kg/m^3)

Fig. 3 Synthetic fiber versus density (kg/m^3)

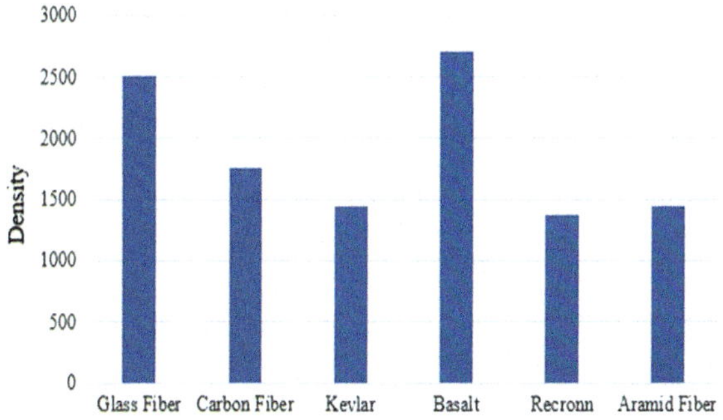

processed products [10]. Characterized by its tough, waxy leaves forming a ground-level rosette, the pineapple plant thrives in tropical weather with consistent warmth, humidity, and ample sunlight. The pineapple plant, a unique herbaceous perennial and a ground-dwelling member of the bromeliad family, typically takes 24–36 months to bear fruit from seed. However, in commercial cultivation, the use of suckers and slips reduces this time to about 16–24 months. The pineapple itself is not a single fruit but a cluster of individual berries that fuse around the central core, topped with a tuft of stiff leaves. This distinctive growth and fruiting process underscores the plant's unique place in the plant kingdom and its agricultural cultivation. Pineapple growth and yield can significantly decline due to different non-living environmental challenges, impacting its development and output [11]. Beyond its culinary uses, the pineapple plant holds cultural and historical significance in many societies, symbolizing hospitality and welcome. Moreover, various parts of the plant are utilized in traditional medicine, and it's a source of important enzymes like bromelain [12, 13], which has various health and industrial applications, According to a 2020 report from the Costa Rican National Electricity Institute, Costa Rica has impressively generated over 98% of its electricity from renewable resources like hydroelectricity, geothermal, wind, and solar power over the last five years, although this primarily pertains to electricity use. This achievement not only emphasizes the country's commitment to renewable energy but also showcases its potential for growth in the biorefinery industry, aiming for sustainable production of gas, fuel, and chemicals. Additionally, during the 1993–1994 period, key agricultural exports included bananas, coffee, milk, and beef, marking significant sectors in the nation's economy. Nonetheless, they also produced approximately 1.41 million tons of solid agricultural waste, with coffee, banana, sugarcane, and oil palm industries being the primary sources, each generating about 421,000 tons of waste annually [14, 15]. The dehydrated waste of Pineapple is used as feeding food for cattle, chickens, and pigs [16].

2.1 History

The pineapple's journey from its origins to a globally recognized symbol of luxury and indulgence highlights its significance in history and culture. Initially found in South America, it was brought to Europe by Christopher Columbus in 1493, quickly becoming a prized fruit among the European elite. Its cultivation required specialized environments to simulate its native tropical conditions, making it a symbol of wealth and status. Today, countries like the Philippines, Thailand, and Brazil lead in pineapple production, showcasing its widespread appeal and importance in both historical and contemporary contexts. The Portuguese made a vast contribution to the plantation of the pineapple in tropical regions like India, China, Africa, the Philippines, and Malaysia [17].

2.2 Pineapple Fiber

Pineapple leaf fiber (PALF) has a low density, is lightweight, and is also reasonably priced. It has a lot of cellulose, nearly 80%. It is easy to use and non-toxic. It has characteristics and exceptional mechanical capabilities that are ecologically friendly. High specific strength, stiffness, flexural, and torsional rigidity are all characteristics of PALF. It has fantastic chemical makeup. Pineapple leaf fibers (PALF) are notable for their superior mechanical strength, yet they remain underutilized due to limited awareness of their potential applications, including use in synthetic fibers, sound absorption, and thermal insulation, among others. Various techniques exist for PALF extraction from pineapple leaves [18], such as the scrapping method, which employs a machine comprising three specific rollers: a feed roller for leaf insertion, a scratching roller to eliminate the waxy surface layer, and a serrated roller with densely positioned blades to crush the leaves, facilitating the entry of retting microbes [19].

Another method, retting, involves submerging scratched leaf bundles in a solution with a substrate-to-liquor ratio of 1:2 supplemented with 0.5% urea or diammonium phosphate (DAP) to accelerate the retting reactions. This process necessitates regular manual checks to confirm fiber loosening and the extraction of various chemical components like pentosans, lignin, and others. Post-retting, fibers are mechanically separated and washed before air drying [20]. PALF has a cellulose content of (70–82%) and also seems to be equal to the arrangement in the cotton. [21–23]. The extraction of PALF can also be achieved using mechanical grinding methods such as ball or disc milling from fresh pineapple leaves, with wet ball milling yielding a higher number of elementary fibers despite being a slower process.

These methods stand out for their simplicity, efficiency, and the higher fiber yield they offer compared to traditional techniques [24]. In making yarn it shows that PALF has better mechanical strength than jute [25, 26]. Industries are using pineapple fiber as a perfect replacement raw material in the prospect of matrix composites [27].

2.3 PALF-Based Composites

PALF is a perfect compensation for glass fiber in every research and development [28]. PALF has been studied in thermoset, thermoplastic, biodegradable plastics, and natural rubber [29–32]. Mechanical properties of PALF include tensile strength, tensile modulus, elongation at break and impact strength [33].

2.4 PALF with Epoxy Resin

Epoxy resins, when mixed with a hardener like Araldite HY951, form a durable adhesive that's capable of bonding fibers together, distributing loads evenly. This mixture is known for its excellent adhesion, corrosion resistance, minimal shrinkage, and robust chemical resilience, making it ideal for aerospace composites and other applications. Araldite HY951, an aliphatic primary amine, can cure at room temperature or faster with heat, offering good mechanical strength and chemical degradation resistance [33]. Meanwhile, Pineapple Leaf Fiber (PALF) has gained recognition for its cost-effectiveness and environmental friendliness compared to synthetic fibers, due to its strong mechanical properties derived from its lignocellulosic structure. However, its integration into polyethylene and polypropylene composites reveals challenges like fiber dispersion, necessitating further research to improve its performance and compatibility in composite materials.

Presently, the research spotlight is on natural fibers as viable replacements for glass fibers in fiber-reinforced plastics, attributable to their lower density, cost-efficiency, minimal energy consumption during manufacturing, reduced equipment wear, and non-toxicity [34]. Yet, despite these advantages, PALF's potential as a reinforcing material in plastics remains underexplored, though interest in this application is growing. The focus is now shifting toward employing PALF in polymer composites to devise value-added applications. However, PALF's drawbacks, such as inadequate fiber-matrix adhesion and a propensity for moisture uptake, persist. Despite the dearth of literature on PALF-vinyl composites, it's acknowledged that vinyl esters are durable, pliable, and exhibit lower hydrophilicity [35]. Natural fibers like PALF, when integrated with vinyl ester matrices, often show enhanced interfacial shear stress (IFSS), indicating stronger fiber-matrix adhesion compared to other materials. The hand lay-up method is predominantly used for manufacturing PALF-reinforced thermoset composites, though techniques like liquid compression molding are also explored. Utilized in both raw and treated forms, PALF can be organized in various orientations within composites to assess their suitability for environmentally friendly applications, focusing on attributes such as mechanical strength, water absorption, and thermal resilience. The efficacy of these fibers is deeply influenced by their extraction process, the maturity of the source plants, environmental exposures, and moisture content, all of which contribute to the fibers' overall performance and application potential in different domains and the specific techniques used in their harvesting, retting, and decortication stages [36, 37]. Das et al. [38] made a study on pineapple and banana fiber in both physical and mechanical properties in the ratios 3:1, 1:1 and 1:3 ratios and they found 3:1 has higher properties than the other combinations.

Zulkafli et al. and Feng et al. [39, 40] did two experiments using the pineapple fiber with Kevlar and glass fiber and he found that it is better to use a hybrid than using the neat ones since neat ones break easily but the hybrid doesn't. Nitayaphat et al. [41] hydrogen peroxide concentration increases in the bleaching process, and there's a noticeable decrease in the PALF's strength, toughness, and stretchability, in comparison to the untreated PALF. This trend suggests that higher hydrogen peroxide levels more effectively strip away non-cellulosic components within the PALF, components that are crucial for binding the cellulose fibers and preserving the fiber's tensile integrity. At a lower hydrogen peroxide concentration (5%), the removal of these non-cellulosic materials is minimal, resulting in better mechanical properties, though the achieved level of whiteness is less than ideal. Conversely, a higher concentration (15%) results in excessive removal of these materials, leading to significantly compromised mechanical properties, albeit with an improved whiteness level. Study by Lopattananon and Cheirmakani showed that treating pineapple leaf fibers with solutions like NaOH and DGEBA significantly enhances the mechanical properties of epoxy resin-based composites, making them suitable for various industrial applications [28, 42]. Jain's et al [43] discovered optimal mechanical properties in epoxy-based composites with 10% pineapple leaf fiber, while Indro Reddy's et al. [44] studies found that composites with jute, pineapple, and glass fibers show superior mechanical strength when using epoxy resin compared to polyester resin. PALF contains 70–82% cellulose, 5–12% lignin and 1.1% ash [45].

PALF that are fabricated using the resins are updated in the following Table 3.

Table 3 PALF with Resin

Natural fiber	Resin	References
Palf	Epoxy	[46]
Palf	Vinyl Ester	[47]
Palf + Banana	Epoxy Resin	[48]
Palf + Glass fiber	Polyester	[49]

3 Banana Plant

The banana plant, scientifically referred to as Musa sap, stands as a crucial crop in tropical and subtropical areas, offering vast nutritional, economic, and cultural benefits. This large herb, often mistaken for a tree, features a stout, succulent pseudostem and grows from an underground rhizome. They produce edible fruits, technically berries, that come in various sizes, colors, and textures, commonly elongated and curved with a starchy, soft interior and a rind that can be green, yellow, or even purple when ripe. Bananas are a key source of vitamins C and B6, and minerals like potassium and magnesium, and are both low in calories and high in dietary fiber, making them a fundamental food source globally. Their economic impact is significant, with extensive commercial farming in regions such as Latin America, the Caribbean, and parts of Asia and Africa, underscoring their importance as one of the world's leading food crops due to their quick growth and high yield.

3.1 History

The origins of banana cultivation are steeped in mystery, but it's widely believed to have started in present-day Malaysia, spreading to the Philippines and India through ancient trade routes. Historical records, including Buddhist texts from 600 BCE, highlight the banana's longstanding significance in Asian cultures. Alexander the Great is credited with introducing bananas to the West in 327 BCE, while Arab traders later spread them across Africa, giving rise to the name "banana," derived from an Arabic word for "finger." In Europe, bananas were a rare luxury, often consumed and dried by the wealthy, until advancements in shipping in the late 19th and early twentieth centuries allowed for fresh bananas to be transported from the Caribbean to North America and Europe, transforming them from an exotic treat to a household staple. Despite challenges like political strife, economic difficulties, and diseases like the Panama disease, bananas have remained a global dietary mainstay. Today, they are grown in over 107 countries, with India, China, the Philippines, and Ecuador leading production. The journey of the banana illustrates the impact of human innovation and global trade on spreading this fruit from regional delicacy to a worldwide commodity, reflecting broader changes in agriculture, commerce, and cultural practices.

3.2 Banana Fiber

In addition to having excellent mechanical, dimensional stability, and mechanical properties, banana fiber is also environmental friendly, practical, affordable, biodegradable, renewable, and sustainable. It can withstand heat, spin well, and has a high tensile strength. Other fibers may be mixed with it.

3.3 Banana-Based Composites

Banana fiber, traditionally coarse and stiff, has been innovatively blended with softer fibers like cotton to improve its textile applications. Companies such as Japan's Nisshin Textile pioneered this, creating a blend that capitalizes on the softness of cotton and the luster and absorption of banana fiber [50, 51]. India, a leading producer of banana products, has been instrumental in advancing banana fiber research. Studies have focused on its blending techniques, spinning performance, and strength characteristics. Beyond textiles, the eco-friendly trend has ushered in a rise in hybrid fiber composites, combining materials like banana and carbon fibers for enhanced mechanical strength [52, 53]. Additionally, banana fiber's potential in polymer-based composites is significant. These composites, using matrices like unsaturated polyesters and epoxy resins, tap into banana fiber's strength-enhancing attributes [54]. Notably, such integrations have shown improved elastic properties, tensile strength, and thermal stability. Kumar et al. [55] used banana fiber and pineapple fiber and they found that the specimen PPBP specimen has very high tensile strength, toughness and can withstand very high loads without breaking.

The study by Rai et al. [56] explored the use of euphorbia coagulum as a binder in polyester banana fiber composites, demonstrating significant enhancements in mechanical properties and water resistance. By replacing 40% of polyester resin with euphorbia coagulum, the composite exhibited a 50% increase in bending strength and a 68% decrease in water absorption, alongside improved fire resistance. This research highlights the potential of renewable materials in creating environmentally friendly composites, offering an alternative to petroleum-based products. Bhoopathi et al. [57] conducted an experimental study on the mechanical properties of composites made from glass fiber combined with treated banana and hemp fibers, using hand layup process and enzymatic extraction for fiber preparation. The findings revealed that the glass/banana/hemp fiber composite exhibited superior tensile strength, with the hemp-glass fiber composite leading in flexural strength, suggesting their potential application in fields requiring robust mechanical properties. Jagadeesh et al. [58], and others have highlighted the growing interest in eco-friendly materials in engineering industries, emphasizing the advantages of natural fibers like banana, jute, and hemp over synthetic ones due to their high strength, light weight, and biodegradability. Their research delves into the extraction and analysis of banana fibers, exploring their mechanical, chemical, and physical properties, along with their potential in hybrid composites.

Rao et al. [59] reviewed the rising interest in natural fiber composites for their environmental benefits, ease of availability, and cost-effectiveness, with a special focus on the promising mechanical properties of banana fiber- reinforced composites. Mukhopadhyay et al. [60] studied the diameter variation of banana fibers, finding a range from 0.08 mm to 0.32 mm, with more uniformity in coarser fibers. They conducted tensile testing on fibers in the 0.17–0.19 mm range, noting the significant impact of strain rates on the fibers' strength and failure nature. Prabu's [61] study focused on the impact of red mud on banana/polyester hybrid composites, observing enhanced mechanical, vibration, and chemical resistance properties. They found a 50% increase in mechanical strength with 4 lm particle size red mud at 8 wt%, also noting improved damping characteristics and chemical resistance. Jústiz-Smith et al. [62] explored the potential of Jamaican banana, coconut coir, and bagasse fibers in composites. They found that banana fibers exhibited the highest ash, carbon, and cellulose content, along with superior hardness and tensile strength, while coconut had the highest lignin content Herrera-Estrada et al. [63] investigated banana fiber-reinforced composites for automotive and transportation applications, focusing on fiber surface treatments and environmental exposure. They reported higher flexural and compressive strengths in alkaline-pretreated banana fiber/eco-polyester composites, demonstrating their enhanced durability and resistance to environmental conditions.

4 Conclusion

Pineapple leaf fiber (PALF) and banana fiber, both sourced from tropical regions, present innovative and sustainable alternatives to synthetic fibers, favored for their environmental benefits and recyclability. PALF's utility in composites is hindered by its moisture-absorbing nature, requiring modifications for better polymer compatibility, yet its high cellulose content offers great promise for applications in construction, automotive, and furniture. Research into its compatibility with various resins and its properties in hybrid composites is ongoing. Banana fiber, meanwhile, emerges from India's vast banana waste, offering economic and environmental advantages with its strength, light weight, and biodegradability, holding the potential to transform industries like automotive and construction through improved composite technologies. Despite the need for further study on their mechanical and environmental behaviors, both fibers are recognized for their potential to replace more traditional, less sustainable materials, marking a significant shift toward greener alternatives in material science.

Conflict of Interests The authors declare that there is no conflict of interests regarding the publication of this paper.

References

1. Dejene BK (2024) False banana (Enset Ventricosum) fibers: an emerging natural fiber with distinct properties, promising potentials, challenges and future prospects—A critical review. J Nat Fibers 21(1):2311303
2. Demir M, Ekici R (2024) Investigation of low-velocity impact behaviors of polymer composites reinforced with different natural fiber fabrics. Polym Compos
3. Taj S, Munawar MA, Khan S (2007) Natural fiber-reinforced polymer composites. Proc-Pakistan Acad Sci 44(2):129
4. Faruk O, Bledzki AK, Fink HP, Sain M (2012) Biocomposites reinforced with natural fibers: 2000–2010. Prog Polym Sci 37(11):1552–1596
5. John MJ, Thomas S (2008) Biofibres and biocomposites. Carbohydr Polym 71(3):343–364
6. Foulk JA, Akin DE, Dodd RB (2000) New low cost flax fibers for composites (No. 2000-01-1133). SAE Technical paper
7. Satyanarayana KG, Arizaga GG, Wypych F (2009) Biodegradable composites based on lignocellulosic fibers—An overview. Prog Polym Sci 34(9):982–1021
8. Wambua P, Ivens J, Verpoest I (2003) Natural fibres: can they replace glass in fibre reinforced plastics? Compos Sci Technol 63(9):1259–1264
9. Van Tran A (2006) Chemical analysis and pulping study of pineapple crown leaves. Ind Crops Prod 24(1):66–74
10. Muti S, Ndiso J, Abdallah B, Monthe E, Nzau S, Kinoti D, Amukhoye R (2024) Status and challenges facing pineapple (Ananas comosus) production in Magarini sub-county of Kilifi county in coastal Kenya. J Agric Ecol Res Int 25(1):1–18
11. Sinaga AOY, Marpaung DSS (2024) Abiotic stress-induced gene expression in pineapple as a potential genetic marker. Adv Agrochem
12. Ketnawa S, Rawdkuen S, Chaiwut P (2010) Two phase partitioning and collagen hydrolysis of bromelain from pineapple peel Nang Lae cultivar. Biochem Eng J 52(2–3):205–211
13. Schieber A, Stintzing FC, Carle R (2001) By-products of plant food processing as a source of functional compounds—recent developments. Trends Food Sci Technol 12(11):401–413
14. Eixenberger D, Carballo-Arce AF, Vega-Baudrit JR, Trimino-Vazquez H, Villegas-Peñaranda LR, Stöbener A, ... Liese A (2024) Tropical agroindustrial biowaste revalorization through integrative biorefineries—Review part II: pineapple, sugarcane and banana by-products in Costa Rica. Biomass Convers Bioref 14(4):4391–4418
15. Ulloa JB, Van Weerd JH, Huisman EA, Verreth JAJ (2004) Tropical agricultural residues and their potential uses in fish feeds: the Costa Rican situation. Waste Manag 24(1):87–97
16. Cathcart DJ (1995) The importance of maintaining bromeliad imports. Fla Entomol 16–21
17. Heinicke RM, Gortner WA (1957) Stem bromelain—a new protease preparation from pineapple plants. Econ Bot 11(3):225–234
18. Debasis N, Sanjoy D (2007) A pineapple leaf fibre decorticator assembly. India Patents, 2334
19. Kannojiya R, Gaurav K, Ranjan R, Tiyer NK, Pandey KM (2013) Extraction of pineapple fibres for making commercial products. J Environ Res Dev 7(4):1385–1390
20. Banik S, Basak MK, Paul D, Nayak P, Sardar D, Sil SC, ... Ghosh A (2003) Ribbon retting of jute—a prospective and eco-friendly method for improvement of fibre quality. Ind Crops Prod 17(3):183–190
21. Fakirov DBS (2007) Wood fibre thermoplastic composite: processing properties and future developments
22. Mohanty AK, Misra MA, Hinrichsen GI (2000) Biofibres, biodegradable polymers and biocomposites: an overview. Macromol Mater Eng 276(1):1–24
23. Omojasola PF, Jilani OP, Ibiyemi SA (2008) Cellulase production by some fungi cultured on pineapple waste. Nat Sci 6(2):64–79
24. Dey SK, Bhattacharyya GK, Bhattacharyya SK (2005) Magic yarns from ramie and pineapple—a new dimension in 21st century. In: Proceedings of the 20th Indian engineering congress, vol 69, pp 129–141

25. Sinha MK (1982) A review of processing technology for the utilisation of agro-waste fibres. Agric Wastes 4(6):461–475
26. Lopattananon N, Panawarangkul K, Sahakaro K, Ellis B (2006) Performance of pineapple leaf fiber–natural rubber composites: the effect of fiber surface treatments. J Appl Polym Sci 102(2):1974–1984
27. Kengkhetkit N, Amornsakchai T (2012) Utilisation of pineapple leaf waste for plastic reinforcement: 1. A novel extraction method for short pineapple leaf fiber. Ind Crops Prod 40:55–61
28. Lopattananon N, Payae Y, Seadan M (2008) Influence of fiber modification on interfacial adhesion and mechanical properties of pineapple leaf fiber-epoxy composites. J Appl Polym Sci 110(1):433–443
29. George J, Joseph K, Bhagawan SS, Thomas S (1993) Influence of short pineapple fiber on the viscoelastic properties of low-density polyethylene. Mater Lett 18(3):163–170
30. Luo S, Netravali AN (1999) Interfacial and mechanical properties of environment-friendly "green" composites made from pineapple fibers and poly (hydroxybutyrate-co-valerate) resin. J Mater Sci 34:3709–3719
31. Liu W, Misra M, Askeland P, Drzal LT, Mohanty AK (2005) 'Green' composites from soy based plastic and pineapple leaf fiber: fabrication and properties evaluation. Polymer 46(8):2710–2721
32. Bhattacharyya TB, Biswas AK, Chatterjee J, Pramanick D (1986) Short pineapple leaf fibre reinforced rubber composites. Plast Rubber Process Appl 6(2):119–125
33. Ellis B, Found MS, Bell JR (1998) Effect of cure on the impact behaviour of an epoxy resin. In: ECF 12: fractures from defects, vol 3, pp 1459–1464
34. Arib RMN, Sapuan SM, Ahmad MMHM, Paridah MT, Zaman HK (2006) Mechanical properties of pineapple leaf fibre reinforced polypropylene composites. Mater Des 27(5):391–396
35. Mallick PK (2007) Fiber-reinforced composites: materials, manufacturing, and design. CRC Press
36. Joffe R, Andersons J, Wallström L (2003) Strength and adhesion characteristics of elementary flax fibres with different surface treatments. Compos A Appl Sci Manuf 34(7):603–612
37. Kiruthika AV (2024) A review of leaf fiber reinforced polymer composites. J Eng Appl Sci 71(1):24
38. Das S, Rahman M, Hasan M (2018) Physico-mechanical properties of pineapple leaf and banana fiber reinforced hybrid polypropylene composites: effect of fiber ratio and sodium hydroxide treatment. In IOP conference series: materials science and engineering, vol 438, No 1, p 012027. IOP Publishing
39. Zulkafli N, Malingam SD, Fadzullah SHSM, Razali N (2019) Quasi and dynamic impact performance of hybrid cross-ply banana/glass fibre reinforced polypropylene composites. Mater Res Exp 6(12):125344
40. Ng LF, Dhar Malingam S, Subramaniam K, Selamat M, Ali MB, Bapokutty O (2019) The influence of fibre stacking configurations on the indentation behaviour of pineapple leaf/glass fibre reinforced hybrid composites. Def S T Tech Bull 12:113–123
41. Nitayaphat W, Jintakosol T (2024) Dyeing of pineapple leaf fibers using various natural dye extracts and mordants. J Nat Fibers 21(1):2313867
42. Cheirmakani BM, Subburaj B, Balasubramanian V (2022) Exploring the properties of pineapple leaf fiber and prosopis julifora powder reinforced epoxy composite. J Nat Fibers 19(6):2065–2076
43. Jain J, Jain S, Sinha S (2019) Characterization and thermal kinetic analysis of pineapple leaf fibers and their reinforcement in epoxy. J Elastomers Plast 51(3):224–243
44. Reddy MI, Kumar MA, Raju CRB (2018) Tensile and flexural properties of jute, pineapple leaf and glass fiber reinforced polymer matrix hybrid composites. Mater Today: Proc 5(1):458–462
45. Hoque MB, Mollah MZI, Faruque MRI, Hannan MA, Khan RA (2021) Review on the mechanical properties of pineapple leaf fiber (PALF) reinforced epoxy resin based composites. Polymer 14(15):16–17

46. Mukherjee PS, Satyanarayana KG (1986) Structure and properties of some vegetable fibres: Part 2 pineapple fibre (Anannus comosus). J Mater Sci 21:51–56
47. Mohamed AR, Sapuan SM, Khalina A (2010) Selected properties of hand-laid and compression molded vinyl ester and pineapple leaf fiber (PALF)-reinforced vinyl ester composites. Int J Mech Mater Eng 5(1):68–73
48. Payae Y, Lopattananon N (2009) Adhesion of pineapple-leaf fiber to epoxy matrix: the role of surface treatments. Songklanakarin J Sci Technol 31(2)
49. Madhukiran J, Rao SS, Madhusudan S (2013) Tensile and hardness properties of banana/ pineapple natural fibre reinforced hybrid composites. Int J Eng Res Technol 2(7):1260–1264
50. Preetham, Tukaram PR, Chittappa H (2023) A review of the pineapple leaf fiber variants, structure, physical properties and chemical composition 12. https://doi.org/10.17577/IJERTV 12IS110127
51. Fan Y (2023) Study on preparation and application of banana fiber-based composites. J Phys: Conf Series 2539(1):012093). IOP Publishing
52. Dang M (2001) Banana fiber and products thereof. Text Technol Overseas 12:11–13
53. Sivaranjana P, Arumugaprabu V (2021) A brief review on mechanical and thermal properties of banana fiber based hybrid composites. SN Appl Sci 3:1–8
54. Dinesh Kumar S, Ponraj Sankar L, Sathish T, Vijayan V, Parthiban A, Kamalakannan R, Rajkumar S (2021) Improving the mechanical properties of natural fiber composites of hemp fiber with ramie and banana fiber through compression molding method. Adv Mater Sci Eng 2021:1–8
55. Kumar VA, Neeraj TS, Meghana Y (2022) Mechanical characterization and fabrication of banana and pineapple fibers. IOP Conf Series: Mater Sci Eng 1248(1):012061. IOP Publishing
56. Rai B, Kumar G, Diwan RK, Khandal RK (2011) Study on effect of euphorbia coagulum on physico-mechanical and fire retardant properties of polyester-banana fiber composite. Indian J Sci Technol 443–446
57. Bhoopathi R, Ramesh M, Rajaprasanna R, Sasikala G, Deepa C (2017) Physical properties of glass-hemp-banana hybrid fiber reinforced polymer composites. Indian J Sci Technol
58. Jagadeesh D, Venkatachalam R, Nallakumarasamy G (2015) Characterisation of banana fiber-a review. J Environ Nanotechnol 4(2):23–26
59. Rao P, Rao D, Lakshumu Naidu A, Bahubalendruni MVAR (2017) Mechanical properties of banana fiber reinforced composites and manufacturing techniques: a review 8
60. Mukhopadhyay S, Fangueiro R, Arpac Y, Şentürk Ü (2008) Banana fibers–variability and fracture behaviour. J Eng Fibers Fabr 3(2):155892500800300200
61. Uthayakumar M, Manikandan V, Rajini N, Jeyaraj P (2014) Influence of redmud on the mechanical, damping and chemical resistance properties of banana/polyester hybrid composites. Mater Des 64:270–279
62. Jústiz-Smith NG, Virgo GJ, Buchann VE (2008) Potential of Jamaican banana, coconut coir and bagasse fibres as composite materials. Mater Charact 59(9):1273–1278
63. Bhatnagar R, Gupta G, Yadav S (2015) A review on composition and properties of banana fibers. Cellulose 60:65

A Study on Surface Treatment Methods of Metal Surface for Metal Fiber Laminates

Renu Sahu, Gaurav K. Gupta, Pradeep Kumar Ghosh, and Mohit Jaiswal

Abstract Fiber Metal Laminate (FML) is a lightweight material that combines the advantages of fiber-reinforced composites and metal laminates with good mechanical properties. It is extensively utilized in the transportation and aerospace sectors, especially as a structural component in the skins, wings, and tails of aircraft. However, under challenging service conditions, the interfacial properties and structure between the layers are crucial to load performance and, as such, require industrial applicability research. In FMLs, interlaminar failure drastically decreased mechanical properties, even to the point of catastrophic consequences for safety and the economy. People are put in danger when vital structural parts of airplanes and other manned vehicles fail or are destroyed. Therefore, it is essential to describe the interlaminar failure behavior of FMLs and provide preventative measures for these flaws. For the metal layers and fiber to form a strong bond, the surface treatment of the metal sheets is crucial. This review paper explores the various studies conducted by numerous groups that systematically discuss the effect of surface treatment on interlaminar behavior and other mechanical behavior. Readers can obtain research progress on the surface treatment method, and failure behavior of FMLs through this review article more efficiently.

Keywords Fiber Metal Laminate (FML) · Surface treatment · Interlaminar behavior

R. Sahu (✉) · P. K. Ghosh
University Teaching Department (UTD), Chhattisgarh Swami Vivekanand Technical University, Bhilai 491107, India
e-mail: rsahu05.se2019@gmail.com

G. K. Gupta
CSIR-Advanced Materials and Processes Research Institute, Bhopal 462026, India

M. Jaiswal
National Institute of Technology Raipur, Rajpur 492010, India

P. Bhingole et al. (eds.), *Advances in Materials Engineering*, Lecture Notes in Mechanical Engineering, https://doi.org/10.1007/978-981-97-7114-1_4

1 Introduction

Day by day Requirement for exceptionally lightweight materials is rising having high mechanical properties in the aircraft and aerospace industries. In this regard, FMLs (Fiber Metal Laminates) are one of the choices for such applications. A modern hybrid composite made of alternately organized fibers and metal alloys [1, 2]. The FML materials have continuously improved by adding additional metal alloys like titanium and aluminum-magnesium [3, 4], and combinations like woven fabric and prepreg of carbon fibers [5]. Such types of laminates have been used as portions of aircraft, including the wings, fuselage, and other sections. Additionally, FMLs are being used in other industrial sectors like athletic goods and automotive engineering thanks to the optimization of the production process of materials and the reduction of production costs [6, 7].

One of the most important variables for increasing the endurance of fiber metal laminates is a strong bonding interface [9–11]. Surface treatment is a useful technique for bolstering the bond between metal sheets and a fiber composite, in addition to increasing the surface energy and wettability of metal substrates [12]. By eliminating inefficient surface contaminants and oxides, surface treatment can enhance the metal substrate's surface composition and microstructure. As a result, fiber metal laminates have improved mechanical properties and fiber bridging mechanisms, and the rate of fracture propagation at the bi-material interface has also been greatly slowed [8, 13].

In addition to the significance of the matrix and reinforcement in polymer composites, the adherence of the composite laminae to the aluminum foil is a critical issue for the overall performance of the metal-fiber laminate. A sufficient amount of surface preparation of the aluminum alloy is required to ensure a solid mechanical binding between the aluminum surface and epoxy. Numerous studies have been conducted in this field. Chromic acid anodization (CAA) as a surface treatment is the method that is most frequently used to increase the surface activity of aluminum alloys. This process produces a fantastic interface between the fiber and metal because of the toxicity of the chromium, but it has serious environmental consequences as well. As a solution to this problem, sulfuric-boric-oxalic acid anodization (SBOA), a new Anodization treatment was initiated [14–17].

To evaluate the effectiveness of a metal surface modification's surface treatment, the most common techniques for calculating contact angle and roughness are used [10–12].

This review article examines the numerous research carried out by numerous organizations that methodically evaluate the impact of surface treatment on interlaminar behavior and other mechanical behavior. Through this review article, readers can more effectively learn about research developments on the surface treatment method and failure behavior of FMLs.

2 Types of Surface Treatments

To enhance the interlaminar bond between metal and fibers, surface roughness is needed for better adhesive properties. There are many types of surface treatments occurs which change the surface properties of either metal or fiber. The following types of surface treatments are cure.

2.1 *Mechanical Treatments*

Using mechanics medium metal surfaces are treated. In mechanical treatments, Scotch-Brite abrasion, Sandblasting, and emery paper rubbing are defined as:

Scotch-Brite abrasion technique was carried out manually and abrasively using a Scotch-Brite disc pad 07447+ from 3M, Saint Paul, Minnesota, USA. It was done once more. To get a cross coat, the Scotch-Brite pad was rotated 90° in the other direction after being moved side to side. Using lint-free tissue and acetone, the surface was then painstakingly cleaned.

Sandblasting Using alumina (Al_2O_3) powder, sandblasting was carried out. The granularity of the powder was 180 m. After that, the surface was painstakingly cleaned with acetone and lint-free tissue.

Emery Paper Rubbing Different-sized emery paper is used to rub metal surfaces in two perpendicular directions. Numerous grooves are created as a result of rubbing, aiding in the formation of a grip between layers.

Plasma Treatments The plasma process was conducted at 140 W of power, 30 L/min of He flow, and 0.1 L/min of O_2 striking flow. Plasma activation was accomplished using Surfx Technologies' Atomflo 500 plasma devices (Fig. 1).

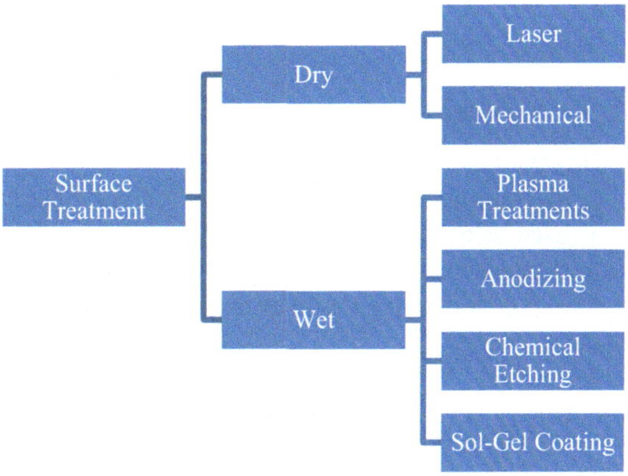

Fig. 1 Category of surface treatment process

2.2 Anodization

Chromic acid anodization a sulfuric bath. Chrmic acid anodizing was made using CrO_3 chromic acid anhydride. The procedure ran for 45 min at a temperature of 38–42 °C with a voltage of 20 V.

Sulphuric acid anodizing Drying, also known as $K_2Cr_3O_7$. The process temperature for the sulfuric acid anodizing was 10–15 °C, the voltage ranged from 13 to 24 V, and the process time was 24 min.

2.3 Chemical Etching

Chemical etching is a type of engraving that removes material from metal while also etching a permanent image using a high-pressure, high-temperature chemical spray. To generate the desired picture, the material's surface is covered with a mask or resist, which is then carefully peeled off, revealing the metal. The aluminum surface was etched for 11 min at temperatures ranging from 63 to 65 °C, then rinsed in a water tank and drained.

2.4 Phosphate-Fluoride Method

The procedure included degreasing, rinsing, and digestion using hydrofluoric acid, nitric acid, and sodium sulfate for two to three minutes at room temperature (24 °C). In addition, a 1.5–2.5 min phosphate–fluoride treatment using potassium fluoride, sodium phosphate, and hydrofluoric acid was carried out. 24 °C was the procedure temperature.

2.5 Sol–gel Coating

The two-component 3MTM Surface Pre-Treatment AC 130-2 (3MTM) formulation was used to make the sol–gel coatings. The combination of two components was introduced for 30 min. Using a lint-free tissue, the mixture was then applied to the surface and left to dry for 60 min at room temperature after draining for 10 min.

2.6 *Primer*

The 3MTM Scotch-WeldTM EC-3924B corrosion-resistant structural adhesive primer was applied to the aluminum surface. EC-3960 (3MTM Scotch-WeldTM) was applied to the titanium surface to coat it. A thin primer coating is sprayed onto the surface as part of the procedure.

3 Review of Research

Magda Drózdziel-Jurkiewicz and Jarosław Bienia have investigated the use of surface treatments on titanium Grade 2 and aluminum 2024-T3 in fiber metal laminates (FML) to enhance adhesion at the metal-composite interface. The most robust mechanical interlocking effect occurs on the bonding surface when using medium-sized (180#) sandpaper. Aluminum surface treatment techniques that work well are P2 etching and sol–gel coating. Surface homogeneity is produced via chromic acid anodizing. A moderate roughness is found to be ideal for binding the materials. Too high roughness or too high smoothness both have negative effects on metal-fiber interlayer joining. The study is based on SEM analysis and sample preparation using ion etching and impact tests [14].

In 1981, L. G Stringer the constraints of pre-treating aluminum alloy surfaces for bonding and the application of straightforward approaches to retain durability have been investigated. When the typical sodium dichromate/sulfuric acid etching technique was tried for a longer duration of time at room temperature, the outcomes were superior to those obtained at the suggested temperature of 60–65 °C. The findings demonstrated that when the sodium dichromate/sulfuric acid etch pretreatment was utilized at room temperature for a longer period, adhesively bonded aluminum (4% Zn-2% Mg) joints performed better in warm, humid situations. Additionally, testing with 120 °C-curing epoxy adhesives has shown progress [15].

Laser surface treatment can increase the shear strength of single-lap bonded joints made of CFRP as understood by Yingxi Xie et al. The joint performance was improved using a two-stage laser surface treatment, where step one involved removing resin from the surface to reveal the carbon fiber fabric, and step two involved making minute grooves on the surface. The shear strength of the CFRP laminate bonding was increased by crossing grooves, according to the findings of the experiments. The 150m groove distance, 6 reprocessing cycles, and 0/90° grooves were the best processing variables for optimum shear strength. Compared to smooth CFRP, the grooved-CFRP bonded joint's shear strength increased by 40.8% [16].

Hong et al. have evaluated the impact of three surface treatments on the aluminum alloy AA6061 composite Al/CFRP/Al laminates' shear strength. Shear strength after sanding was 38 MPa, while shear strength after anodizing and surface modification (with a silane coupling agent) was around 50 MPa. According to SEM and XPS research, the anodized layer strengthened the mechanical anchoring and chemical bonds between AA6061 and CFRP, increasing shear strength. The silane alteration enhanced the mechanical stability of the laminates by allowing the thermal expansion mismatch to be controlled. Anodizing yielded a shear strength of 51.4 MPa, which is the highest [17].

Lee et al. looked into how aluminum sample samples' surface roughness affected their mechanical characteristics. Tensile, bending, and shear lap tests were performed on a sample made of a hybrid material consisting of Al5052 and carbon fiber-reinforced plastic. According to the findings, the surface roughness value with the highest shear stress and best adhesion was 1.2 mm (Ra). To increase adhesion, surface preparation was essential [18].

A study by Muhammad Yasir Khalid and coworkers examined the interlaminar shear strength (ILSS) of fiber-metal laminates (FMLs) consisting of sheets of aluminum 7075-T6 reinforced with a variety of fibers. Short beam shear tests were used to analyze the ILSS, and it was discovered that carbon fiber-reinforced aluminum laminates (CARALL), as opposed to glass and aramid-reinforced. The displacement rates did not affect the ILSS; instead, the adherence of the fiber-reinforced epoxy to the aluminum sheet was the significant factor. Future research might examine the ILSS utilizing various surface treatment techniques on the aluminum matrix and at various strain rates, frequencies, and temperatures [19].

A modified study was conducted on the interface of carbon-reinforced aluminum laminates (CARE) employing amine silane A-1387 A-187 and epoxy silane as the silane coupling agents. On the mechanical properties and microstructure, the impact of coupling agent type and content was examined. According to the findings, increasing the coupling agent content initially enhanced and then decreased wettability. The best moisture content and mechanical characteristics were found at 1% mass fraction. In comparison to the epoxy silane A-187, the amine silane A-1387 had greater wettability and greatly improved the interface structure and characteristics, increasing interlaminar shear and flexural strengths by 19.1% and 8.9%, respectively [20].

Selim Mrzljak studied the hybrid laminate constructed of AA6082 aluminum alloy sheets, glass, and polyamide 6 with carbon fiber reinforcement and was the subject of a study to determine how aluminum surface treatment affected its tensile and fatigue behavior. After the laminate was cemented, double-notch shear tests and tensile testing were carried out to ascertain the interfacial strength. The surface treatments were characterized using laser scanning microscopy. Digital image correlation, thermography, and hysteresis measurement techniques were utilized to determine the impact of the aluminum surface treatment on fatigue-induced damage. The fatigue performance was evaluated utilizing short-time instrumented multiple and constant amplitude tests [21].

Monika Ostapiuk examined the microstructures of glass or carbon fiber-reinforced aluminum-epoxy composites. The effect of pressure and aluminum surface treatment on the interface between metal and polymer composites was investigated during the autoclave process. The outcomes demonstrate that the metal-polymer interface is significantly impacted by aluminum pretreatment and that there is direct contact between the anodizing layer and the resin there. Low pressure during the autoclave process causes the polymer layer to become porous and delaminate. It also causes the epoxy composite's and the untreated aluminum surface's adhesion to be weak, especially when carbon reinforcing is used [22].

Mehra et al. study, Al2024-T3 sheets, basalt fibers, and epoxy fiber metal laminates' flexural and impact behavior were examined about the effects of various surface treatments (FPL etching, SAA, sandblasting, sandblasting + FPL, and sandblasting + SAA). Al 2024-T3, basalt fiber, and epoxy were used to bend and prepare the laminates, and profilometry was used to measure the surface roughness. Results revealed that the surface treatments significantly affected how well Al2024-T3 adhered to the polymeric layers. In contrast to sandblasting, which produced the maximum strain to failure, SAA treatment produced the highest bending strength. Impact test findings demonstrated that the FMLs' low-velocity impact behavior was improved by combining FPL or SAA treatment with sandblasting. It was determined what each sample's flexural strength and strain were. After being treated with SAA (or FPL), the sandblasted A12024-T3's adhesion to epoxy increased and was on par with that of the FMLs that had also been treated with SAA [23].

Gonzalez-Canche et al. study investigated the impact of surface treatments on the Al5052-H32 aluminum alloy utilized in the polypropylene (FML)-based composite material reinforced with aramid fibers. To describe the surface treatments, contact angle, SEM, profilometry, and XPS were used. Shear adhesion strength testing and uniaxial tensile testing were used to assess the adhesion and mechanical performance of the FML. The findings demonstrated that the FML's adhesion and tensile strength were enhanced by pickling the metal layers in nitric and sodium hydroxide, producing superior mechanical qualities in terms of stiffness and strength [24].

Zakaria et al. examined how aluminum substrates in carbon fiber metal laminates interacted with acid and alkaline surface treatments. When treated with acid or alkaline, respectively, AFM pictures demonstrated roughness and a porous oxide layer, both of which were maintained due to the mutual influence of the treatments. Double cantilever beam specimens were used to assess the interfacial fracture toughness, and plate theory and the compliance calibration method were used to determine the energy release rates. The outcomes demonstrated that both the acid and the alkaline treatments increased the interfacial fracture toughness, resulting in a higher rate of energy release than with other treatments. Finite element analysis was used to assess the accuracy of the calculated fracture energy, and the results showed that the derived cohesive laws from the plate theory were more in line with tests than those from the compliance calibration approach [25].

The results showed that wettability was highly influenced by surface roughness, with rough surfaces having higher surface energy and better wettability. On the substrate with the highest weight and surface energy, a double-layer phosphating film formed after the treatment period reached five minutes. The magnesium alloys' wettability and surface energy were enhanced by phosphating treatment. Following abrasion with #120 and #3000 grit abrasive paper, the phosphate substrate's surface energies were 84.31 mJ/m^2 and 83.65 mJ/m^2, respectively, and its wettability was significantly higher than that of the abraded magnesium. In terms of corrosion resistance, the phosphate AZ31B sheet outperformed the abraded AZ31B sheet very quickly. By mixing magnesium alloy with glass fiber/epoxy prepregs, its corrosion resistance was increased even more [26].

For usage in medical implants, Trauth investigated the use of laser-based surface pre-treatment to strengthen the binding between titanium and a self-reinforced biocompatible polylactide (srPLA). Through a parameter investigation utilizing shear edge testing and scanning electron microscopy, the optimal laser structures for pre-treatment were identified. The results showed that the laser-induced pre-treatment significantly increased the interface strength when compared to an untreated surface and that this pre-treatment technique is suitable for titanium and srPLA pregs together [27].

A study was conducted on an FML composed of epoxy Epon 828, basalt fibers, and Al2024 to investigate the effects of four different surface treatments on its flexural behavior. SEM, optical microscopy, and profilometry were employed to investigate the surface morphologies and fracture surfaces. The results demonstrated that FPL-etching and anodizing treatments improved the flexural characteristics, while mechanical abrasion and alkaline etching treatments did not, due to the formation of surface oxide layers with microscale pits that facilitated solid interfacial bonding.

The major kind of failure in the FPL-etched and anodized samples was a cohesive failure, as shown by SEM and OM pictures [28].

Almeida examined the behavior of metal/fiber laminates (MFLs) consisting of carbon fiber and aluminium foil under tension-tension fatigue. To encourage adhesive bonding, two surface treatments—chromic acid anodization (CAA) and sulfuric-boric-oxalic acid anodization (SBOA)—were used. The effectiveness of the surface treatments was assessed using roughness measures and scanning electron microscopy. The findings demonstrated that both treatments can be employed to achieve strong structural adhesion between the continuous fiber/epoxy composites and aluminum 2024-T3 alloy in MFL specimens made with SBOA and CAA treatments. A crack in 0/90 was discovered to be the main damage mode in plain weave carbon fiber reinforced plastic laminates. According to SEM examination, the initial fracture in high-cycle (low-stress) CARALL specimens happened on the aluminum alloy and not the CF/E fracture [29].

Karunagaran et al. investigate how surface treatments affect the mechanical characteristics of hybrid composites comprised of epoxy-reinforced stainless steel wire mesh and glass fiber. While the stainless steel wire mesh is treated with electro-dissolution or sandblasting, the glass fiber is either treated with sulfuric acid or sodium hydroxide. The findings demonstrate that compared to composites without surface treatments, the hybrid composite made of sandblasted stainless steel wire mesh and acid-treated glass fiber has better tensile, flexural, inter-laminar shear, and impact properties. The surface alterations strengthen the bond between the reinforcement and resin [30].

Shanmugam et al. By merging metal with fiber-reinforced polymer layers, a brand-new class of hybrid material known as T-FML (thermoplastic fiber metal laminates) has been developed. T-FML was created using a Ti6Al4V metal alloy, UHMWPE, and Elium thermoplastic resin. The metal-thermoplastic composite interface and the interlaminar interface, which control damage resistance of T-FML, can be improved to increase stiffness and decrease crack propagation and delamination under impact loading. With polydopamine deposition with MWCNT on the fiber surface and electrochemical treatment on the metal surface, the surfaces of the fibers and metals were treated to improve the interfaces. Impact tests at low speeds revealed that T-FML with enhanced interfaces had a decreased rate of single-crack development [31].

Based on the above review following findings are constructed (Table 1).

Most of the authors have performed chemical treatments. Because chemical-etching and sulfuric acid anodizing treatments markedly enhanced strain to failure and toughness compared to mechanical abrasion and alkaline etching. Microscale pits, resulting from the etching treatment on the metal surface, played a key role in these improvements by facilitating epoxy resin infiltration and strong mechanical interlocking between laminates. Fracture analyses and mechanical tests revealed a combination of adhesive and cohesive failure in chemical-etching and anodized samples.

4 Conclusion

The paper presents the different issues of metal treatments using mechanical, anodization, and chemical treatments, depending on different experiments conducted on the surface treatment of metals the following conclusions have been drawn:

1. Every surface treatment of metal surfaces enhances the adhesive property of metal fiber interlayer joining with different efficiency.
2. In mechanical treatments, sandblasting provides a high value of roughness while emery pepper rubbing gives a pleasant value of surface roughness.
3. Plasma treatment is good for cleaning the surfaces but is unable to form proper grooves so not that much good method for enhancing the layer joining.
4. During the anodizing process, is oxide layer is formed which helps the metal to create a good bond with interfaces.
5. In the case of chemical etching, moderate values of surface roughness are created.
6. In combination with mechanical and chemical etching surface treatments provide a more suitable surface roughness value.

Table 1 Review chart based on different surface treatment methods

SN	Author	Treatment	Finding
1	Drózdziel-Jurkiewicz and Jarosław Bienia [14]	Mechanical	These methods enhance surface area but lack the ability to achieve the specific shape needed for mechanical interlocking and adhesive anchoring
		Plasma treatment	Clean the surface with low roughness value. So having limited application to maintain high adhesion
		Anodizing	The surface exhibits a uniform structure with significant micro-irregularities. So it is advantageous methods
		Chemical etching	The surface was shaped with micro-irregularities and porosity to achieve suitable topography and morphology. Also a good method
2	Stringer [15]	Chemical etching at warm condition	Employing the sodium dichromate/ sulfuric acid etch pretreatment at room temperature for an extended duration enhanced the performance of adhesively bonded aluminum joints under warm, humid conditions
3	Xie et al. [16]	Laser surface treatment	Suggesting a 40.8% improvement when compared to laminates with a smooth surface
4	Hong et al. [17]	Mechanical	The most robust mechanical interlocking effect occurs on the bonding surface when using medium-sized (180#) sandpaper
		Anodizing	C–O–Al chemical bond was formed by the reaction of epoxy resin with alumina, which increases bond shear strength
5	Lee et al. [18]	Mechanical	Three times shrear strength increases and 50% tensile strength increases due to sandblasting in which Ra value varies between 0.33 um to 0.65 um
6	Khalid [19]	Chemical	Provide a better adhesion between interfaces which increases ILSS
7	Mrzljak [21]	Mechanical	Interlaminar shear strength increases but tensile strength does not depend on surface treatments

(continued)

Table 1 (continued)

SN	Author	Treatment	Finding
		Anodizing and phosphating	Attaining the longest fatigue life, the adhesion promoter resulted in a notably diminished fatigue life when compared to Anodizing and phosphating. Phosphating allows for the prolonged preservation of stiffness with minimal crack formation but induces more extensive delamination propagation and the greatest overall stiffness reduction
8	Ostapiuk [22]	Anodizing	The untreated surface of an aluminum sheet exhibits insufficient adhesion to epoxy composite, particularly when reinforced with carbon, whereas treated signifies good adhesion at interface
9	Mehra et al. [23]	Mechanical, chemical anodization	Out of all the samples, the sandblasted sample had the lowest S.A.E. value, which was roughly 17.83% less than the SAA sample. On the other hand, the SAA sample's S.A.E. value was about 2.94% lower than that of the sandblast + SAA samples. In conclusion, out of all the samples, the one that underwent sandblasting and SAA treatment showed the best impact behavior, making it the most likely candidate for use in aircraft applications
10	Gonzalez-Canche et al. [24]	Mechanical chemical	The chemical treatment produced a surface with enhanced wettability and deeper roughness compared to the surface achieved through sanding and degreasing. Additionally, the chemical treatment provided a superior cleaning effect, contributing to a superior interfacial contact between the metal and the adhesive-composite material, in conjunction with the increased roughness area
11	Zakaria et al. [25]	Chemical	The aluminum surface's finely patterned peaks and valleys created by acid washing improve the interfacial area and encourage the mechanical interlocking of the metal and composite layers. But aluminum substrates may not receive enough surface oxide from an acid treatment to provide sufficient bond durability

(continued)

Table 1 (continued)

SN	Author	Treatment	Finding
12	Zhang et al. [26]	Phosphating	The magnesium alloy's wettability was significantly impacted by surface roughness. High surface energy and good wettability were demonstrated by a rough surface. The magnesium alloy's wettability and surface energy were both improved by the phosphating treatment, which also significantly increased the surface roughness
13	Trauth et al. [27]	Laser surface treatments	In order to achieve the best mechanical interface properties within the given parameter range, the titanium sheet must be pre-treated using a laser with 90% power and a 60 kHz frequency
14	Aghamohammadi et al. [28]	Mechanical chemical anodization	Layers are not suitable for adhesion because of inadequate layer bonding caused by grinding and alkaline treatments. Adhesive failure in both abrasion and etching samples was confirmed by fracture analyses. Sulfuric acid anodizing and FPL etching significantly increased toughness and strain to failure
15	Karunagaran [30]	Chemical	The surface layer is changed by the acid treatment, adding tiny rings and helical features that improve bonding with the resin. This ultimately results in a strength factor that falls between 80 and 93

References

1. Ostapiuk M, Bieniaś J, Surowska B (2018) Analysis of the bending and failure of fiber metal laminates based on glass and carbon fibers. Sci Eng Compos Mater 25(6):1095–1106
2. Rajkumar GR, Krishna M, Narasimhamurthy HN, Keshavmurthy YC, Nataraj JR (2014) Investigation of tensile and bending behavior of aluminum based hybrid fiber metal laminates. Procedia Mater Sci 5:60–68
3. Nassir NA, Birch RS, Cantwell WJ, Sierra DR, Edwardson SP, Dearden G, Guan ZW (2020) Experimental and numerical characterization of titanium-based fiber metal laminates. Compos Struct 245:112398
4. Hu YB, Li HG, Cai L, Zhu JP, Pan L, Xu J, Tao J (2015) Preparation and properties of Fibre-Metal Laminates based on carbon fiber reinforced PMR polyimide. Compos Part B Eng 69:587–591
5. Li H, Xu Y, Hua X, Liu C, Tao J (2018) Bending failure mechanism and flexural properties of GLARE laminates with different stacking sequences. Compos Struct 187:354–363
6. Vermeeren C (2003) A historic overview of the development of fiber metal laminates. Appl Compos Mater 10:189–205
7. Sinmazcelik T, Avcu E, Bora MO, Coban O (2011) A review: fibre metal laminates, background, bonding types, and applied test methods. Mater Des 32:3671–3685

8. Múgica JI et al (2014) Impact characterization of thermoformable fiber metal laminates of 2024–T3 aluminum and AZ31B-H24 magnesium based on self-reinforced polypropylene. Compos A Appl Sci Manuf 61:67–75
9. Wei R et al (2017) Effect of surface treatment on the interfacial adhesion performance of aluminum foil/CFRP laminates for cryogenic propellant tanks. Mater Des 116:188–198
10. Zain NM, Ahmad SH, Ali ES (2014) Effect of surface treatments on the durability of green polyurethane adhesive bonded aluminum alloy. Int J Adhes Adhes 55:43–55
11. Li H et al (2016) Effect of adhesive quantity on failure behavior and mechanical properties of fiber metal laminates based on the aluminum–lithium alloy. Compos Struct 152:687–692
12. He P, Chen K, Yang J (2015) Surface modifications of Ti alloy with tunable hierarchical structures and chemistry for improved metal–polymer interface used in the deepwater composite riser. Appl Surf Sci 328:614–622
13. Pan Y et al (2017) Effects of surface pre-treatments on mode I and mode II interlaminar strength of CFRP/Mg laminates. Surf Coat Technol 319:309–317
14. Droździel-Jurkiewicz M, Bieniaś J (2022) Evaluation of surface treatment for enhancing adhesion at the metal-composite interface in fibre metal-laminates. Materials 15(17):6118
15. Stringer LG (1981) Effect of a room temperature sodium dichromate/sulphuric acid etching pretreatment on the durability of aluminum alloy bonded joints. Int J Adhes Adhes 1(5):228–232
16. Xie Y, Yang B, Lu L, Wan Z, Liu X (2020) Shear strength of bonded joints of carbon fiber reinforced plastic (CFRP) laminates enhanced by a two-step laser surface treatment. Compos Struct 232:111559
17. Zhu W et al (2019) Characterization and properties of AA6061-based fiber metal laminates with different aluminum-surface pretreatments. Compos Struct 227:111321
18. Lee MS et al (2020) A study on mechanical properties of Al5052/CFRP/Al5052 composite through three-point bending tests and shear lap tests according to surface roughness. J Compos Mater 54(12):1549–1559
19. Khalid MY et al (2021) Interlaminar shear strength (ILSS) characterization of fiber metal laminates (FMLs) manufactured through VARTM process. Forces Mech 4:100038
20. Zhu W et al (2021) Effect of coupling agent quantity on composite interface structure and properties of fiber metal laminates. Polym Compos 42(7):3195–3205
21. Mrzljak S et al (2020) Influence of aluminum surface treatment on tensile and fatigue behavior of thermoplastic-based hybrid laminates. Materials 13(14):3080
22. Ostapiuk M, Surowska B, Bieniaś J (2014) Interface analysis of fiber metal laminates. Compos Interf 21(4):309–318
23. Mehr ME et al (2019) Effects of applying a combination of surface treatments on the mechanical behavior of basalt fiber metal laminates. Int J Adhes Adhes 92:133–141
24. Gonzalez-Canche NG et al (2018) Evaluation of surface treatments on 5052–H32 aluminum alloy for enhancing the interfacial adhesion of thermoplastic-based fiber metal laminates. Int J Adhes Adhes 82:90–99
25. Zakaria AZ et al (2017) Effects of aluminum surface treatments on the interfacial fracture toughness of carbon-fiber aluminum laminates. Eng Fracture Mech 172:139–151
26. Zhang X et al (2017) Effect of surface treatment on the corrosion properties of magnesium-based fiber metal laminate. Appl Surf Sci 396:1264–1272
27. Trauth A et al (2022) Interface characterization of hybrid biocompatible fiber-metal laminates after laser-based surface treatment. Compos Struct 281:115054
28. Aghamohammadi H et al (2018) Effects of various aluminum surface treatments on the basalt fiber metal laminates interlaminar adhesion. Int J Adhes Adhes 84:184–193
29. Almeida RS et al (2008) Effect of surface treatment on fatigue behavior of metal/carbon fiber laminates. J Mater Sci 43:3173–3179
30. Karunagaran N, Rajadurai A (2016) Effect of surface treatment on mechanical properties of glass fiber/stainless steel wire mesh reinforced epoxy hybrid composites. J Mech Sci Technol 30:2475–2482

31. Shanmugam L et al (2020) Influence of UHMWPE fiber and Ti6Al4V metal surface treatments on the low-velocity impact behavior of thermoplastic fiber metal laminates. Adv Compos Hybrid Mater 3:508–521

Comparative Study of Fe_3O_4 and Calcium-Doped Fe_3O_4

S. D. Raut and S. G. Dahotre

Abstract Sustaining relation between Spintronics and Quantum Computing and a new generation of 5G and 6G devices is a measure challenge in front of researchers and scientist. Half metal ferromagnetic nanoparticles play a very energetic role in the development of nanoscale-devices and spintronics devices. In this work, we synthesized successfully Fe_3O_4 and $Ca_{0.8}Fe_{2.2}O_4$ nanoparticles via sol–gel auto-combustion method for the study of specific morphological and magnetic properties for the application of spintronics devices. Field Emission Scanning Electron Microscope (FESEM) image shows the crystalline nature of given samples. Electron Diffraction X-ray (EDX) confirms the presence of Fe, Ca, and O atoms in the prepared sample. Fourier Transform Infra-Red (FTIR) confirms the presence of metal bond M–O in a given sample. Vibrating Sample Magnetometer (VSM) measures the magnetic field at room temperature which confirms the existence of a superparamagnetic behavior of Fe_3O_4 and $Ca_{0.8}Fe_{2.2}O_4$ nanoparticles which plays an essential role in nanoscale devices such as spin-MRAM, DRAM, spin valve, spin filter, MTJ, and bio-sensing devices applications.

Keywords $Ca_{0.8}Fe_{2.2}O_4$ · FESEM · VSM

1 Introduction

In recent years vast development in research field of spintronics and quantum computing have a relationship between the physical and imaginary behavior of compact nanoscale devices such as sensors, supercapacitors, SRAM, DRAM, spin-FET, MRI Nanoscale devices. Artificial Intelligence gives the faster speed to store the data as well as work functions performed by these devices [1–5]. The world

S. D. Raut (✉) · S. G. Dahotre
Department of Physics, Dr. Babasaheb Ambedkar Technological University, lonere, Mangaon, Raigad, Maharashtra 402103, India
e-mail: Supriya@dbatu.ac.in

S. G. Dahotre
e-mail: sgdahotre@dbatu.ac.in

P. Bhingole et al. (eds.), *Advances in Materials Engineering*, Lecture Notes in Mechanical Engineering, https://doi.org/10.1007/978-981-97-7114-1_5

gets revolutionize due to enormous changes in day today life like android phones, smart watches, with spintronics devices. Due to quantum computing number of things can be done within milliseconds, hence to design the devices at nanoscale we require ultrafine range of nanoparticles, which can be synthesized through top down approach. Here we used sol-gel auto-combustion synthesis method [6–8].

The influence of quantum computing, related to the electron magnetic spin investigated power technologies having source GMR at nanoscale level. Modern technology has concentrated on manipulating and controlling both an electron's spin (magnetic moment) and charge in order to achieve important applications like memory and sensing. For many applications, including fuel cells, smart gadgets, environmental protection, supercapacitors, and gas leak alarms, metal oxide nanoparticles are essential [9, 10].

Fe_3O_4 is one of the best metal oxide, half-metallic ferromagnetic material, and the best elusive quantum material which is highly spin-polarized and plays vital role in non-volatile and highly magnetic storage devices [11–13]. Hence to enhance the magnetic and electrical properties at nanoscale devices, doping of elements such as Co, Zn, Ni, La, Sr, Mn, Mg in Fe_3O_4 has been done by many researchers [14–17].

Calcium is an amphoteric ion and is a member of the alkaline earth metals group and is the fifth most plentiful element in the Earth's crust, also it is non-hazardous and less radioactive. Its Face-Centered Cubic (FCC) structure and characteristics are comparable to those of strontium, but it has better mass, volume, and specific conductivity than that of strontium. Thus, calcium produces remarkable results from its reaction with Fe_3O_4 nanoparticles, which makes it easier to reduce the density of materials used in spintronics. This improvement results in more effective applications for the next nanoscale gadgets [18–20].

In this research article, we discussed structural and magnetic characterizations of magnetite and calcium-doped magnetite and analyzed a comparative study of these two samples.

2 Literature Survey

For the formation of spintronics or nanoscale devices the structural and magnetic properties plays unique role. Magnetite is half metal which gives highly magnetic anisotropy and nano-size which helps to design inductor core and coils. Magnetite is superparamagnetic in nature because of its higher magnetic saturation (M_s) and lower coercivity (H_c) whereas due to the inequivalence crystallographic structure of calcium ferrite, it shows two phases as ferrimagnetic and antiferromagnetic [21–24]. It has been seen that magnetite and calcium ferrite are used in the production of permanent magnets, microwave devices as well as in sensors [25, 26].

Thapa et al. observed the crystalline size of Fe_3O_4 is below 10 nm, hence surface to volume ratio increases which causes a decrease in saturation magnetization, hence magnetite material changes from ferromagnetic to superparamagnetic in nature and it is used for flexible spintronics devices [27]. Lemine et al. synthesized magnetite

NP using a sol–gel method and characterized by XRD, giving ultrafine particle size of 8 nm and also confirming the single phase and good crystallanity of magnetite NP [28, 29]. Parmar et al. synthesized magnetite NPs using sol–gel auto-combustion method at different temperature as 270 °C, 320 °C, 370 °C. The crystalline size obtained is in the range of 37–43 nm and the synthesis method also concludes single phase of magnetite NP at low temperature [30]. Thavrani et al. studied Ca^{2+} doped $ZnFe_2O_4$ synthesized via solvothermal method and observed that lattice parameter increases whereas x-ray density decreases, this is due to Zn^{2+} ion replacing Ca^{2+} ion. This replacement affects the change in magnetic properties (M_s) from 31.103 emu/g to 36.355 emu/g with an increase in the concentration of calcium ion doped $ZnFe_2O_4$ for high frequency applications [31].

To the best of our literature survey, a comparative study of Fe_3O_4 and calcium-doped magnetite ($Ca_{0.8}Fe_{2.2}O_4$) via sol–gel auto-combustion method has been not reported yet. Thus, exploring dramatic changes in magnetite and calcium-doped magnetite behavior is an experimentally and theoretically interesting issue, which might be helpful in designing nanoscale devices.

3 Materials and Methodology

In this research, chemical materials were used such as citric acid, Merck's chemical reagents (AR grade) calcium nitrate, ferric nitrate, and deionized water. The process of sol-gel auto combustion is shown in Fig. 1.

A simple sol-gel auto-combustion method was adopted for the preparation of Fe_3O_4 and $Ca_{0.8}Fe_{2.2}O_4$ nanoparticles. In the synthesis procedure, Ferric Nitrate and Calcium nitrate were taken as sources of Fe^{3+}/Fe^{2+} and Ca^{2+}. The pH was adjusted to 7 by drop-wise addition of NH_4OH [32–34]. The reaction mixture was kept on hot plate at 80 °C and after 1 h mixture was converted into homogeneous sol with brown slurry. After continuously heating brown slurry gets converted into gel form and within 5 sec auto-combustion takes place and gel is converted into fluppy powder. This fluppy powder is crushed into pestle and mortar, and it is characterized with the help of FESEM with EDX, FTIR, and VSM characterization [8, 33, 35, 36].

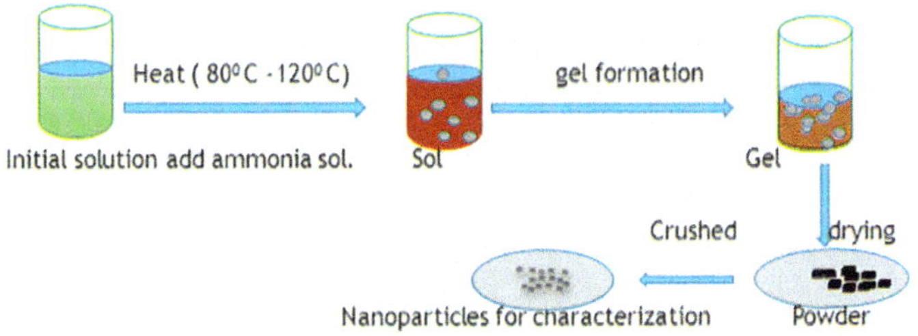

Fig. 1 Sol–gel auto combustion synthesis process

4 Results and Discussion

4.1 FESEM with EDXA

The morphological and elemental structure studies were carried out using a Field Emission Scanning Electron Microscope (FESEM) that can be operated at voltage 20 kV with an Energy Dispersive X-ray Analyzer (EDXA) [37]. Fe_3O_4, and $Ca_{0.8}Fe_{2.2}O_4$ nanoparticles magnifying at 60,000 X so we can observe at 500 nm scale. The FESEM micrograph is shown in Figs. 2 and 3. The FESEM image reveals the structure and surface morphology of the synthesized Fe_3O_4 and $Ca_{0.8}Fe_{2.2}O_4$. From the image, it is evident that the formation of crystalline form of nature of Fe_3O_4, and $Ca_{0.8}Fe_{2.2}O_4$ nanoparticles. Average particle size of Fe_3O_4 is nearly 150 nm calculated by using ImageJ software and $Ca_{0.8}Fe_{2.2}O_4$ is 12 nm to 18 nm directly observed from the image. It concludes that the particle size of magnetite is higher than calcium-substituted magnetite. Pure magnetite shows cubic particle nature of sample and as Ca^{2+} was added, agglomeration of the particles was observed. Ca^{2+} ions (0.99 Å) have larger ionic radii than Fe^{2+} (0.77 Å), resulting in decreased grain size or particle size due to reduced solubility in the spinel lattice [18]. At the same time due to high ionic radius of Ca^{2+} than that of Fe the bond length between Ca-Fe increases, hence crystallite size decreases.

From Fig. 2 it is confirmed that the EDXA of Fe_3O_4 shows the existence of Fe and O ions in Fe_3O_4 sample and Fig. 3 confirms that $Ca_{0.8}Fe_{2.2}O_4$ reveals the attendance of Ca, Fe, and O elements.

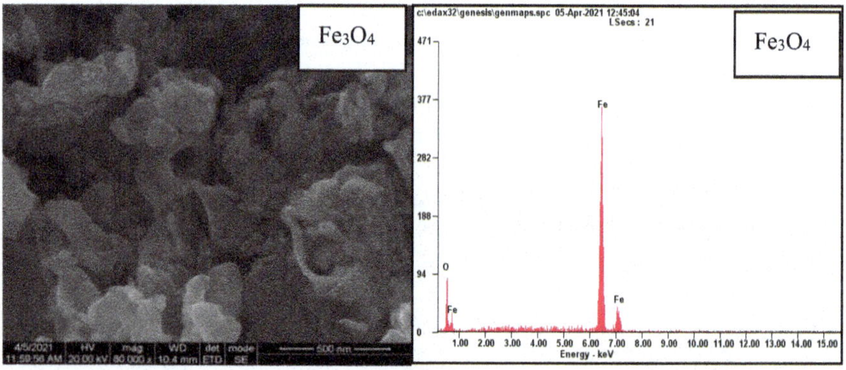

Fig. 2 FESEM micrograph with EDXA of Fe_3O_4

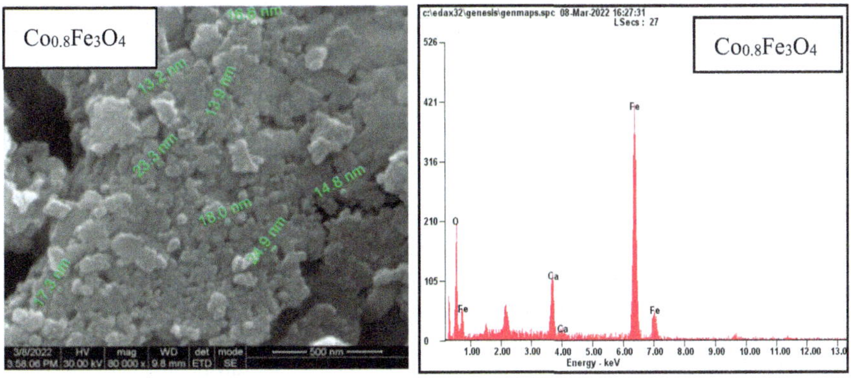

Fig. 3 FESEM micrograph with EDXA of $Ca_{0.8}Fe_{2.2}.O_4$

4.2 *FTIR*

The Fourier Transform Infra-Red (FTIR) is shown in Fig. 4. FTIR absorption peak confirms the M–O bond is present in the region 400–800 cm^{-1} for both sample. It has been noticed that the functional group O–H is observed at 3435 cm^{-1} due to presence of oxygen in prepared sample. Due to the presence of moisture, the absorption peak is observed at frequency range nearly 1454 cm^{-1} [25, 38–40].

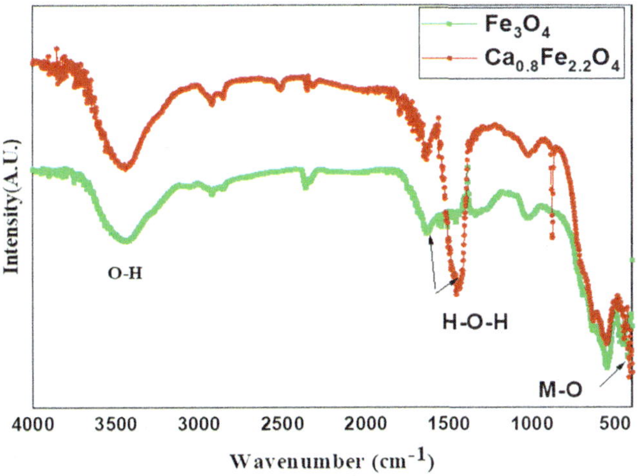

Fig. 4 FTIR of Fe_3O_4 and $Ca_{0.8}Fe_{2.2}O_4$

4.3 Magnetic Study via VSM Characterization

Magnetic characterization has been done on Vibrating Sample Magnetometer (VSM) which is as shown in Figs. 5 and 6. From VSM graph we calculated Coercive Field (H_c), Saturation Magnetization (M_s), and Remenant Magnetization (M_r) which is given in Table 1. The lower value of remenant magnetization and approximately zero value of coercivity (H_c) show superparamagnetic nature of Fe_3O_4 and $Ca_{0.8}Fe_{2.2}O_4$. Eventhuogh Calcium is non-magnetic element, when it is doped in pure Fe_3O_4 at nano-level it gives higher saturation magnetization (M_s) because Ca^{2+} ion goes to A-site resulting in an increase in Fe-O bond length and reduction in Fe-O-Fe bond angle which weakens double exchange effect which is an interesting behavior shown in Fig. 6 [29, 41–43].

Since crystalline structure of Fe_3O_4, is FCC type which has lower saturation magnetization whereas, calcium is alkaline earth metal which is less atomic number, it easily react with water giving higher saturation magnetization due to hexagonal spinel crystalline ferrite. This drastic change of saturation magnetization in two samples might be due to particle size variation observed from the FESEM images which are shown in Fig. 2 and Fig. 3, at the same time there may be super-exchange interaction between Ca^{2+} ion and Fe^{3+} ion in the synthesis process [13, 24, 34, 44, 45].

From Table 1 it is observed that the squareness ratio is less than 0.5 which confirms both samples are polycrystalline in nature [1, 23, 46] whereas from Figs. 5 and 6, it has been observed that calcium-doped magnetite has lower particle size and higher saturation magnetization than that of magnetite, Ca^{2+} ions prefer to occupy A-site by pushing Fe^{3+} ion to go to B-site and hence magnetization of B-site will

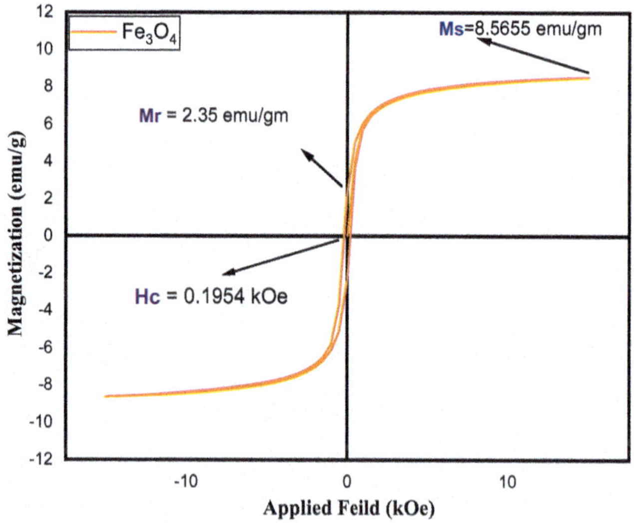

Fig. 5 VSM of sample Fe_3O_4

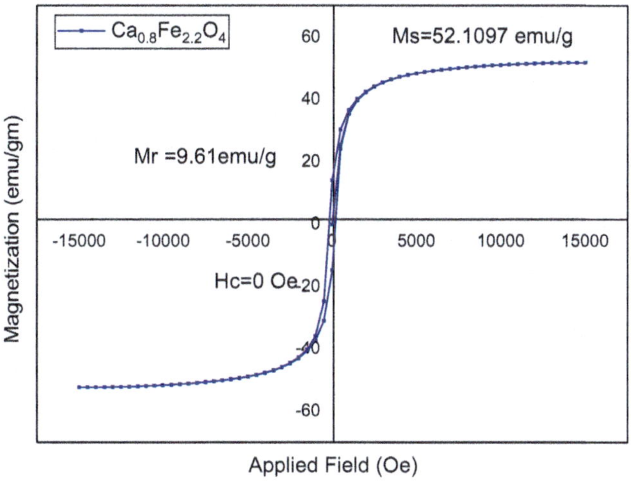

Fig. 6 VSM of sample $Ca_{0.8}Fe_{2.2}O_4$

Table 1 Saturation magnetization (M_s), Remenant magnetization (M_r) Coercive Field (H_c), and squareness ratio of magnetite and Calcium-doped magnetite

Composition	Saturation magnetization (M_s) (emu/g)	Remenant magnetization (M_r) (emu/g)	Coercive field (H_c)kOe	Squareness ratio (M_r/M_s)
Fe_3O_4	8.5655	2.89	0.1954	0.337
$Ca_{0.8}Fe_{2.2}O_4$	52.1097	9.61	0.00	0.184

increases and total magnetization increases with increase in Ca^{2+} doping hence, it plays a vital role in spintronics and nanoscale devices. High saturation magnetization gives miniaturization of devices. This also reveals that calcium-doped magnetite is worthwhile for compact devices and its magnetic properties reveal a faster speed of work function for new quantum computing generation.

5 Conclusion

Fe_3O_4 and calcium-doped Fe_3O_4 are successfully synthesized via sol–gel auto-combustion method. FESEM shows higher particle size of Fe_3O_4 than that of $Ca_{0.8}Fe_{2.2}O_4$, which confirms that $Ca_{0.8}Fe_{2.2}O_4$ is more effective than pure magnetite to design compact device. FTIR confirms the presence of M–O bond at frequency range of 400–800 cm^{-1} in prepared sample. Calcium-doped magnetite has higher saturation magnetization than that of magnetite, which reveals faster speed of work function for new quantum computing generation. Currently, for the development of electronic information, industry for quantum computing devices, high frequency, miniaturization, and integration are required for electronic devices, therefore to obtain superior soft magnetic properties at high frequency, magnetic materials should have high saturation magnetization and low coercivity. The high temperature sintering could be a feasible method to improve soft magnetic properties of nanoparticle sample. Practically the thickness of sample is required to reduce at nanolevel to get lower coercive field and higher saturation magnetization. As the saturation magnetization is high and coercivity is low, these nanoparticles are ideal candidate for high frequency electronic devices.

Acknowledgements We would like to express sincere appreciation to the Department of Physics, Dr. Babasaheb Ambedkar Technological University Lonere, Mangaon, Raigad, Maharashtra for the experimental work and also thanks to the IIT madras for sample characterizations.

References

1. Hirohata A, Sukegawa H, Yanagihara H, Zutic I, Seki T, Mizukami S, Swaminathan R (2015) Roadmap for emerging materials for spintronic device applications. IEEE Trans Magn 51:1–11. https://doi.org/10.1109/TMAG.2015.2457393
2. Jayakumar OD, Tyagi AK (2012) Spintronic materials, synthesis, processing and applications. Funct Mater 193–228. https://doi.org/10.1016/B978-0-12-385142-0.00005-2
3. Dua P, Parmar KPS, Mishra AK, Kumar S, Chauhan A, Kuchhal P (2021) Selection criteria for spintronic materials: role of spin dependent band-width narrowing/broadening. J Phys Chem Solids 149:135907. https://doi.org/10.1016/j.jpcs.2020.109786
4. Kotnala RK, Shah J (2015) Ferrite materials: nano to spintronics regime. Elsevier. https://doi.org/10.1016/B978-0-444-63528-0.00004-8
5. Raut SD, Dahotre SG, Singh LN, Jadhav SN (2021) Synthesis and characterization of magnetite and cobalt ferrite nanoparticles by sol-gel auto combustion technique. Int J Innov Eng Sci 6:17. https://doi.org/10.46335/ijies.2021.6.10.4
6. Zegeye SM (2019) A review paper on spintronics and its role to improve electronic devices. Am J Quantum Chem Mol Spectrosc 3:41–47. https://doi.org/10.11648/j.ajqcms.20190302.13
7. Liu W, Bryan MT, Xu Y (2019) Introduction to spintronics and 2D materials. LTD. https://doi.org/10.1016/B978-0-08-102154-5.00001-1
8. Shinde VS, Dahotre SG, Singh LN (2020) Synthesis and characterization of aluminium substituted calcium hexaferrite. Heliyon 6:4–9. https://doi.org/10.1016/j.heliyon.2020.e03186
9. Ponkumar R, Nesaraj AS, Arunkumar M (2023) Facile chemical synthesis and fundamental characterization of calcium-doped bismuth oxide nanoparticles for technological applications. Integr Ferroelectr 232:105–113. https://doi.org/10.1080/10584587.2023.2173445
10. Korichi K, Doumi B, Mokaddem A, Sayede A, Tadjer A (2020) Ferromagnetism, half-metallicity and spin-polarised electronic structures characterisation insights in Ca1−xTixO. Philos Mag 100:1172–1190. https://doi.org/10.1080/14786435.2020.1723812
11. Liu X, Chang CF, Rata AD, Komarek AC, Tjeng LH (2016) Fe_3O_4 thin films: controlling and manipulating an elusive quantum material. NPJ Quantum Mater 1:1–22. https://doi.org/10.1038/npjquantmats.2016.27
12. Ansar MZ, Atiq S, Riaz S, Naseem S, Ramay SM, Mahmood A (2015) Synthesis and investigating the magnetic properties of magnetite nanocrystallites for drug delivery. Mod Phys Lett B 29:1–8. https://doi.org/10.1142/S0217984915501225

13. Wang X, Liao Y, Zhang D, Wen T, Zhong Z (2018) A review of Fe_3O_4 thin film: synthesis, modification and applications. J Mater Sci Technol 34:1259–1272. https://doi.org/10.1016/j.jmst.2018.01.011
14. Bohra M, Agarwal N (2019) Nanostructured magnetite thin films: an avenue for spintronics. Front Nanosci 14:121–140. https://doi.org/10.1016/B978-0-08-102572-7.00005-2
15. Son S, Taheri M, Carpenter E, Harris VG, McHenry ME (2002) Synthesis of ferrite and nickel ferrite nanoparticles using radio-frequency thermal plasma torch. J Appl Phys 91:7589–7591. https://doi.org/10.1063/1.1452705
16. Pirouzfar A, Seyyed Ebrahimi SA (2014) Optimization of sol-gel synthesis of $CoFe_2O_4$ nanowires using template assisted vacuum suction method. J Magn Magn Mater 370:1–5. https://doi.org/10.1016/j.jmmm.2014.06.058
17. Jeseentharani V, George M, Jeyaraj B, Dayalan A, Nagaraja KS (2013) Synthesis of metal ferrite (MFe_2O_4, M = Co, Cu, Mg, Ni, Zn) nanoparticles as humidity sensor materials. J Exp Nanosci 8:358–370. https://doi.org/10.1080/17458080.2012.690893
18. Raut SD, Dahotre SG, Singh LN, Shinde VS, Jadhav SN (2023) Materials today: proceedings exploring the structural and magnetic characteristics of ferrous ferric oxide nanoparticles doped with Ca2+ ions for nanoscale devices. Mater Today Proc. https://doi.org/10.1016/j.matpr.2023.09.124
19. Khan S, Riaz S, Arshad F, Azhar M, Ahmad N, Noor H, Atiq S, Naseem S (2022) Role of Ca doping on oxygen vacancy production in modulating dielectric, ferroelectric and magnetic polarization in BaTiO3 thin films. J Mater Res Technol 16:993–1007. https://doi.org/10.1016/j.jmrt.2021.11.148
20. Serraon ACF, Del Rosario JAD, Abel Chuang PY, Chong MN, Morikawa Y, Padama AAB, Ocon JD (2021) Alkaline earth atom doping-induced changes in the electronic and magnetic properties of graphene: a density functional theory study. RSC Adv 11:6268–6283. https://doi.org/10.1039/d0ra08115a
21. Devi SM, Nivetha A, Prabha I (2019) Superparamagnetic properties and significant applications of iron oxide nanoparticles for astonishing efficacy—A review. J Supercond Nov Magn 32:127–144. https://doi.org/10.1007/s10948-018-4929-8
22. Kushwaha N, Kumar S, Kushwaha R, Jain P (2013) Spintronics: combination of nanotechnology & superconductivity 3:17–22
23. Damerio S, Kaverzin AA, Ocelík V, Hoogeboom GR, van Wees BJ, Noheda B (2022) Antiferromagnetic ordering and uncoupled spins in $CaFe_2O_4$ thin films probed by spin hall magnetoresistance. Adv Electron Mater 8. https://doi.org/10.1002/aelm.202100963
24. Zia J, Farhat SM, Aazam ES, Riaz U (2021) Highly efficient degradation of metronidazole drug using $CaFe_2O_4$/PNA nanohybrids as metal-organic catalysts under microwave irradiation. Environ Sci Pollut Res 28:4125–4135. https://doi.org/10.1007/s11356-020-10694-3
25. Yunas J, Sulaiman NH, Ghazali MJ (2018) Comparative study of the calcium ferrite nanoparticles ($CaFe_2O_4$-NPs) synthesis process. In: IEEE international conference semiconductor electronics. Proceedings, ICSE. 2018-August, pp 101–103. https://doi.org/10.1109/SMELEC.2018.8481301
26. Aravazhi RSS, Selva C, Senthil S (2019) Tuning of ferrites—(CoxFe3–xO4) nanoparticles by co-precipitation technique. SN Appl Sci 1:1–11. https://doi.org/10.1007/s42452-019-0244-7
27. Thapa D, Palkar VR, Kurup MB, Malik SK (2004) Properties of magnetite nanoparticles synthesized through a novel chemical route. Mater Lett 58:2692–2694. https://doi.org/10.1016/j.matlet.2004.03.045
28. Zhao FY, Li YL, Li LH (2014) Preparation and characterization of magnetite nanoparticles. Appl Mech Mater 618:24–27. https://doi.org/10.4028/www.scientific.net/AMM.618.24
29. Rashid H, Mansoor MA, Haider B, Nasir R, Abd Hamid SB, Abdulrahman A (2020) Synthesis and characterization of magnetite nano particles with high selectivity using in-situ precipitation method. Sep Sci Technol 55:1207–1215. https://doi.org/10.1080/01496395.2019.1585876
30. Parmar C, Verma R, Ghosh A, Modak SS, Kane SN (2019) Sol-gel auto-combustion synthesis of magnetite and its characterization via X-ray diffraction. AIP Conf Proc 2142:0–5. https://doi.org/10.1063/1.5122595

31. Thavarani M, Robert MC, Pavithra N, Saravanan R, Kannan YB, Prasath SB (2022) Effect of Ca2+ doping on the electronic charge density and magnetic properties of $ZnFe_2O_4$ spinel ferrites. J Mater Sci Mater Electron 33:4116–4131. https://doi.org/10.1007/s10854-021-07605-8
32. Senthil VP, Gajendiran J, Raj SG, Shanmugavel T, Ramesh Kumar G, Parthasaradhi Reddy C (2018) Study of structural and magnetic properties of cobalt ferrite ($CoFe_2O_4$) nanostructures. Chem Phys Lett 695:19–23. https://doi.org/10.1016/j.cplett.2018.01.057
33. Jumril Y, Noor Humam S, Mariyam Jameelah G (2018) Synthesis of calcium ferrite nanoparticles ($CaFe_2O_4$-NPs) using auto-combustion method for targeted drug delivery. Key Eng Mater 775 KEM:115–119. https://doi.org/10.4028/www.scientific.net/KEM.775.115
34. Fahlepy MR, Tiwow VA (2018) Subaer: characterization of magnetite (Fe_3O_4) minerals from natural iron sand of Bonto Kanang Village Takalar for ink powder (toner) application. J Phys Conf Ser 997. https://doi.org/10.1088/1742-6596/997/1/012036
35. Venkatesan K, Rajan Babu D, Kavya Bai MP, Supriya R, Vidya R, Madeswaran S, Anandan P, Arivanandhan M, Hayakawa Y (2015) Structural and magnetic properties of cobalt doped iron oxide nanoparticles prepared by solution combustion method for biomedical applications. Int J Nanomed 10:189–198. https://doi.org/10.2147/IJN.S82210
36. Tadesse A, Ramadevi D, Hagos M, Battu G, Basavaiah K (2018) Synthesis of nitrogen doped carbon quantum dots/magnetite nanocomposites for efficient removal of methyl blue dye pollutant from contaminated water. RSC Adv 8:8528–8536. https://doi.org/10.1039/c8ra00158h
37. Thummer KP, Pandya MP, Jani KH, Modi KB, Joshi HH (2004) Microscopic and macroscopic magnetic properties of the MgAl xCr xFe 2–2xO 4 spinel ferrite system. Mater Sci 40:102–112. https://doi.org/10.1023/B:MASC.0000042792.25335.14
38. Slavov L, Abrashev MV, Merodiiska T, Gelev C, Vandenberghe RE, Markova-Deneva I, Nedkov I (2010) Raman spectroscopy investigation of magnetite nanoparticles in ferrofluids. J Magn Magn Mater 322:1904–1911. https://doi.org/10.1016/j.jmmm.2010.01.005
39. Javed Iqbal M, Naeem Ashiq M, Hussain Gul I (2010) Physical, electrical and dielectric properties of Ca-substituted strontium hexaferrite (SrFe12O19) nanoparticles synthesized by co-precipitation method. J Magn Magn Mater 322:1720–1726. https://doi.org/10.1016/j.jmmm.2009.12.013
40. Sulaiman NH, Ghazali MJ, Majlis BY, Yunas J, Razali M (2015) Superparamagnetic calcium ferrite nanoparticles synthesized using a simple solgel method for targeted drug delivery. Biomed Mater Eng 26:S103–S110. https://doi.org/10.3233/BME-151295
41. Tomar MS, Singh SP, Perales-perez O, Guzman RP (2005) Synthesis and magnetic behavior of nanostructured ferrites for spintronics. Microelectronics J 36:475–479. https://doi.org/10.1016/j.mejo.2005.02.086
42. Kershi RM, Ali FM, Sayed MA (2018) Influence of rare earth ion substitutions on the structural, optical, transport, dielectric, and magnetic properties of superparamagnetic iron oxide nanoparticles. J Adv Ceram 7:218–228. https://doi.org/10.1007/s40145-018-0273-5
43. Zhang W, He B, Zou Z, Xie Z, Jiang X, Mao Z, Xu C, Feng M (2023) Role of Ca2+ doping on the structure, magnetic, and magnetocaloric properties of La0.75Gd0.05Sr0.2−xCaxMnO3 (x = 0, 0.05, 0.075, and 0.10) perovskite manganites. J Mater Sci Mater Electron 34:1–13. https://doi.org/10.1007/s10854-023-10311-2
44. Li Q, Kartikowati CW, Horie S, Ogi T, Iwaki T, Okuyama K (2017) Correlation between particle size/domain structure and magnetic properties of highly crystalline Fe_3O_4 nanoparticles. Sci Rep 7:1–4. https://doi.org/10.1038/s41598-017-09897-5
45. Sun S, Zeng H, Robinson DB, Raoux S, Rice PM, Wang SX, Li G (2004) Monodisperse MFe_2O_4 (M = Fe Co, Mn) nanoparticles. J Am Chem Soc 126:273–279. https://doi.org/10.1021/ja0380852
46. Ismail MM, Jaber NA (2018) Structural analysis and magnetic properties of lithium-doped Ni-Zn ferrite nanoparticle. J Supercond Nov Magn 31:1917–1923. https://doi.org/10.1007/s10948-017-4428-3

Sound Absorption Performance of 3D-Printed PLA/Wood Composites

Hiral H. Parikh, Hiren Rana, and Axit Raj

Abstract This article delves into the acoustic properties of Wood-filled PLA composite, specifically filament made using an extrusion process and specimen prepared with 3D printing techniques. To assess the sound absorption capabilities of this novel material, the study utilized the impedance tube method. The results of the tests indicated that the newly developed material exhibited subpar sound absorption characteristics at low and high frequency range when compared with medium frequency range. This deficiency can be attributed to the material's dense and closely packed structure. It is worth highlighting that the infill percentage and surface roughness emerged as the primary determinant influencing the acoustic properties.

Keywords Sound absorption coefficient · Fiber reinforced polymer matrix composite · Extrusion process · 3D printing

1 Introduction

In recent times, there has been a growing emphasis on the utilization of natural fibers as a sustainable alternative to mineral and synthetic fiber composites. These fibers have gained recognition in diverse industries, including automotive, construction, and various engineering applications, where they serve as reinforcement materials. While it is true that their mechanical properties may not match those of synthetic fibers, natural fibers offer several compensatory advantages. They are inherently biodegradable, cost-effective, readily available, possess a high strength-to-weight ratio, and contribute to reduced CO_2 emissions [1]. Numerous researchers have undertaken investigations into the mechanical and tribological properties of various natural fibers and their composites [2].

3D printing, also known as additive manufacturing (AM), creates complex parts layer by layer using polymers, metals, ceramics, and composites [3]. It's widely used in various industries like aerospace, automotive, and biomedical due to its

H. H. Parikh (✉) · H. Rana · A. Raj
School of Engineering and Technology, Navrachana University, Vadodara, Gujarat 391410, India
e-mail: hiralp@nuv.ac.in

P. Bhingole et al. (eds.), *Advances in Materials Engineering*, Lecture Notes in Mechanical Engineering, https://doi.org/10.1007/978-981-97-7114-1_6

accuracy, speed, and cost-effectiveness. Polymer parts are common in 3D printing, using techniques like SLA, SLS, and FDM [4, 5]. However, pure polymer parts often lack strength and stiffness. To improve this, researchers have explored adding fibers for reinforcement, particularly continuous fibers, which enhance mechanical properties [6, 7]. This resolves the issue of poor mechanical performance in 3D-printed materials.

Li et al. [8], Ahn et al. [9], undertook the creation of 3D-printed components using pure polymer, specifically Acrylonitrile Butadiene Styrene (ABS). Their primary focus was on assessing the material's mechanical strength through rigorous testing. The outcomes of these tests unveiled that the tensile strength of the ABS material fell within the range of 65–72 MPa. Hiral et al. [10] have developed a wood PLA composite with the filament extrusion process identified its flexural properties and compared this property with the neat PLA. The test results revealed that the wood dust improve the flexural strength of the material.

In a notable study, Devadiya Bhat and Mahesh [11] integrated sugarcane fibers as a filler material within the concrete and conducted an investigation into the thermal stability of the resulting composite when subjected to elevated temperatures. In a parallel effort, El-Tayeb [12] pioneered the development of a composite material using sugarcane fibers combined with polyester resin. This composite underwent an assessment of its tribological properties, with a particular focus on comparing them to traditional glass fiber composites. The findings from this study revealed that sugarcane fiber composites exhibit promising potential as a substitute for glass fiber composites, showcasing notable improvements in friction and wear properties. Furthermore, Mahapatra and Chaturvedi [13] executed a strategic approach by aligning sugarcane fibers within a composite material. They subsequently conducted a rigorous evaluation of its resistance to abrasive wear, shedding light on yet another advantageous aspect of utilizing sugarcane fibers in composite applications.

Noise, often described as unwelcome and bothersome sound, presents a considerable threat to our well-being. With the continuous advancement of technology, urban expansion, and industrial growth, completely avoiding noise in our daily lives has become nearly impossible [14]. Excessive noise can lead to a diminished quality of life and adverse effects on human health. To address this issue, researchers are diligently working on various technologies, strategies, and materials. One way to control the noise is to use sound absorbing materials. Researchers are increasingly shifting their attention toward an alternative approach—the development of sound absorption panels. Yilmaz et al. have conducted research on various factors affecting sound absorption, including fiber porosity, size, and layer sequence [15], as well as the impact of heat treatment on multi-fiber nonwoven composites [16] and surface treatment on material sound absorption capabilities [17]. Several studies have also assessed the sound absorption properties of materials developed from recycled resources like tea leaf fibers, rubber waste, waste wool, and discarded polyester fibers, among others [18–20]. In line with this shift, many researchers are exploring the potential of natural fibers as materials for sound absorption.

Traditionally, materials like glass wool, cotton, polyethylene foam, and rock wool have been used for sound absorption. However, these substances come with drawbacks such as skin and eye irritation, as well as respiratory issues [21]. There is a demand for the new material which can replace these conventional materials. In line with this in the present article, Wood/PLA composite is developed with the 3D printing technique and checked for its possible sound absorption properties.

2 Materials and Testing

This study employed wood dust particles as filler materials in conjunction with PLA polymer. The filament production process utilized a single screw extrusion method. To enhance the flexibility of the filament and mitigate its brittleness, glycerin was incorporated as a plasticizer. The result of this process was the successful production of a PLA matrix filament reinforced with 20 wt. % wood dust. Fused deposited 3D printing technique has been used to print the material. Design software is used to create the geometry of the material, .stl file is used to print the material on 3D printer. In this process, material is deposited on the bed layer by layer till the final shape is achieved. The parameters set for the printing are listed in Table 1. For the acoustic testing, specimens were manufactured using the fused deposition 3D printing technique. As the glass transition temperature of the PLA is 220 °C, to ensure the proper material flow from nozzle 270 °C has been set. Bed temperate set at 70 °C to ensure the proper adhesion of first later with the bed. Also to reduce the chances of voids and porosity full infill with honeycomb structure is used. Specimen for the sound absorption has been printed using ASTM E 1050-12 standards as shown in the Fig. 1 Specimens measuring 29.5 mm for the lower frequency range and 99.5 mm for the higher frequency range.

The printing and the process parameters were set on the 3D printer for printing is shown in Table 1.

Table 1 3D printing parameters for specimen development

Printer bed temperature	70 °C
Nozzle temperature	270 °C
Infill pattern	Full honey comb
Printing speed	60 mm/min
Infill percentage	100%
Top and bottom layer thickness	2 mm

Fig. 1 3D-printed wood/PLA specimen for acoustic testing

3 Results and Discussions

The samples were tested for the sound absorption coefficient (SAC). When the sound waves pass through the porous material, it travels through the air cavities and lose its energy as heat due to vibration. The experimental setup involved the use of an impedance tube system, as illustrated in Fig. 2 To carry out investigations, both the standard large tube setup (Type 4206) and the standard small tube setup were used.

Several factors, including surface roughness, infill percentage, and material thickness, influence the acoustic coefficient. Figure 3 illustrates that the developed material tested for low frequency range starting with 125 Hz to high frequency range of 6300 Hz. Figure 4 shows that the developed material exhibits subpar acoustic characteristics at both lower frequencies (125–500 Hz) and higher frequencies (2500–6300 Hz). Conversely, it demonstrates favorable acoustic performance within the medium frequency range (500–2000 Hz).

Increasing the infill percentage results in a denser material, enabling it to absorb more sound waves. The denser structure enhances its ability to absorb sound energy compared to materials with lower infill percentages, which create porous structures that reduce sound absorption. Infill pattern affect the acoustic properties. Honeycomb structure increases the air pockets which may lead to the more absorption coefficient.

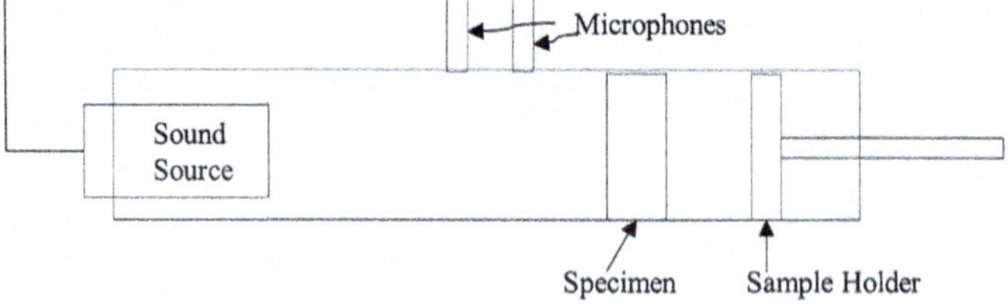

Fig. 2 Impedance tube setup for SAC measurement

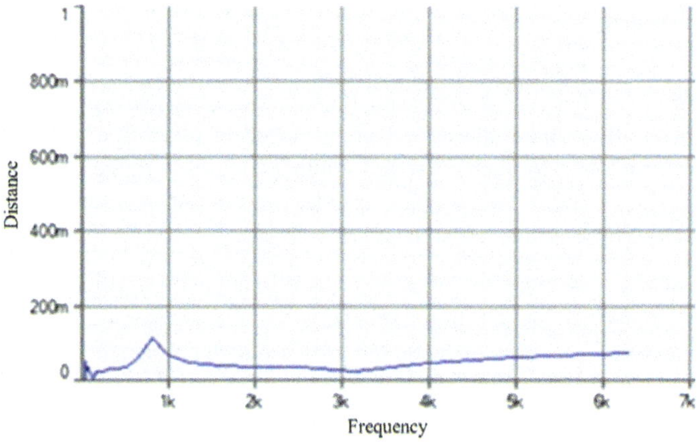

Fig. 3 Variation in sound absorption coefficient along the impedance tube length

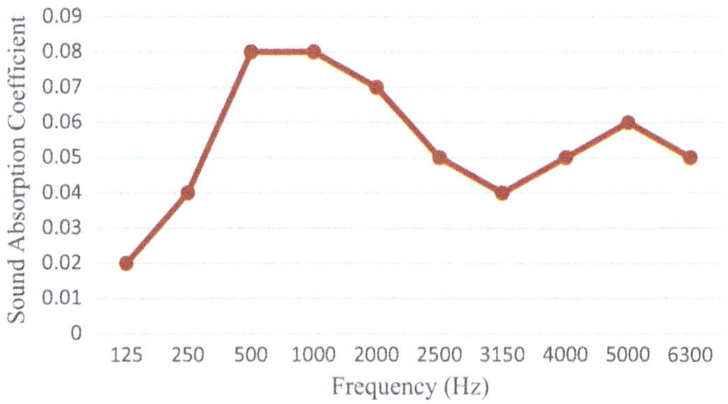

Fig. 4 Sound absorption coefficient for wood/PLA composite

Experimenting with different infill patterns can help optimize acoustic properties. This result is in accordance with the research studies conducted by Sebastian et al. [22] and Vignesh [23], which evaluated the effect of infill percentage on the SAC of composite material. The material's smooth surface finish causes sound waves to reflect, resulting in a reduced absorption coefficient. Conversely, a rough outer surface allows more sound waves to pass through the air cavities. Printing speed and temperature settings can influence the bonding between layers and the overall density of the printed object. Adjusting these parameters can impact sound absorption. The thickness of the 3D-printed object can also play a role. Thicker objects may offer better sound absorption, especially at lower frequencies.

4 Conclusion

An effort was made to develop composite filament using extrusion process and printed by 3D printer. 3D-printed structure was produced to check the sound absorption coefficient from lower frequency range to higher frequency range with the full infill percentage, infill pattern honeycomb structure, and top and bottom layer thickness 2 mm. The findings from test results are as follows:

- The wood/PLA composite developed is well-suited for applications within the medium frequency range (500–2000 Hz).
- The specimen shows poor acoustic properties at lower range, this can be enhanced by increasing the specimen's thickness.
- The produced specimen shows a mirror surface finish which reflects the sound waves. Conversely, for improved properties in the higher frequency range, it is advisable to augment the outer surface roughness.

Funding This work was in part supported by the Khoj Summer Grant of Navrachana University (Grant No: KHOJ/2022-23/111).

References

1. Adil A, Hiral P, Mohamed R (2022) A review on chicken feather fiber (CFF) and its application in composites. J Nat Fiber 19(16):12565–12585
2. Hiral P (2023) Tribology of plant-based natural fiber reinforced polymer matrix composites—A short review. J Nat Fibers 20(1):1–15
3. Levy G, Schindel R, Kruth J (2003) Rapid manufacturing and rapid tooling with layer manufacturing (LM) technologies, state of the art and future perspectives. CIRP Annals- Manuf Technol 52(2):589–609
4. Ahn S, Montero M, Odell D (2002) Anisotropic material properties of fused deposition modeling ABS. Rapid Prototyp J 8:248–257
5. Tymrak B, Kreiger M, Pearce J (2014) Mechanical properties of components fabricated with open-source 3-D printers under realistic environmental conditions. Mater Des 58:242–246
6. Goodridge R, Shofner M, Hague R (2011) McClelland: processing of a polyamide- 12/carbon nanofibre composite by laser sintering. Polym Test 30(1):94–100
7. Subramanian K, Vail N, Barlow J, Marcus H (1995) Selective laser sintering of alumina with polymer binders. Rapid Prototyp J 1(2):24–35
8. Li L, Sun Q, Bellehumeur C, Gu P (2002) Composite modeling and analysis for fabrication of FDM prototypes with locally controlled properties. J Manuf Process 4(2):129–141
9. Ahn SH, Lee C, Jeong W (2004) Development of translucent FDM parts by post- processing. Rapid Prototyp J 10(4):218–224
10. Hiral H, Sagar C, Vijay C, Adil K, Jitendra M (2023) Flexural response of 3D printed wood dust reinforced polymer composite, Mater Today: Proc (2023). (In Printing)
11. Devadiga D, Bhat K, Mahesha G (2020) Sugarcane bagasse fiber-reinforced composites: recent advances and applications. Cogent Eng 7(1):1823159
12. El-Tayeb N (2008) A study on the potential of sugarcane fibers/polyester composite for tribological applications. Wear 265(1–2):223–235
13. Mahapatra S, Chaturvedi V (2009) Modeling and analysis of abrasive wear performance of composites using Taguchi approach. Int J Eng, Sci Technol 1(1):123–35

14. Zhu X, Kim B, Wang Q, Wu Q (2013) Recent advances in the sound insulation properties of bio based materials. Bioresources 9(1):1764–1786
15. Yilmaz ND, Banks-Lee P, Powell NB (2011) Effects of porosity, fiber size, and layering sequence on sound absorption performance of needle-punched nonwovens. J Apply Polym Sci 121:3056–3069
16. Yilmaz ND, Powell NB, Banks-Lee P (2013) Multi-fiber needle-punched nonwoven composites: effects of heat treatment on sound absorption performance. J Ind Text 43:231–246
17. Yilmaz ND, Powell NB, Banks-Lee P (2012) Hemp-fiber based nonwoven composites: effects of alkalization on sound absorption performance. Fibers Polym 13:915–922
18. Ersoy S, Küçük H (2009) Investigation of industrial tea-leaf-fibre waste material for its sound absorption properties. Appl Acoust 70:215–220
19. Patnaik A, Mvubu M, Muniyasamy S (2015) Thermal and sound insulation materials from waste wool and recycled polyester fibers and their biodegradation studies. Energy Build 92:161–169
20. Zhou X, Jang S, Yan X (2014) Damping acoustic properties of reclaimed rubber/sevenhole hollow polyester fibers composite materials. J Compos Mater 48:3719–3726
21. Lipworth L, La Vecchia C, Bosetti C (2009) Occupational exposure to rock wool and glass wool and risk of cancers of the lung and the head and neck: a systematic review and Meta-analysis. J Occup Environ Med 51:1075–1087
22. Zaharia S, Pop M, Cosnita M, Croitoru C, Matei S, Spîrchez C (2023) Sound absorption performance and mechanical properties of the 3D-printed bio-degradable panels. Polymers 15(18):3695
23. Vignesh S, Se Yong E, Sivakumar S, Azma P, Desmond D, Hafizah K (2022) Effect of thickness and infill density on acoustic performance of 3D printed panels made of natural fiber reinforced composites. J Nat Fibers 19(13):7132–7140

A Review of Biomedical Applications of Zirconia-Based Nanomaterials

Mukund Kumar, Shibashis Halder, and Arvind Kumar

Abstract Over the past decade, significant advancements and growth have been observed within the scientific field across various domains of Nano research. Zirconia NPs garnered significant interest across various disciplines owing to their diverse array of uses. Zirconia (ZrO_2) is a ceramic compound composed of the transition metal zirconium. Due to their effective curative biomedical domains, zirconium nanoparticles are used in tissue engineering, antibacterial, anticancer, and antioxidant medication. Zirconium oxide is a metal oxide that exhibits the characteristics of economic feasibility, nonhazardous properties, and sustainability, hence providing a wide range of prospective uses. The ZrO_2 nanocomposite demonstrates a notable capacity for biosensing, particularly in the detection of glucose and other living entities, displaying a high level of sensitivity. This review article encompasses an examination of the biological applications of nanoparticles based on zirconia.

Keywords Nanoscience · Zirconia · Nanomaterial · Biomedical applications

1 Introduction

Zirconium is an extremely corrosion-resistant transition metal with superior thermal, mechanical, and catalytic properties [1]. Zirconium dioxide, or zirconia is a ceramic material made from zirconium (Zr). Zirconia is sometimes referred to as ceramic steel due to its high toughness, low abrasiveness, low friction resistance, and high melting

M. Kumar (✉)
Department of Mechanical Engineering, National Institute of Technology, Jamshedpur, Jharkhand 831014, India
e-mail: 2021rsme018@nitjsr.ac.in

S. Halder
Department of Chemistry, T.N.B. College, T.M. Bhagalpur University, Bhagalpur, Bihar 812007, India

A. Kumar
Department of Botany, T.N.B. College, T.M. Bhagalpur University, Bhagalpur, Bihar 812007, India

P. Bhingole et al. (eds.), *Advances in Materials Engineering*, Lecture Notes in Mechanical Engineering, https://doi.org/10.1007/978-981-97-7114-1_7

point. ZrO_2 has substantial antibacterial, anti-oxidant, and oxidative capabilities; it also has substantial specific area; is biocompatible; and resistant to high temperatures; all of these qualities make it a material of tremendous technological importance [2]. It exhibits a diverse array of potential applications and is financially viable, safe, and sustainable [3]. Because of its great strength, chemical inertness, and capacity to withstand extreme temperatures, ZrO_2 is a promising material for use in cutting-edge ceramics [4]. ZrO_2's transformation-toughening mechanism gives exceptional mechanical properties, including fracture toughness and fracture strength [5].

The benefits of zirconia-based materials have various cutting-edge applications, such as photocatalysts, adsorption, anti-corrosion coatings, supercapacitors, lithium-ion batteries, sensors, water splitting, solar cells, and more [6–9]. There is constantly more work being done on nanoparticles based on ZrO_2. ZrO_2-based materials are gaining popularity because of their excellent features such as high refractive index, chemical and photoelectron stability, and high hardness [10]. Because it is more biocompatible and has greater mechanical strength than traditional ceramics, ZrO_2 ceramic is increasingly being used in the medical engineering industry [11].

2 Biomedical Applications of ZrO_2 Nanoparticles

Zirconium dioxide nanoparticles are being used for several purposes, including their use as anti-microbial substances, nanopowder fillers, raw materials for sintering, nanocoatings, and agents with potential anticancer and antioxidant properties. The utilization of ZrO_2 nanoparticles as hybrid materials has gained significant interest in biomedical applications particularly in areas such as bone prostheses, tissue-engineering scaffolds, and drug delivery systems, among others. This is primarily due to their advantageous bionics, biocompatibility, and mechanical properties [12]. Materials based on ZrO_2 have found successful applications in a variety of biomedical contexts, including packaging and preserving meat, dental work, artificial scaffolds, and more [13–16]. The bio-compatibility of ZrO_2 at the tissue level was observed similar to that of titanium. The proliferation and differentiation of osteoblasts in culture on zirconia are unaffected. ZrO_2 exhibits bioinert characteristics as it does not form any biological bonds with the tissues it is implanted into after implantation.

Zirconium dioxide nanocrystals (ZrO_2 NCs) are commonly synthesized in the context of biomedical applications with the aim of enhancing the inherent capabilities of ZrO_2 and incorporating supplementary functionalities, such as anti-cancer, anti-diabetic, and antibacterial properties. As a result, ZrO_2 NPs have found extensive utilization in a wide variety of applications (Fig. 1). Antioxidants are a group of substances that work together to neutralize the harmful effects of free radicals on the body by removing them from circulation [17]. Antioxidant free radical scavenging with 2,2-azino-bis-3-ethylbenzothiazoline-6-sulfonic acid (ABTS) provided evidence of the ZnO–ZrO_2 heterojunction's antioxidant activity, which was shown to be greater than that of ascorbic acid.

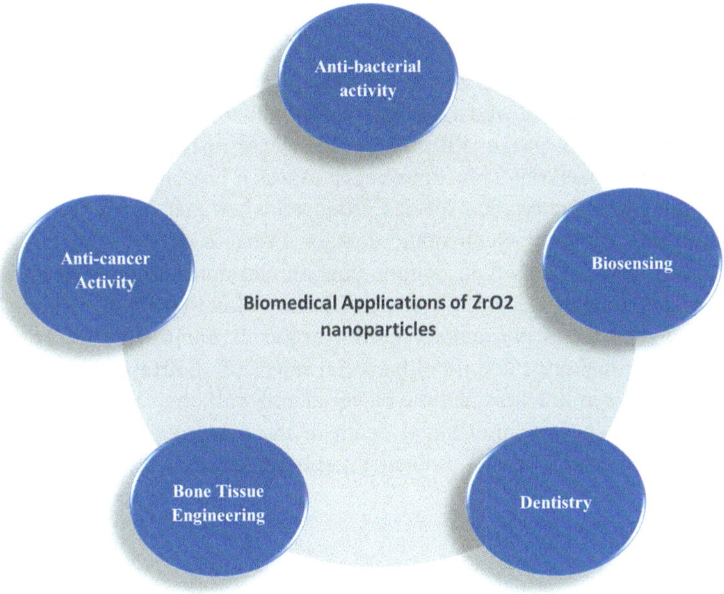

Fig. 1 Various biomedical applications of ZrO_2 NPs

2.1 Anti-Bacterial Activity

As a result of their low cost and less harmful nature, ZrO_2-based nanomaterials have been identified as a novel technique in certain biomedical applications, including dentistry and drug delivery. Zirconia-based constructs have been studied for their potential biological applications due to their composition [18–21]. Zirconia nanoparticles between 50 and 100 nm in size were generated from *aloe vera* leaves and their antibacterial and antifungal capabilities were studied by Gowri et al. [20]. ZrO_2 Nanoparticles have been shown to be effective against *Staphylococcus aureus, Escherichia coli, Fusarium moniliforme, and Fusarium graminearum* [22]. Zirconia Nanoparticles obstruct the *Escherichia coli, Bacillus subtilis,* and *Salmonella typhi* growth [23]. The researchers employed the *Sargassum wightii* as a means to synthesize zirconia nanoparticles. For treating wastewater, a polyaniline zirconia (PANI-ZrO_2) composite was applied [24]. The resulting composite was effective as a phosphate adsorbent and antibacterial and corrosion agents. Zirconia nanoparticles were found to have antibacterial action against Gram-negative and Gram-positive bacteria by the researchers [25].In their investigation, Derbalah et al. [26] documented the presence of antifungal properties targeting *Rhizoctonia solani*, a well-known pathogen associated with the development of root rot in cucumber plants. The manufacture of biogenic ZrO_2 nanoparticles through the utilization of Enterobacter sp. [27]. ZrO_2 nanoparticles, synthesized using the sol-gel technique, were tested for their antimicrobial property toward *S. aureus* and *E. coli* with the findings showing that the nanoparticles hindered both bacterial growth with inhibition regions of 25 mm and 27 mm, respectively [28]. Both gram+ and gram-bacteria are significantly inhibited by ZrO_2-containing polymer nanocomposites [29]. At a concentration of 200 ng/mL, ZrO_2 NPs showed a maximal inhibition zone of 18 mm against *A. niger* and 19 mm against *S. aureus* [30]. Zr4+ ion release restricts the development and dissemination of bacteria. Zr4+ ion release causes a breach in the cell membrane. The oxidizing reactive oxygen species produced when metal ions are allowed into a bacterial cell stimulate electrostatic contact. In bacteria, this kind of damage to protein and DNA disrupts their communication and nutrition networks, ultimately leading to cell death [31]. ZrO_2 nanoparticles are effective antimicrobials because reactive oxygen species, particularly hydroxyl radicals, damage the outer membranes of bacteria, leading to the demise of bacterial cells [32]. Because of the electrostatic interaction between Zr+ ions and the bacterial cell wall, the bacterial cell wall may be ruptured, leading to the eventual death of the bacterium. *S. aureus* and other Gram-positive bacteria have a protective peptidoglycan coating that prevents NPs from entering the cell [33].

2.2 Anticancer Activity

ZrO_2 NPs are outstanding materials for use in biological systems because of many advantages like high thermal stability, biocompatibility, and economical production costs. Several cancer cell lines have been tested for susceptibility to ZrO_2 NPs' anticancer effects [34]. A green method of manufacturing a $ZnO/ZrO_2/rGO$ nanocomposite with ginger rhizome extract showed promising antitumor efficacy in humans. The cytocompatibility of $ZnO/ZrO_2/rGO$ nanocomposites was found superior in human lung fibroblast cells, breast epithelial cells, and normal cell lines. Additionally, these nanocomposites also showed improved anticancer activity in human breast cancer cells and lung cancer cells [35]. Using a one-pot solvothermal green synthesis technique, ZrO2/rGO NCs were produced from *Andrographis paniculate* aqueous leaf extract; these nanocomposites exhibited outstanding anti- cancer properties against A549 and HCT 116 cancer cell lines without damaging healthy cells [36]. The efficacy of ZrO_2 nanoparticles in inhibiting cancer growth has been established through their production using leaf extract of Eucalyptus globulus as a highly effective reducing and capping agent. These nanoparticles have shown promising results in suppressing the growth of several cell lines, including HCT-116 colon cancer and A-549 lung cancer [37].

2.3 Bone Tissue Engineering

Zirconia (ZrO_2) has attracted significant attention as a bioceramic material for biomimetic scaffolds due to its notable mechanical strength, remarkable chemical stability, and extraordinary biocompatibility. Zirconium dioxide has found extensive application in the domain of tissue engineering for bone due to its exceptional characteristics. It is commonly employed in various forms, such as films or coatings on various metallic implants, bone cement, and bone graft substitutes of bone graft. ZrO_2 is therefore widely regarded as a superior bio-ceramic for use in repairing bones. Due to their outstanding toughness, low sintering temperature, and resistance to wear, zirconia-based nanocomposites are increasingly being used as bone tissue substitutes [38]. After being implanted, bioinert ceramic zirconium oxide exhibits only physical connection with the adjacent tissue, lacking any chemical or biological bonding [39]. Due to its non-degradable nature, ZrO_2 exhibits potential as a durable scaffold material for persistent prostheses. This material offers mechanical reinforcement for tissues and creates a conducive environment for cellular infiltration, waste elimination, nutrient transportation, and the formation of new tissue, among other functions [40]. Yttria-stabilized zirconia (Y-ZrO_2) is a biocompatible, and anti-corrosive, and stable material that is ideal for bone regeneration-based applications in vivo [41]. The incorporation of ZrO_2 in the ZrO_2/b-TCP scaffold significantly enhanced the adhesion and proliferation of osteocyte cells [42]. Composite scaffolds exhibited biocompatibility with bone marrow stem cells, as they promote cell development and proliferation and showed no signs of toxicity when in touch with these cells [43].

2.4 Dentistry

Due to its superior biomechanical qualities, such as high durability, toughness, fatigue resistance, and fracture strength, zirconium dioxide has found widespread use in the dental industry. Additionally, it is highly regarded for its biological compatibility, outstanding resistance to wear, and natural coloration, making it a valuable material in dental applications. With the growing demand for cosmetic dental procedures, ZrO_2 has gained significant popularity as a material choice. ZrO_2 has been the subject of investigation for its potential applications in dentistry due to its ability to securely attach to tissues around it without forming biological or chemical bonds [44]. It is commonly utilized as a biomaterial for dental implants and dental crowns. Zirconia-based ceramics have been found to exhibit excellent chemical inertness, making them highly desirable for promoting strong cell adhesion. Furthermore, no adverse systemic reactions occurred with the use of these materials. Additionally, it is worth noting that there have been no reported instances of allergic reactions associated with ZrO_2. The growing interest in utilizing AI in ZrO_2 therapy and research stems from its capacity to evaluate data and uncover connections between events that appear unrelated [45]. The Ti–Zr alloy demonstrates notable corrosion resistance as a result of the development of ZrO_2 and TiO_2 on its surface, which collectively contribute to this effect [46]. ZrO_2 nanoparticles coatings have the potential to enhance the external strength of teeth and potentially contribute to an extended lifespan. It has been observed that ZrO_2 nanoparticles have the ability to interact with bacterial surface, leading to a reduction in their metabolic activity related to food consumption [47]. This interaction may contribute to the prevention of acid synthesis and enamel corrosion, which are known to be factors in tooth decay and cavity formation. The excellent mechanical properties of a yttria-stabilized t-zirconia polycrystal have led to its use in implants and dental applications in recent years [48]. Polymethyl methacrylate (PMMA), commonly used for repairing dentures, can really be enhanced by using zirconia-based nanoparticles. Zirconia nanoparticles have potent antimicrobial and anticarcinogenic effects on teeth [25]. In the course of their metabolic processes, bacteria may produce acids that erode tooth enamel. Coating with zirconia NPs, on the other hand, has shown exceptional effectiveness against oral infections with negatively charged cell walls and resilience to acid assaults. Yamada et al. [49] synthesized silver nanoparticles coated yttria-stabilized ZrO_2 and examined its antibacterial effectiveness against common mouth infections.

2.5 *Biosensing*

Nanomaterials based on zirconia are being researched for use in sensing because their advantageous electrochemical and physicochemical characteristics make them a good option. Because of its substantial bioactivity for biomolecules, zirconia is a promising material. ZrO_2 shows possibilities as a biosensing material because of its resistance in a variety of environmental situations [50]. Zirconium dioxide (ZrO_2) possesses several advantageous characteristics that make it a highly desirable material for electrode applications in substance sensing. These properties include environmental friendliness, thermal resistance, economical production, biological compatibility, and electrochemical reactivity [51]. Zirconium dioxide nanoparticles have received considerable attention in the field of bio-analytical purposes because of non-toxic nature, chemical stability, and strong affinity toward O_2-containing functional groups. Consequently, these nanoparticles have been employed in the development of modified electrodes, as demonstrated by studies conducted by many researchers [52, 53]. In addition to possessing notable characteristics, ZrO_2 nanoparticles have been the subject of research for the advancement of biosensor technology [54]. Nevertheless, ZrO_2 nanoparticles (NPs) do possess several drawbacks that limit their effectiveness in biosensing and electrochemical applications. These constraints include a propensity for excessive aggregation, reduced conductivity, and a lack of required functional groups. Hence, it is strongly advised to employ hybrid systems that effectively harness the whole capabilities of ZrO_2 nanoparticles thereby enhancing both biosensing and electrochemical properties [55].

3 Challenges

Pure ZrO_2 is unusable because to polymorphism, which causes it to expand in volume (by around 4.5–5%) upon cooling. To avoid this, stabilizing oxides including MgO, CaO, Sc_2O_3, CeO_2, and Y_2O_3 are utilized to preserve the oxides' metastable tetragonal phase [56].In addition, doping ZrO_2 with a noble metal or oxides of metal is a practical method for reducing the negative effects of ZrO_2 [57]. Because of their widespread application, NPs can be discharged into the environment in several forms, posing risks to a wide range of animals and ecosystems [58]. When ingested by humans, Nanoparticles have the potential to travel throughout the body, eventually settling in the kidneys, liver, and lungs. This could cause mutations in DNA, damage to mitochondria, and eventually cell death [59].

4 Future Perspectives

Although ZrO_2-based nanomaterials (NMs) have been widely used in recent years, little information is available on the evaluation of their hazardous behaviors in terms of causing cytotoxicity biological activity, and antioxidant properties. It is important to think about the surroundings prior to using ZrO_2 nanomaterials for any biomedical purpose since they can harm living things and the environment. It is advised that various biological functions and uses of ZrO_2 nanomaterials be investigated. This will allow them to do more than one thing, which will make them more useful in medicine. ZrO_2-based biosensors depend substantially on their selectivity for reliable and precise detection in biosensing applications. Because of the strong correlation between ZrO_2 nanoparticles' activity and their size, shape, and morphology, optimizing their activities in a wide range of biological applications may be possible through careful engineering. Adjusting these factors can help bring about the desired effects in biological contexts.

5 Conclusions

ZrO_2 is an excellent potential material for use in the biomedical and pharmaceutical industries since it poses minimal safety risks, has superior qualities, and can be put to a wide variety of uses. Nanomaterials based on ZrO_2 Nanoparticles have been exploited in the field of biomedicine recently due to their exceptional properties. These include non-toxicity, good biocompatibility, and low reactivity. Due to their exceptional biological features, nanomaterial based on ZrO_2 nanoparticles showed remarkable antimicrobial, anti-cancer, and antioxidant activity. ZrO_2 can be employed for precise biosensing of glucose and other biological species because of its great durability under environmental circumstances like pH, temperature, moisture, etc., and its excellent bioactivity for biomolecules. Future research into the biological applications of ZrO_2 NPs should include a thorough assessment of their toxicity.

References

1. Kaushal S, Kumari V, Singh PP (2023) Sunlight-driven photocatalytic degradation of ciprofloxacin and organic dyes by biosynthesized RGO–ZrO2 nanocomposites. Environ Sci Pollut Res 30:65602–65617
2. Keiteb AS, Saion E, Zakaria A, Soltani N (2016) Structural and optical properties of zirconia nanoparticles by thermal treatment synthesis. J Nanomater 1913609
3. Alagarsamy A, Chandrasekaran S, Manikandan A (2022) Green synthesis and characterization studies of biogenic zirconium oxide (ZrO2) nanoparticles for adsorptive removal of methylene blue dye. J Mol Struct (1247):131275
4. Han Z, Liu S, Qiu K, Liu J, Zou R, Wang Y, Zhao J, Liu F, Wang Y, Li L (2023) The enhanced ZrO2 produced by DLP via a reliable plasticizer and its dental application. journal of the mechanical behavior of biomedical materials (141):105751
5. Palmero P (2019) Zirconia-based composites for biomedical applications. In: Antoniac I (ed) Bioceramics and biocomposites: from research to clinical practice; Wiley-American Ceramic Society, Hoboken, NJ, USA, pp 57–85
6. Saeed K, Sadiq M, Khan I, Ullah S, Ali N, Khan A (2018) Synthesis, characterization, and photocatalytic application of Pd/ZrO2 and Pt/ZrO2. Appl Water Sci (8):60
7. Khan I, Zada N, Khan I, Sadiq M, Saeed K (2020) Enhancement of photocatalytic potential and recoverability of Fe3O4 nanoparticles by decorating over monoclinic zirconia. J Environ Health Sci Eng 18:1473–1489
8. Seynnaeve B, Folens K, Krishnaraj C, Ilic IK, Liedel C, Schmidt J, Verberckmoes A, Du Laing G, Leus K, Van Der Voort P (2021) Oxygen-rich poly-bisvanillonitrile embedded amorphous zirconium oxide nanoparticles as reusable and porous adsorbent for removal of arsenic species from water. J Hazard Mater (413):125356
9. Liu X, Cheng W, Yu Y, Jiang S, Xu Y, Zong E (2022) Magnetic ZrO2/PEI/Fe3O4 functionalized MWCNTs composite with enhanced phosphate removal performance and easy separability. Compos Part B: Eng (237):109861
10. Chakraborty D, Devi M, Das B, Dhar SS (2023) Core-shell assembly of ZrO2 nanoparticles with ionic liquid: a novel and highly efficient heterogeneous catalysts for Biginelli and esterification reactions. Environ Sci Pollut Res 30:13846–13861
11. Heng L, Kim JS, Tu JF, Mun SD (2020) Fabrication of precision meso-scale diameter ZrO2 ceramic bars using new magnetic pole designs in ultra-precision magnetic abrasive finishing. Ceram Int 46:17335–17346
12. Yuan Y, Wu Y, Suganthy N, Shanmugam S, Brindhadevi K, Sabour A, Alshiekheid M, Lan Chi NT, Pugazhendhi A, Shanmuganathan R (2022) Biosynthesis of zirconium nanoparticles (ZrO2 NPs) by phyllanthus niruri extract: characterization and its photocatalytic dye degradation activity. Food Chem Toxicol (168):113340
13. Sani IK, Geshlaghi SP, Pirsa S, Asdagh A (2021) Composite film based on potato starch/apple peel pectin/ZrO2 nanoparticles/microencapsulated zataria multiflora essential oil; investigation of physicochemical properties and use in quail meat packaging. Food Hydrocoll (117):106719
14. Zhang C, Jiang Z, Zhao L, Guo W, Gao X (2021) Stability, rheological behaviors, and curing properties of 3Y–ZrO2 and 3Y–ZrO2/GO ceramic suspensions in stereolithography applied for dental implants. Ceram Int 47:3344–13350

15. Shadianlou F, Foorginejad A, Yaghoubinezhad Y (2022) Hydrothermal synthesis ofzirconia-based nanocomposite powder reinforced by graphene and its application for bone scaffold with 3D printing. Adv Powder Technol (33):103406
16. JayakumarR, Ramachandran R, Sudheesh Kumar PT, Divyarani VV, Srinivasan S, Chennazhi KP, Tamura H, Nair SV (2011) Fabrication of Chitin–Chitosan/Nano ZrO2 composite scaffolds for tissue engineering applications. Int J Biol Macromol 274–280
17. Gul T, Saeed K, Ahmad S, Almehmadi M, Alsaiari AA, Alsharif A, Khan I (2023) Investigation of the photocatalytic and biological applications of iron oxide-indium oxide nanocomposite. Chem Pap 1:4547–4558
18. Tang S, Huang X, Chen X, Zheng N (2010) Hollow mesoporous zirconia nanocapsules for drug delivery. Adv Func Mater 20(15):2442–2447
19. Saraswathi VS, Santhakumar K (2017) Photocatalytic activity against azo dye and cytotoxicity on MCF-7 cell lines of zirconium oxide nanoparticle mediated using leaves of Lagerstroemia speciosa. J Photochem Photobiol B 169:47–55
20. Gowri S, Gandhi RR, Sundrarajan M (2014) Structural, optical, antibacterial and antifungal properties of zirconia nanoparticles by biobased protocol. J Mater Sci Technol 30(8):782–790
21. Balaji S, Mandal BK, Ranjan S, Dasgupta N, Chidambaram R (2017) Nano-zirconia-evaluation of its antioxidant and anticancer activity. J Photochem Photobiol B 170:125–133
22. Jalill RDA, Jawad MMHM, Abd AN (2017) Plants extracts as green synthesis of zirconium oxide nanoparticles. J Gen Environ Resour Conserv 5(1):6–23
23. Kumaresan M, Anand KV, Govindaraju K, Tamilselvan S, Kumar VG (2018) Seaweed Sargassum wightii mediated preparation of zirconia (ZrO2) nanoparticles and their antibacterial activity against gram positive and gram negative bacteria. Microb Pathog 124:311–315
24. Masim FCP, Tsai CH, Lin YF, Fu ML, Liu M, Kang F, Wang YF (2019) Synergistic effect of PANI–ZrO2 composite as antibacterial, anti-corrosion, and phosphate adsorbent material: synthesis, characterization and applications. Environ Technol 40(2):226–238
25. Derbalah A, Elsharkawy MM, Hamza A, El-Shaer A (2019) Resistance induction in cucumber and direct antifungal activity of zirconium oxide nanoparticles against Rhizoctonia solani. Pestic Biochem Physiol (157):230–236
26. Ahmed T, Ren H, Noman M, Shahid M, Liu M, Ali MA, Zhang J, Tian Y, Qi X, Li B (2021) Green synthesis and characterization of zirconium oxide nanoparticles by using a native Enterobacter sp. and its antifungal activity against bayberry twig blight disease pathogen Pestalotiopsis versicolor. NanoImpact (21):100281
27. Salih R, Al-Jadiri F, Rahma NM, Mershed K, Odeh AO, Osifo PO Neomagus HJPW, Ishihara A, Tominaka S, Nagai T (2021) Antimicrobial activity of zirconium oxide nanoparticles prepared by the sol-gel method. J Phys: Conf Ser (2114):012058
28. Kadhim KJ, Agool IR, Hashim A (2017) Effect of zirconium oxide nanoparticles on dielectric properties of (PVA-PEG-PVP) blend for medical application. J Adv Phys 6:187–190
29. Chau TP, Veeraragavan GR, Narayanan M, Chinnathambi A, Alharbi SA, Subramani B, Pikulkaew S (2022) Green synthesis of Zirconium nanoparticles using Punica granatum (pomegranate) peel extract and their antimicrobial and antioxidant potency. Environ Res (209):112771
30. Chelliah P, Wabaidur SM, Sharma HP, Majdi HS, Smait DA, Najm MA, Lai WC (2023) Photocatalytic organic contaminant degradation of green synthesized ZrO2 NPs and their antibacterial activities. Separations 10(3):156
31. Korde SA, Thombre PB, Dipake SS, Sangshetti JN, Rajbhoj AS, Gaikwad ST (2023) Neem gum (Azadirachta indicia) facilitated green synthesis of TiO2 and ZrO2 nanoparticles as antimicrobial agents. Inorgan Chem Commun (153):110777
32. Tabassum N, Kumar D, Verma D, Bohara RA, Singh MP (2021) Zirconium oxide (ZrO2) nanoparticles from antibacterial activity to cytotoxicity: a next-generation of multifunctional nanoparticles. Mater Today Commun (26):102156
33. Sumathi P, Renuka N, Subramanian R, Periyasami G, Rahaman M, Karthikeyan P (2023) Prospective in vitro A431 cell line anticancer efficacy of zirconia nanoflakes derived from Enicostemma littorale aqueous extract. Cell Biochem Funct

34. Ahamed M, Lateef R, Khan MM, Rajanahalli P, Akhtar MJ (2023) Biosynthesis, characterization, and augmented anticancer activity of ZrO2 doped ZnO/rGO nanocomposite. J Funct Biomater 14(1):38
35. Kanth Kadiyala N, Mandal BK, Kumar Reddy LV, Barnes CH, De Los Santos Valladares L, Sen D (2023) Efficient one-pot solvothermal synthesis and characterization of zirconia nanoparticle-decorated reduced graphene oxide nanocomposites: evaluation of their enhanced anticancer activity toward human cancer cell lines. ACS Omega 8(2):2406–2420
36. Balaji S, Mandal BK, Ranjan S, Dasgupta N, Chidambaram R (2017) Nano-zirconia—Evaluation of its antioxidant and anticancer activity. J Photochem Photobiol B: Biol (170):125–133
37. Weng W, Wu W, Hou M, Liu T, Wang T, Yang H (2021) Review of zirconia-based biomimetic scaffolds for bone tissue engineering. J Mater Sci 56:8309–8333
38. Almalki AH, Belal A, Farghali AA, Mahmoud R, Mustafa FM, Abd El-Mageed HR (2023) Electronic, mechanical, and thermal properties of zirconium dioxide nanotube interacting with poly lactic-co-glycolic acid and chitosan as potential agents in bone tissue engineering: insights from computational approaches. J Biomol Struct Dyn 1–13
39. Jin M, Sun N, Weng W, Sang Z, Liu T, Xia W, Yang H (2023) The effect of GelMA/alginate interpenetrating polymeric network hydrogel on the performance of porous zirconia matrix for bone regeneration applications. Int J Biol Macromol (242):124820
40. Sakthiabirami K, Kang JH, Jang JG, Soundharrajan V, Lim HP, Yun K. D, Park, SW (2021) Hybrid porous zirconia scaffolds fabricated using additive manufacturing for bone tissue engineering applications. Mater Sci Eng C (123):111950
41. Alizadeh A, Moztarzadeh F, Ostad SN, Azami M, Geramizadeh B, Hatam G, Ai J (2016) Synthesis of calcium phosphate-zirconia scaffold and human endometrial adult stem cells for bone tissue engineering. Artif Cells Nanomed Biotechnol 44(1):66–73
42. Jasemi A, Moghadas BK, Khandan A, Saber-Samandari S (2022) A porous calcium-zirconia scaffolds composed of magnetic nanoparticles for bone cancer treatment: fabrication, characterization and FEM analysis. Ceram Int 48(1):1314–1325
43. Teimouri A, Ebrahimi R, Emadi R, Beni BH, Chermahini AN (2015) Nano-composite of silk fibroin–chitosan/nano ZrO2 for tissue engineering applications: fabrication and morphology. Int J Biol Macromol 76:292–302
44. Singh J, Singh S, Verma A (2023) Artificial intelligence in use of ZrO2 material in biomedical science. J Electrochem Sci Eng 13(1):83–97
45. Fathima JB, Pugazhendhi A, Venis R (2017) Synthesis and characterization of ZrO2 nanoparticles-antimicrobial activity and their prospective role in dental care. Microb Pathog 110:245–251
46. Shahmohammadi M, Sun Y, Yuan JCC, Mathew MT, Sukotjo C, Takoudis CG (2022) In vitro corrosion behavior of coated Ti6Al4V with TiO2, ZrO2, and TiO2/ZrO2 mixed nanofilms using atomic layer deposition for dental implants. Surf Coat Technol (444):128686
47. Camposilvan E, Marro FG, Mestra A, Anglada M (2015) Enhanced reliability of yttria-stabilized zirconia for dental applications. Acta Biomater 17:36–46
48. Gad MM, Abualsaud R, Rahoma A, Al-Thobity AM, Al-Abidi KS, Akhtar S (2018) Effect of zirconium oxide nanoparticles addition on the optical and tensile properties of polymethyl methacrylate denture base material. Int J Nanomed 283–292
49. Yamada R, Nozaki K, Horiuchi N, Yamashita K, Nemoto R, Miura H, Nagai A (2017) Ag nanoparticle–coated zirconia for antibacterial prosthesis. Mater Sci Eng C 78:1054–1060
50. Xiao K, Meng L, Du C, Zhang Q, Yu Q, Zhang X, Chen J (2021) A label-free photoelectrochemical biosensor with near-zero-background noise for protein kinase A activity assay based on porous ZrO2/CdS octahedra. Sens Actuat B: Chem (328):129096
51. Mogha NK, Sahu V, Sharma M, Sharma RK, Masram DT (2016) Biocompatible ZrO2-reduced graphene oxide immobilized AChE biosensor for chlorpyrifos detection. Mater Des 111:312–320
52. Peng HP, Liang RP, Zhang L, Qiu JD (2011) Sonochemical synthesis of magnetic core–shell Fe3O4@ ZrO2 nanoparticles and their application to the highly effective immobilization of myoglobin for direct electrochemistry. Electrochim Acta 56(11):4231–4236

53. Sun W, Wang X, Sun X, Deng Y, Liu J, Lei B, Sun Z (2013) Simultaneous electrochemical determination of guanosine and adenosine with graphene–ZrO2 nanocomposite modified carbon ionic liquid electrode. Biosens Bioelectron 44:146–151
54. Ferlazzo A, Espro C, Iannazzo D, Bonavita A, Neri G (2023) Yttria-zirconia electrochemical sensor for the detection of tyrosine. Mater Today Commun (35):106036
55. Gupta PK, Chauhan D, Khan ZH, Solanki PR (2020) ZrO2 nanoflowers decorated with graphene quantum dots for electrochemical immunosensing. ACS Applied Nano Mater 3(3):2506–2516
56. Gionea A, Andronescu E, Voicu G, Bleotu C, Surdu VA (2016) Influence of hot isostatic pressing on ZrO2–CaO dental ceramics properties. Int J Pharm 510(2):439–448
57. Zahra T, Ahmad KS, Zequine C, Gupta R, Thomas A, Malik MA, Ali D (2022) Biomimmetic ZrO2@ PdO nanocomposites: fabrication, characterization, and water splitting potential exploration. Int J Energy Res 46(6):8516–8526
58. Liu Y, Wang S, Wang Z, Ye N, Fang H, Wang D (2018) TiO2, SiO2 and ZrO2 nanoparticles synergistically provoke cellular oxidative damage in freshwater microalgae. Nanomaterials 8(2):95
59. Sengul AB, Asmatulu E (2020) Toxicity of metal and metal oxide nanoparticles: a review. Environ Chem Lett 18:1659–1683

Influence of Sustainable Lubrication on the Tribological Behaviour of Phosphorus Bronze

Aman Kumar, Palash Biswas, Anal Ranjan Sengupta, and Thia Paul

Abstract In this present work, experimental analysis has been carried out in tribological tests of phosphorus bronze with or without lubricating conditions. The industrial lubrication market is seeing a surge in the usage of vegetable oil-based lubricants and bio-lubricants due to its high tribological performance, minimum environmental effect, sustainability, renewability, and biodegradability. A Pin-on-disc tribometer has been used to process the entire experiment and using Scanning Electron Microscope (SEM), to determine the surface texture of the material used during this experiment. It has been performed in both dry and lubricating conditions by varying the applied load from 20 to 40 N and rotating the disc speed varying between 50 to 90 rpm. Also, the track diameter varied from 50 to 80 mm. In the dry sliding condition, the coefficient of friction increased with increasing applied load and sliding speed. A low coefficient of friction is observed at the 20 N load and 50 rpm sliding speed. In the lubricating condition, the minimum coefficient of friction is found in the ethanol lubricating condition in all cases. In the mustard oil lubricating state, the wear rate is reduced to above 50% as compared to dry-sliding wear. In ethanol as a lubricant, the wear rate is reduced by 92% and 35% as compared to dry-sliding and mustard oil lubricants respectively. Finally from the SEM images related to the wear mechanism, a combination of adhesive, abrasive, and fatigue wear is visible in both dry-sliding and lubricating conditions.

Keywords Friction · Wear · Lubrication · Dry sliding · Mustard oil · Ethanol · Sustainability

A. Kumar · P. Biswas (✉) · A. R. Sengupta · T. Paul
Department of Mechanical Engineering, JIS College of Engineering, Kalyani, West Bengal 741235, India
e-mail: palashbesus@gmail.com

P. Bhingole et al. (eds.), *Advances in Materials Engineering*, Lecture Notes in Mechanical Engineering, https://doi.org/10.1007/978-981-97-7114-1_8

1 Introduction

Phosphor bronze is an alloy of copper, tin, and phosphorus. The phosphorus content is typically between 0.05 and 0.15%. The phosphorus in phosphor bronze improves its wear resistance and stiffness. In dry sliding conditions, phosphor bronze exhibits high friction and wear rates. This is because the phosphorus in the alloy reacts with oxygen to form a layer of copper oxide on the surface. This layer is soft and easily worn away, exposing the underlying metal to further wear [1]. Phosphor bronze is used in a variety of tribology-based applications due to its excellent wear resistance, corrosion resistance, and machinability. Some of the most common applications include bearings, gears, springs, electrical contacts, etc. Overall, phosphor bronze is a versatile and reliable material with a wide range of tribology-based applications. It is strong, durable, corrosion-resistant, and easy to work with [2, 3]. Lubrication is essential for tribology testing because it helps to reduce friction and wear between the test surfaces. This allows for more accurate measurements of friction and wear, and it also helps to prevent premature failure of the test surfaces [2]. There are several different lubricants that can be used for tribology testing, each with its own advantages and disadvantages. The choice of lubricant will depend on the specific test being conducted and the materials being tested [4–6]. Some of the most common lubricants used for tribology testing include oils, greases, and solid lubricants. Oils are the most common type of lubricant used for tribology testing and are effective at reducing friction, wear, in a cost-effective manner. Greases are thicker than oils and provide a more effective barrier between the test surfaces which are considered ideal for applications where high loads or high temperatures are involved. Solid lubricants have low coefficients of friction and are often used in applications where oil or grease cannot be used, such as in vacuum or high-temperature environments. The correct lubrication is essential for accurate tribology testing. By using the right lubricant, it can be ensured that the test results are reliable and test surfaces are protected from premature failure [7, 8]. Some of the benefits of using lubrication for tribology testing are reduced friction, reduced wear, improved accuracy, and protected surfaces. Lubrication helps to protect the test surfaces from premature failure. This is because it creates a barrier between the surfaces and the environment. Overall, lubrication is an essential part of tribology testing. According to market availability and low cost, two types of lubricants are used for tribology tests like ethanol and mustard oil. Ethanol is used as a fuel in a variety of applications. It is blended with gasoline to produce gas holes, which reduces air pollution. Ethanol is also used as a fuel in some vehicles, such as flexible fuel vehicles. Ethanol is a renewable fuel, and it is a cleaner burning fuel than gasoline. It is used in a variety of products, such as paints, varnishes, and cosmetics. Ethanol has several advantages in tribological applications, including low friction, good lubricating properties, corrosion inhibition, and biodegradability [9]. Overall, ethanol is a promising lubricant that has the potential to improve the performance and durability of tribological systems. Mustard oil is a vegetable oil derived from mustard seeds. It is a popular cooking oil in many parts of the world, especially in India, Bangladesh, and Pakistan. Mustard oil can be used for cooking, frying, and salad dressings. It can also be used as lubricant for tribology tests. Other advantages of mustard oil for tribology testing are non-toxicity and biodegradability, high flash point, low viscosity, cost-effectiveness, versatility, high thermal conductivity, and low coefficient of friction [10–12]. Overall, mustard oil is a good choice for tribology testing because it is a versatile, safe, and effective lubricant due to its several beneficial features as mentioned [13, 14].

From this available literature, it is seen that comprehensive analysis of tribological features evaluation of phosphorus bronze using ethanol and mustard oil lubricants considering different applied loads, rotating speed and track diameter is very less which need to be further explored. Again, the analysis of wear rate and coefficient of friction for phosphorus bronze is very important from its application point of view. This current investigation aims to study the tribological behaviour in various parameters like varying applied load, rotating speed, and track diameter with fixed time duration for dry-sliding and various lubricating conditions using ethanol and mustard oil which has not yet investigated by previous researchers. Also, the wear rate and coefficient of friction are compared here between dry sliding and lubricating conditions. Finally, the wear mechanism is observed by Scanning Electron Microscope (SEM).

2 Experimental Details

2.1 *Material*

Phosphorus bronze has been selected for this investigation. Size of the samples is ϕ 6 $\times$ L 30 mm for tribological test and for subsequent characterizations. The samples were prepared by first cutting them from the cylindrical rod by a lathe machine. For a smooth surface, the samples have been ground by a grinding machine. In the group of copper alloys, phosphor bronze is a member. It is made of copper that has been alloyed with 0.01–0.35% phosphorus, 0.5–11% tin, and sometimes other elements to give it particular qualities (such as lead at 0.5–3.0% to create free-machining phosphor bronze). Copper's corrosion resistance and strength are increased by alloyed tin, while its wear resistance and stiffness are increased by phosphorus [15].

2.2 *Tribology Test Dry Sliding and Lubricating Condition*

The pin-on-disc tribometer apparatus (TR-20LE-CHM-400, Ducom, India) is used to measure the coefficient of friction (COF) and wear rate of phosphorous bronze under dry sliding and lubricating conditions at an ambient temperature 32 °C. Two types of lubricant are used for the test. Ethanol and mustard oil i.e. vegetable oil are used for tribo test. The test parameters are displayed in Table 1. A load sensor measures

Table 1 Design of experiment for tribology test

No. of experiments	Load (N)	Speed (rpm)	Track diameter (mm)
1	20	50	50
2	20	75	60
3	20	90	80
4	30	50	50
5	30	75	60
6	30	90	80
7	40	50	50
8	40	75	60
9	40	90	80

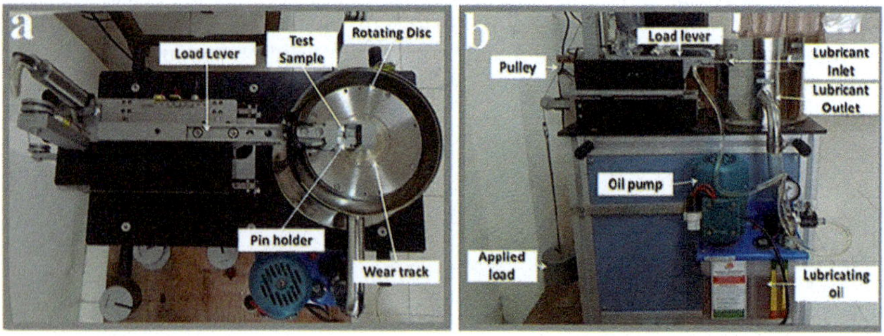

Fig. 1 Pictorial view of pin-on-disc tribometer

the friction force, which is divided by the normal force to get the COF. The COF is then recorded and presented by the "Winducom 2010" software. However, wear rate (K) is calculated by using the following relation: $K = V/(S \times F)$ where V, F, and S are considered as wear volume (mm^3), normal load (N), and sliding distance (m), respectively [9, 16, 17]. Figure 1 shows a pictorial view of the pin-on-disc tribometer.

2.3 *Microstructure Observation of Wear Surface*

As part of a microstructural study of the materials, the surface morphology is observed with the help of a Scanning Electron Microscpe (SEM) by Carl Zeiss, Germany (Model No. EVO LS 10). Everhart-Thornley Detector is used to detect the secondary electrons (SE) [18–21]. The wear surface is observed after a tribology test with or without lubricating conditions.

3 Results and Discussion

3.1 Friction Performance

Phosphor bronze has a low coefficient of friction (COF) in dry sliding, which means that it can reduce the amount of friction between two surfaces. This can lead to improved wear resistance and energy efficiency. The friction performance of phosphorus bronze in dry sliding is affected by several factors, including the surface finish, and the sliding speed, track diameter, and applied load during the test. Table 2 displayed the coefficient of friction in various track diameters, sliding speed, and applied load in constant time duration.

In general, phosphorus bronze has a lower coefficient of friction than other common materials, such as steel and aluminum. This is due to the presence of phosphorus in the alloy, which forms a thin film on the surface of the metal. This film helps to reduce friction by providing a barrier between the two surfaces. The surface finish of phosphorus bronze can also affect its friction performance.

A smooth surface will have lower friction than a rough surface. This is because a smooth surface has fewer irregularities that can catch on the other surface and cause friction. Figure 2 shows the coefficient of friction in various speed, track diameter, and applied load in dry sliding test. In 20 N applied load, the low COF is revealed for low sliding speed with higher track diameter. The higher track diameter does not depend on friction performance. At low speeds, the friction coefficient is relatively constant. In these loads, the higher COF is obtained at 75 rpm with 60 mm track diameter due to high speeds, the friction can increase. This is because the film of phosphorus that forms on the surface of the metal can be worn away at high speeds. The applied load is increased and the COF is also increased (refer to Fig. 2b). This COF is not only affected by the load. It also affected another parameter. The higher COF is obtained by 40 N applied load with 60 mm track diameter at low speed. It

Table 2 Coefficient of friction and wear rate in dry sliding condition

No. of experiments	Load (N)	Speed (rpm)	Track diameter (mm)	Coefficient of friction (μ)	Wear rate $\times 10^{-4}$ $mm^3 N^{-1} m^{-1}$
1	20	50	80	0.10	4.15
2	20	75	60	0.38	8.03
3	20	90	50	0.275	7.54
4	30	50	50	0.55	8.34
5	30	75	80	0.49	2.14
6	30	90	60	0.54	7.9
7	40	50	60	0.58	8.05
8	40	75	50	0.54	8.65
9	40	90	80	0.4292	2.17

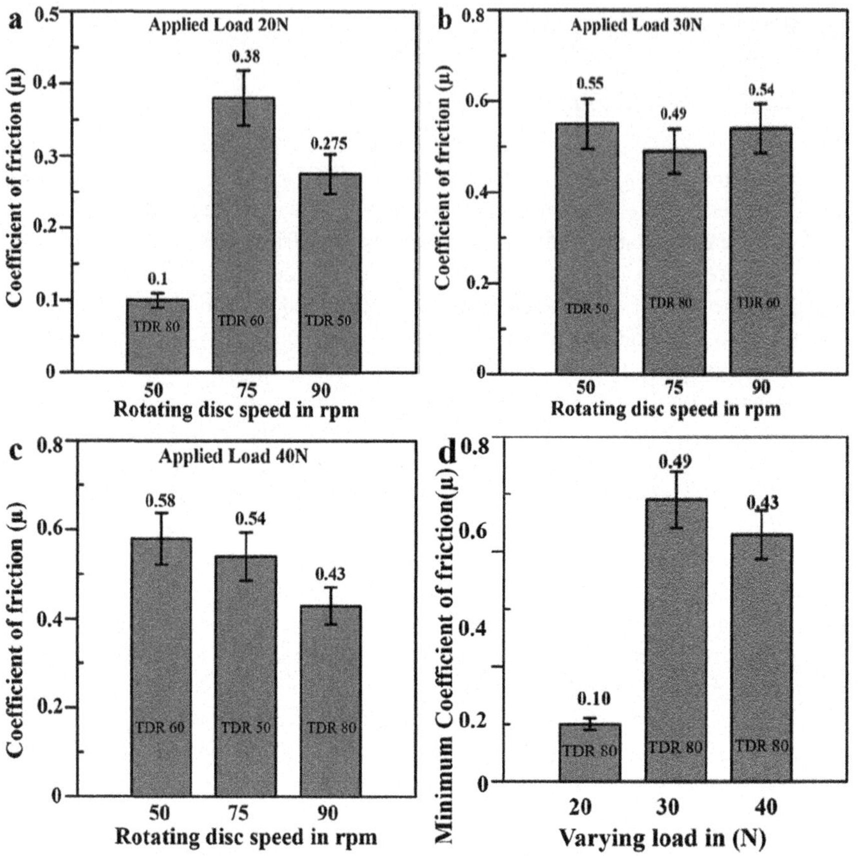

Fig. 2 Coefficient of friction in various speeds and applied load, **a** 20 N, **b** 30 N, **c** 40 N and **d** Minimum coefficient of friction in dry sliding test

is very interesting to know the low COF is obtained in low load and low speed with higher track diameter.

The use of a lubricant can help to reduce friction by providing a thin film between the two surfaces. This film helps to prevent direct contact between the two surfaces and reduces the amount of friction. Table 3 Coefficient of friction and wear rate in used mustard oil as lubricant. Figure 3 displayed the coefficient of friction at various speeds, applied load, and track diameter in the mustard oil lubricant test. The COF is reduced from the dry-sliding test in all aspects due to the mustard oil used (refer to Tables 2 and 3). Mustard oil can have a significant effect on the coefficient of friction in tribo-tests. In general, mustard oil can reduce the coefficient of friction between two surfaces due to mustard oil being a good lubricant and can help to reduce the friction between two surfaces. The effect of mustard oil on the coefficient of friction can vary depending on several factors, including the type of mustard oil, the surface finish of the materials, and the load applied to the surfaces. In a 20 N load with the

Table 3 Coefficient of friction and wear rate in lubricating (Mustard oil) condition

No. of experiments	Load (N)	Speed (rpm)	Track diameter (mm)	Coefficient of friction	Wear rate × 10^{-4} $mm^3\ N^{-1}\ m^{-1}$
1	20	50	80	0.08	2.7
2	20	75	60	0.12	0.36
3	20	90	50	0.27	3.2
4	30	50	50	0.07	0.39
5	30	75	80	0.21	4.7
6	30	90	60	0.12	2.2
7	40	50	60	0.081	1.1
8	40	75	50	0.13	0.86
9	40	90	80	0.08	0.34

same parameter, using the mustard oil, the COF is reduced by 20% with respect to the dry sliding test due to the generated protective thin layer.

Similarly in the dry test, the speed increased in 20 N applied load, the COF increased. But at 30 N load, the maximum COF is observed at 75 rpm with higher (at 80TDR). This COF also depends on high track diameter. High track diameter indicates more distance travelled and repeated area movement can be happening at high COF. It is very interesting to show that the maximum applied load and minimum speed obtained higher COF (Fig. 3c). In the meantime, high speed and higher track diameter obtained minimum COF due to the protective thin layer of mustard oil. Figure 3b displayed the minimum COF with varying loads. This mostly depends on varying speeds and loads. These studies suggest that mustard oil can be an effective way to reduce the coefficient of friction. In addition to reducing the coefficient of friction, mustard oil can also have other beneficial effects in tribo-tests. For example, mustard oil can help to protect the surfaces from wear and tear. Mustard oil can also help to improve the performance of the materials in the tribo-test.

Table 4 displayed the coefficient of friction and wear rate in ethanol lubricant used. Ethanol can have a significant effect on the COF in tribo-tests. In general, ethanol reduces the COF, meaning that it makes it easier for two surfaces to slide against each other. Ethanol is a good solvent, which means that it can help to remove any contaminants that may be present on the surface of the materials being tested.

This can help to create a smoother surface, which reduces friction. Ethanol is a polar molecule, which means that it has a positive and negative end. This polarity can help to form hydrogen bonds with the surface of the materials being tested, which can also help to reduce friction. The COF is reduced in all aspects from the dry-sliding test. Figure 4 displayed the coefficient of friction at various speeds, track diameter, and applied load in the ethanol lubricating test. In 20 N load, the high speed (90 rpm) obtained minimum COF due to applied ethanol (refer to Fig. 4a). It can help to reduce the amount of friction between two surfaces. This is because ethanol can fill in any

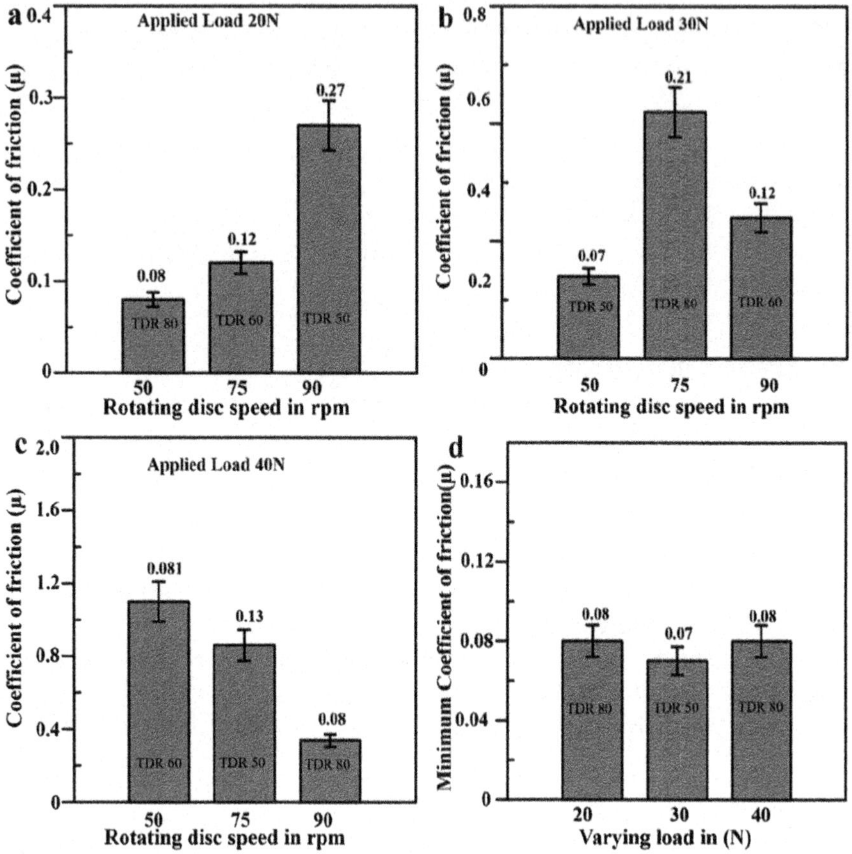

Fig. 3 Coefficient of friction in various speed and applied load, **a** 20 N, **b** 30 N, **c** 40 N and **d** Minimum coefficient of friction mustard oil lubricant test

Table 4 Coefficient of friction and wear rate in ethanol lubricant used

No. of experiments	Load (N)	Speed (rpm)	Track diameter (mm)	Coefficient of friction	Wear rate × 10^{-4} mm^3 N^{-1} m^{-1}
1	20	50	80	0.18	1.8
2	20	75	60	0.22	0.18
3	20	90	50	0.16	0.59
4	30	50	50	0.2	0.7
5	30	75	80	0.17	0.45
6	30	90	60	0.15	0.19
7	40	50	60	0.15	0.29
8	40	75	50	0.23	0.74
9	40	90	80	0.16	2.4

microscopic gaps between the surfaces, which prevents them from rubbing directly against each other.

At this condition, the COF is reduced by around 42% as compared to the dry-sliding tribology test. Also, a similar observation is obtained in the 30 N applied load condition (refer to Fig. 4b). At a high load (40 N), the COF increased with increasing sliding speed. The COF is reduced above 50% to similar test parameters in dry-sliding tests. The effect of ethanol on the coefficient of friction can vary depending on several factors, including the type of materials being tested, the temperature, and the pressure. However, in general, ethanol can be a very effective way to reduce friction in tribo-tests.

Overall, it has been observed that the lubricating effect is prominent for the tribological test. The COF values are reduced in both kinds of lubricant. Out of the two lubricants, mustard oil performs better in most cases. Mustard oil is more viscous

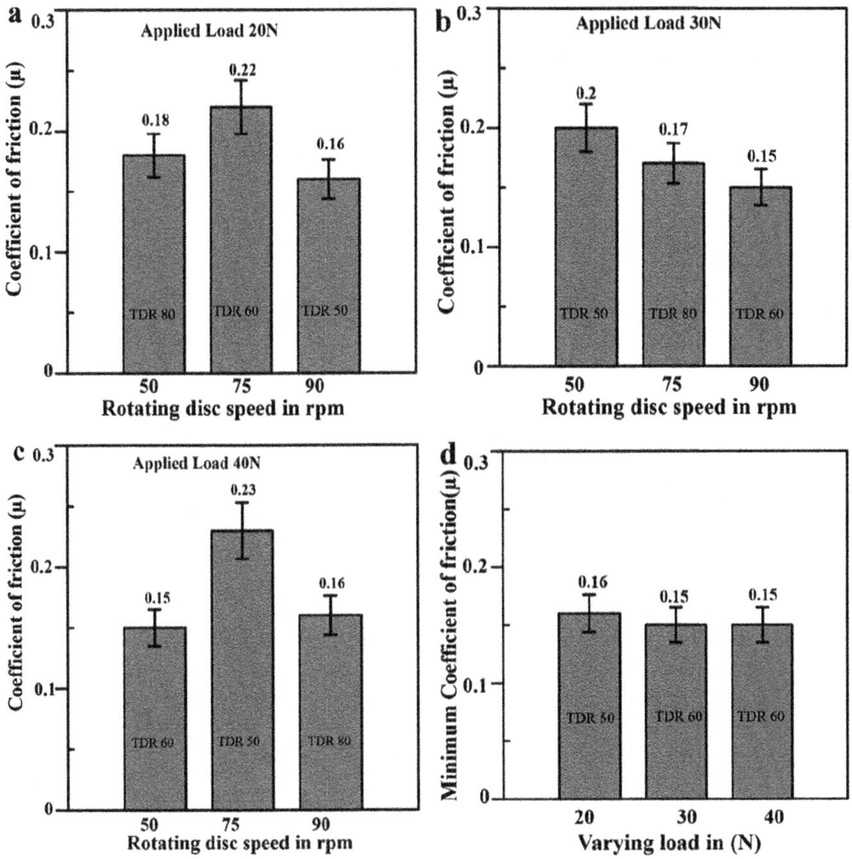

Fig. 4 Coefficient of friction in various speed and applied load, **a** 20 N, **b** 30 N, **c** 40 N and **d** minimum coefficient of friction ethanol lubricating test

and polar than ethanol, while ethanol is more biodegradable and non-toxic. The best choice for a particular tribology test will depend on the specific factors involved in the test.

3.2 Wear Behaviour

Figure 5 shows the wear rate at various speeds, track diameter, and applied load in the dry sliding test. The dry-sliding wear behaviour of phosphorus bronze is generally good. The wear rate of phosphorus bronze is significantly lower than that of other copper alloys, such as pure copper and brass. This is due to the formation of a protective layer of copper phosphate on the surface of the alloy.

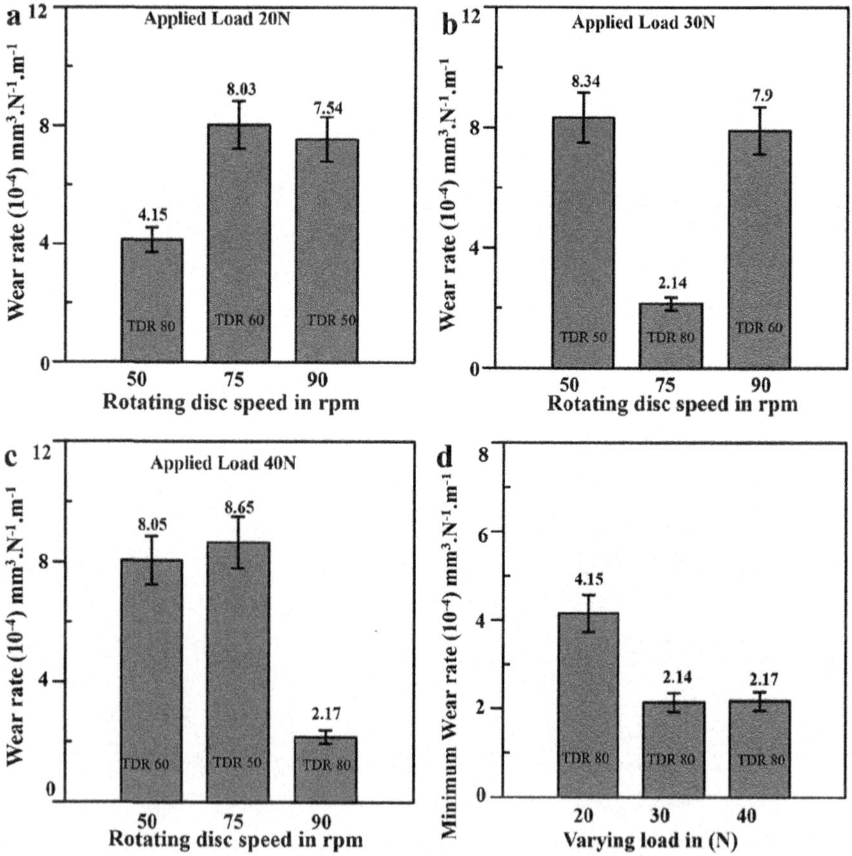

Fig. 5 Wear rate in various speeds and load in dry sliding test

This layer is very hard and resistant to wear and helps to protect the underlying metal from further damage. The dry-sliding wear behaviour of phosphorus bronze is affected by the type of wear, the load, the speed, and the environment. In Fig. 5a–c, the wear rate is increased with increasing load and sliding speed. In general, the wear rate of phosphorus bronze increases with increasing load and speed. The wear rate is also affected by the presence of abrasive particles in the environment. In 20 N applied load, the maximum wear rate is observed at 75 rpm sliding speed with 60 mm TDR. The wear rate not only affected sliding speed and applied load. It also depends on the track diameter or distance travelled and test duration. In the case of a 30 N applied load, the minimum wear rate is observed at 75 rpm sliding distance and 80TDR. It is interesting to display that a high load with a higher sliding speed obtained a minimum wear rate. The wear debris may seem to be welded on the rubbing surface due to high sliding speed and applied load. It is very difficult to understand the optimal choice due to various parameters affecting wear rate (refer to Fig. 5d).

Figure 6 displayed the wear rate at various speeds and load in mustard oil lubricating condition. The wear rate is reduced in all test parameters due to the use of mustard oil. Mustard oil reduces the wear rate in tribology tests in several ways. First, it helps to reduce friction between the two surfaces in contact. This reduces the amount of heat generated, which in turn reduces the wear rate. Second, mustard oil has a few anti-wear properties. These properties help to protect the surfaces from damage, even under high-stress conditions. Third, mustard oil can help to form a protective film on the surfaces in contact. This film helps to reduce the amount of direct contact between the two surfaces, which further reduces the wear rate. The wear rate is reduced above 50% as compared to dry-sliding wear (refer to Fig. 5). In 20 N applied load condition, the minimum wear rate is obtained at 75 rpm rotating speed with 60 mm TDR. In the meantime, the high speed (90 rpm) of the rotating disc obtained a maximum wear rate (refer to Fig. 6a).

Using this lubricant, the wear rate is reduced by around 57% with respect to the dry-sliding state. It is more beneficial for high-speed conditions. In the case of a 30 N load applied, the optimal wear rate is obtained at a low speed with a lower track diameter. Load and track diameter are inversely proportional during the wear test. High rotational speed (90 rpm) reduced wear from median speed (75 rpm) due to mustard oil having several anti-wear properties (Fig. 6b). These properties help to protect the surfaces from damage, even under high-stress conditions. When two surfaces rub against each other, they can experience high stress. This stress can cause the surfaces to deform and eventually break down. When the applied load is 40 N, the minimum wear rate is obtained at high speed with a higher track diameter as shown in Fig. 6c. The wear rate is reduced by around 84% as compared to the dry-sliding wear rate. It can be very useful as a lubricant in engine oil, gear oil, and brake fluid. It can also be used as a rust inhibitor and a protectant for metal surfaces. It is very difficult to understand which test condition is the best choice because the wear rate depends on various parameters like applied load, sliding speed, track diameter, test duration, etc. According to the set test parameters, the best performing parameter is shown in Fig. 6d.

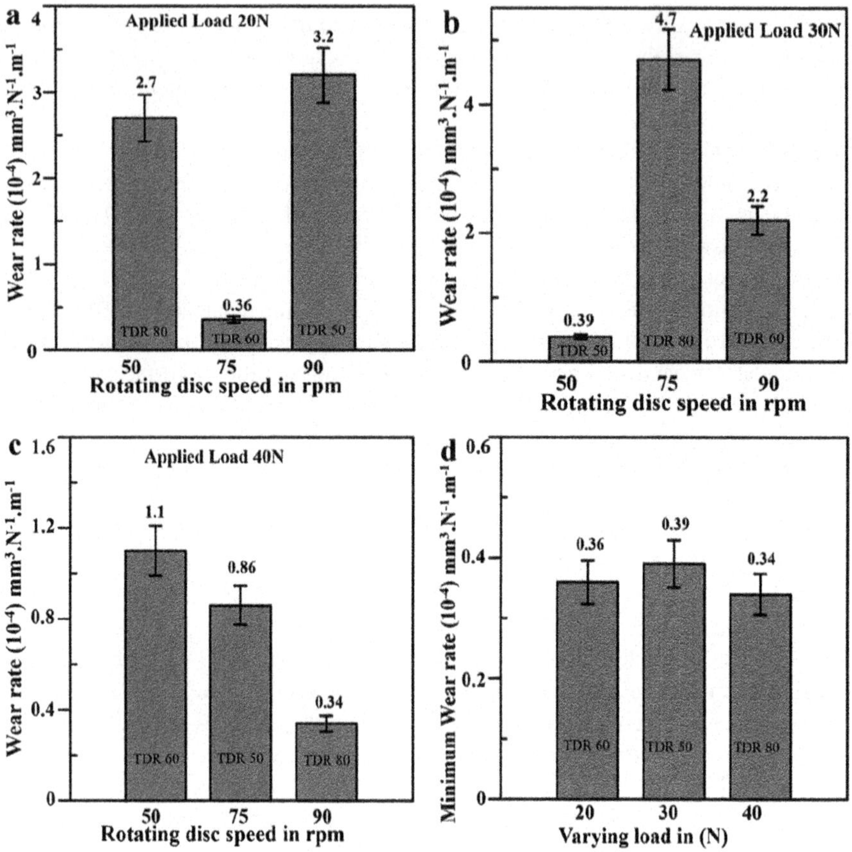

Fig. 6 Wear rate in various speeds and load in mustard oil lubricating condition

Ethanol is used as another kind of lubricant for tribological tests. The wear rate is displayed in Fig. 7 and the detailed wear test condition is shown in Table 4. The wear rate is reduced in all testing conditions. Mostly, the wear rate is reduced by around 92% as compared to the dry-sliding wear rate. When two surfaces rub against each other, friction can cause heat to build up. This heat can damage the surfaces and lead to wear.

Ethanol helps to reduce friction by forming a thin film between the two surfaces. This film acts as a barrier and prevents direct contact between the surfaces. As a result, less heat is generated and wear is reduced. In addition to its lubricating properties, ethanol also has several other properties that can help to reduce wear. For example, ethanol is a good solvent and can help to remove dirt and debris from the surfaces in contact. This can help to prevent the surfaces from becoming damaged and can further reduce wear. When the applied load is 20 N, the low rotational speed with maximum track diameter obtained the highest wear rate as shown in Fig. 7a. The wear track is the sole factor of wear rate. It is interesting to find that a high sliding

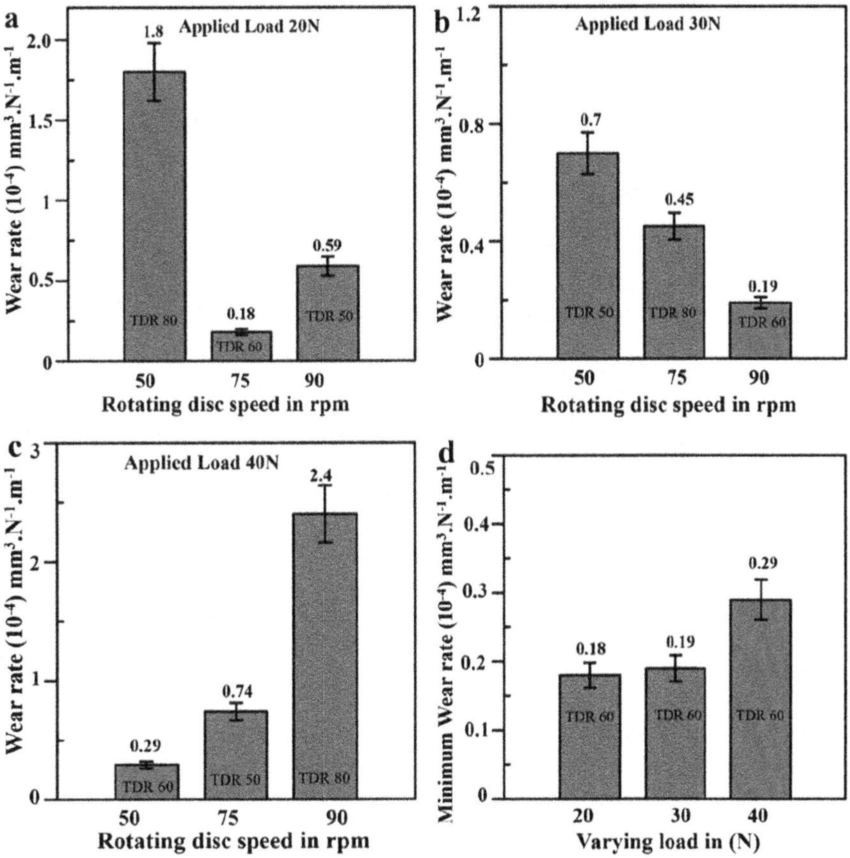

Fig. 7 Wear rate in various speeds and load in ethanol lubricating condition

speed with a lower track diameter reduced the wear rate by 66% with respect to a lower sliding speed with a higher track diameter. The minimum wear rate is obtained at 90 rpm sliding speed with a 60 mm track diameter in a 30 N applied load state (refer to Fig. 7b) whereas at 40 N applied load with 50 rpm sliding speed, the minimum wear rate is obtained for the same track diameter (Fig. 7c). It is very difficult to choose optimal parameters because the wear rate depends on several parameters. According to said set parameters, the minimum wear rate is shown in Fig. 7d.

Overall, ethanol is used as a lubricant and the best wear result is observed in all states. In comparison to mustard oil, the wear rate is reduced by above 50% in low-load conditions. Mostly, the wear rate is reduced by more than 50% with respect to mustard oil lubricant. In the inter-test conditions, ethanol is the optimal choice as a lubricant used for tribology tests.

3.3 Wear Mechanism

Figure 8 shows the SEM image of the sample after dry-sliding wear test in various applied load and sliding speed. The wear mechanism under dry sliding conditions is typically a combination of adhesive wear, abrasive wear, and fatigue wear. Adhesive wear occurs when the asperities on the sliding surfaces adhere to each other and are then torn apart, resulting in the loss of material from both surfaces. Abrasive wear occurs when hard particles, such as dirt or grit, are embedded in one of the sliding surfaces and then scratch or gouge the other surface. Fatigue wear occurs when the surface of the bronze is repeatedly stressed and then relieved, resulting in the formation of small cracks (refer to Fig. 8a) that eventually grow and lead to the loss of material. Delaminating is a common wear mechanism in phosphorus bronze for higher applied load and sliding speed as shown in Fig. 8b, c. It can be caused by several factors like high loads that can cause the surface of the bronze to plastically deform, which can lead to the formation of cracks. High sliding speeds can cause the surface of the bronze to heat up, which can also lead to the formation of cracks. Contaminants, such as dirt or grit, can become embedded in the surface of the bronze and cause it to wear more quickly. The wear tracks and ploughing effects are revealed on all worn surfaces.

Figure 9 displayed the SEM image of the sample after the mustard oil lubricating wear test in various applied loads and sliding speeds. The wear mechanism in mustard oil lubricating conditions is typically a combination of adhesive wear, abrasive wear, and fatigue wear. Adhesive wear occurs when the asperities on the sliding surfaces adhere to each other and are then torn apart, resulting in the loss of material from both surfaces. Abrasive wear occurs when hard particles, such as dirt or grit, are embedded in one of the sliding surfaces and then scratch or gouge the other surface. Fatigue wear occurs when the surface of the material is repeatedly stressed and then relieved, resulting in the formation of small cracks that eventually grow and lead to the loss of material as shown in Fig. 9a. A few ploughing effects are observed at 40 N applied load with 90 rpm sliding speed (refer to Fig. 9b).

The wear mechanism in ethanol lubricating condition is typically a combination of adhesive wear, abrasive wear, and fatigue wear. Similar to Fig. 9, adhesive wear is also visible in Fig. 10a. A few micro-cracks and micro-furrow are observed on the worn surface. Again, abrasive wear and fatigue wear can be observed in Fig. 10b. Ploughing effects are also observed on the worn surface.

4 Conclusions

In this present investigation, the tribology test is conducted on phosphor bronze material considering dry-sliding and lubricating conditions for different test conditions like variation of loads, sliding speeds, and track diameters. The following conclusions are drawn from the present study.

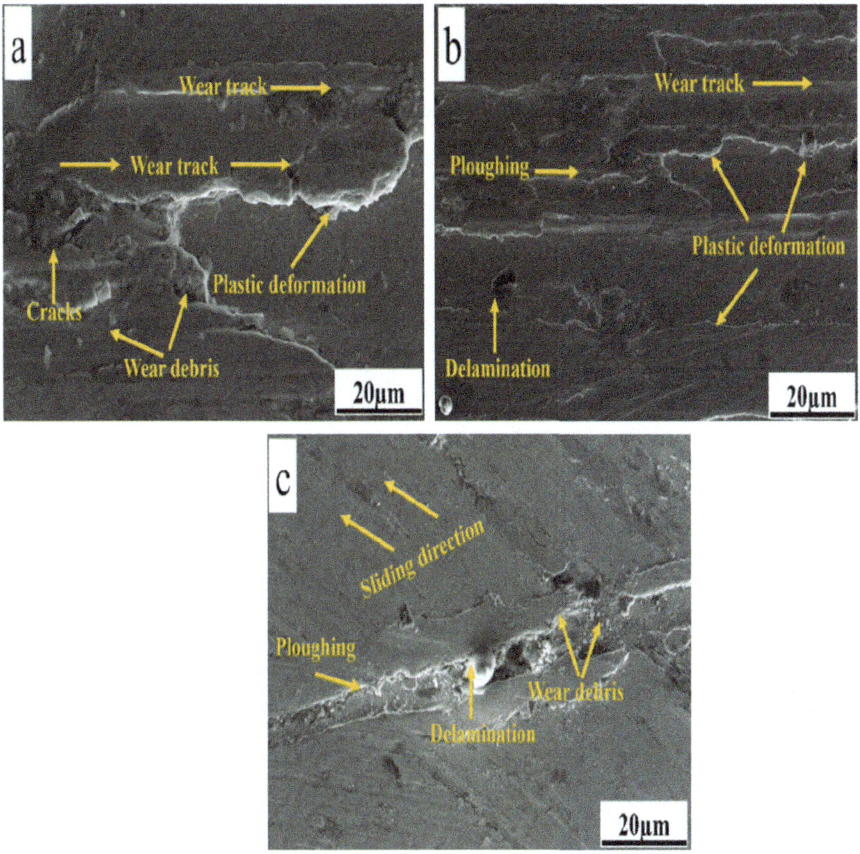

Fig. 8 SEM image of the sample after dry-sliding wear test in various applied load and sliding speed, **a** load 20 N with 50 rpm, **b** load 30 N with 75 rpm, **c** load 40 N with 90 rpm

- The coefficient of friction is increased with increasing applied load and sliding speed in the dry-sliding tribology test. The low coefficient of friction is obtained in 20 N applied load and 50 rpm sliding speed with 80 track diameters.
- The coefficient of friction is reduced by using mustard oil as a lubricant during the tribology test. But the best result is obtained when ethanol is used as a lubricant.
- In wear results in all cases, the best wear resistance is observed in ethanol lubricating conditions. Each lubricant reduces the wear rate from the dry sliding test.
- The wear mechanism is typically a combination of adhesive wear, abrasive wear, and fatigue wear in both dry-sliding and lubricating conditions.

Overall, it is very difficult to conclude in the said test parameter which one is the best. It depends on the various parameters based on the application.

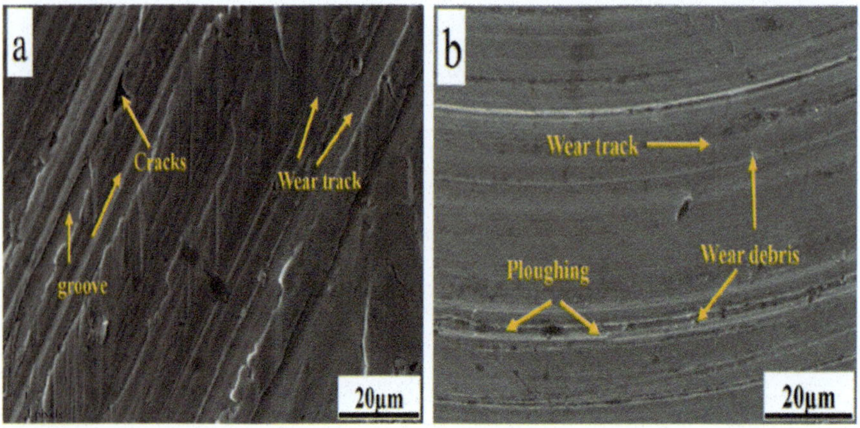

Fig. 9 SEM image of the sample after mustard oil lubricating wear test in various applied load and sliding speed, **a** load 20 N with 50 rpm, **b** load 40 N with 90 rpm

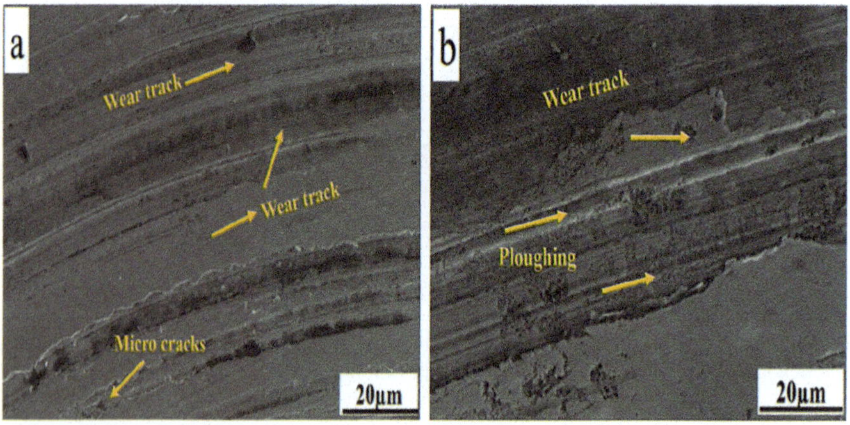

Fig. 10 SEM image of the sample after ethanol lubricating wear test in various applied load and sliding speed, **a** load 20 N with 50 rpm, **b** load 40 N with 90 rpm

References

1. Xia Y, Wang L, Liu X, Qiao Y (2008) Tribological properties of phosphor bronze and nanocrystalline nickel coatings under PAO+ MoDTC and ionic liquid lubricated condition. Tribol Lett 31:149–158
2. Xia YQ, Sasaki S, Murakami T, Nakano M (2007) Tribological behavior of phosphor bronze against SAE52100 steel under different lubricants. Key Eng Mater 353:852–855
3. Van der Heide E, Stam ED, Giraud H, Lovato G, Akdut N, Clarysse F, Heikillä I (2006) Wear of aluminium bronze in sliding contact with lubricated stainless steel sheet material. Wear 261(1):68–73
4. Trivedi HK, Bhatt DV (2017) Effect of lubricating oil on tribological behaviour in pin on disc test rig. Tribol Ind 39(1):90

5. Shahabuddin M, Masjuki HH, Kalam MA (2013) Experimental investigation into tribological characteristics of bio-lubricant formulated from Jatropha oil. Procedia Eng 56:597–606
6. Jain AK, Suhane A (2014) Investigation of tribological characteristics of non edible castor and mahua oils as bio lubricant for maintenance applications. In: 5th international & 26th all India manufacturing technology, design and research conference. Indian Institute of Technology
7. Chourasia SK, Lakdawala AM, Patel RN (2021) Tribological analysis of automotive material under wet lubrication condition using diesel, biodiesel and their blend. IOP Conf Ser: Mater Sci Eng (IOP Publishing) 1146(1):012028
8. Kumar MA (2017) The effect of lubrication on tribological properties of bearing materials using pin on disc: a review. Int J Res Appl Sci Eng Technol 8:267–276
9. Bakshi S, Saha M, Mandal S, Biswas P, Ghosh S (2022) Influence of heat treatment on mechanical and tribological behaviors of brass. Int J Surf Eng Interdiscip Mater Sci 10(1):1–14
10. Sajeeb A, Rajendrakumar PK (2019) Comparative evaluation of lubricant properties of biodegradable blend of coconut and mustard oil. J Clean Prod 240:118255
11. Koszela W, Pawlus P, Galda L (2007) The effect of oil pockets size and distribution on wear in lubricated sliding. Wear 263(7–12):1585–1592
12. Ukonsaari J (2003) Wear and friction of synthetic esters in a boundary lubricated journal bearing. Tribol Int 36(11):821–826
13. Baskar S, Sriram G (2014) Tribological behavior of journal bearing material under different lubricants. Tribol Ind 36(2):127–133
14. Singh Y (2016) Friction and wear characteristics of pongamia oil based blended lubricant at different load and sliding distance. Int J Mechan Mechatron Eng 10(7):1315–1321
15. Prashanth M, Satish N (2018) Effect of brass and silver on mechanical properties of copper. Mater Today: Proc 5(11):25404–25411
16. Kato H, Todaka Y, Umemoto M, Haga M, Sentoku E (2015) Sliding wear behavior of sub-microcrystalline pure iron produced by high-pressure torsion straining. Wear 336:58–68
17. Kabir MA, Higgs III CF, Lovell MR (2018) A pin-on-disk experimental study on a green particulate-fluid lubricant. J Tribol 130(4):041801
18. Saha M, Bakshi S, Mandal S, Biswas P, Ghosh S (2022) Tribologicalbehavior of EN8 and EN24 steel in lubricating condition. Mater Today: Proc 66:3890–3896
19. Biswas P, Das SK, Sahoo P (2022) Duplex electroless Ni-P/Ni-Cu-P coatings: preparation, evaluation of microhardness, friction, wear, and corrosion performance. J Electrochem Sci Eng 12(6):1261–1282
20. Biswas P, Das SK, Sahoo P (2022) Duplex electroless Ni-P/Ni-PW coatings: effect of heat treatment on tribological and corrosion performance. Mater Today: Proc 66:2237–2244
21. Biswas P, Das SK, Sahoo P (2022) Role of heat treatment on the friction and wear behavior of duplex electroless nickel deposits. Mater Today: Proc 66:3902–3909

Material Processing

Calculation of Ni–Cu, Ni–Ag, and Cu–Ag Phase Diagram Using CALPHAD Method

Uttam Kumar Tarai, Swagat Dwibedi, Subhranshu Kumar Singh, and Suyog Jhavar

Abstract MATLAB is a powerful tool for scientific computing, offering seamless integration of computation, representation, and computer programming within an intuitive surrounding. Using well-known mathematical symbols and expressions, users can effectively express and solve complex problems, including calculating phase diagrams. The present work introduces a MATLAB-based approach for constructing collective a tangent between Intersection curves, specifically applied to the Ni–Cu, Ni–Ag, and Cu–Ag phase diagrams. This paper describes the underlying principles and presents a detailed flowchart for implementation, offering a valuable resource for researchers and practitioners in materials science and related fields. The computed melting temperatures for Ni, Cu, and Ag are 1728 K, 1356 K, and 1235 K, respectively, which align well with existing data. Notably, the model identifies the maximum solubility of copper in silver as 14.07%. These results highlight the model's efficacy for predicting phase diagrams in more complex systems, offering valuable materials science and engineering insights.

Keywords Phase diagram · Common tangent method · Gibbs free energy · CALPHAD · TLP bonding

U. K. Tarai
Department of Mechanical Engineering, DRIEMS University, Cuttack, Odisha 752023, India
e-mail: uttamtarai@driems.ac.in

S. Dwibedi
Department of Mechanical Engineering, Indian Institute of Technology Guwahati, Guwahati, Assam 781039, India
e-mail: swagatsudhadwibedi@iitg.ac.in

S. K. Singh
Department of Electronics and Telecommunications Engineering, DRIEMS University, Cuttack, Odisha 752023, India
e-mail: subhranshu@driems.ac.in

S. Jhavar (✉)
School of Mechanical Engineering, VIT-AP University, Amaravati, Andhra Pradesh 522237, India
e-mail: suyog.jhavar@vitap.ac.in

P. Bhingole et al. (eds.), *Advances in Materials Engineering*, Lecture Notes in Mechanical Engineering, https://doi.org/10.1007/978-981-97-7114-1_9

1 Introduction

As Transient Liquid Phase (TLP) bonding is a diffusional controlled process [1–3], therefore the bond characteristics generally depend upon the phase diagram ($P_{\text{diag.}}$) of the multicomponent alloy system and interlayer material. The $P_{\text{diag.}}$ of a multi-component system can be calculated by using theoretical thermodynamic models based on the CALPHAD method. Many software packages (like ChemSage, Lukas programs, MTDATA, and Thermo-Calc) have been developed to calculate the $P_{\text{diag.}}$ of a multicomponent alloy system [4]. Although these software are user-friendly for calculating the $P_{\text{diag.}}$, their capabilities are confined to the software packages [5]. Therefore, in this research, an efficient and simple solution is provided to calculate the $P_{\text{diag.}}$ using the common tangent method based on Gibbs Free Energy (GFE) minimization. MATLAB is used to solve non-linear equations to calculate the equilibrium phases. In this research work the principle of the CALPHAD method is discussed, and then the Ni–Cu, Ni–Ag, and Cu–Ag phase is calculated. CALPHAD method of calculating $P_{\text{diag.}}$ is the minimization of total GFE, (G), of all phases contribute to equilibrium in a multicomponent system. The total GFE is the summation of GFE of the pure elements and GFE due to mixing. It is written below in Eq. (1).

$$G(\text{total}) = \sum_{i=1}^{n} c_i G_i^{\alpha} + \Delta G_{\text{mix}}^{\alpha} \tag{1}$$

where c_i is the number of moles, and G_i^{α} is the GFE (phase i) and $\Delta G_{\text{mix}}^{\alpha}$ is GFE of mixing. Considering a simple system of two-phase equilibrium in binary systems. Assuming both phases are single-phase α and β, the GFE of Gα and Gβ is a combined function C while the temperature is held steady, if α is in equilibrium with β, then GFE of two phases with different compositions is calculated. The minimum total GFE of the mixture is found to form on the common tangent [6], as shown in Fig. 1. These common tangent points are the equilibrium points of different phases and can be found by solving the common tangent equations. The slope of the common tangent line remains the same all over the lines. That gives rise to the following set of equations:

$$\begin{cases} \left(\frac{\partial G^{\alpha}}{\partial c}\right)_a = \left(\frac{\partial G^{\beta}}{\partial c}\right)_b |a \\ \left(\frac{\partial G^{\alpha}}{\partial c}\right)_a * (b - a) = G^{\beta}(b) - G^{\alpha}(a) \end{cases} \tag{2}$$

Solving the above set of equations at different temperatures gives the equilibrium composition α and β phases. MATLAB can solve these nonlinear equations with a proper algorithm or by the Fsolve function (MATLAB). To address the nonlinear equation, we require the starting points of iterations, which is very important in solving the nonlinear equations. The starting point of iteration can be taken as the intersection of the GFE curve of different phases, where the GFE of two phases are equal. Mathematically, it can be written as follows:

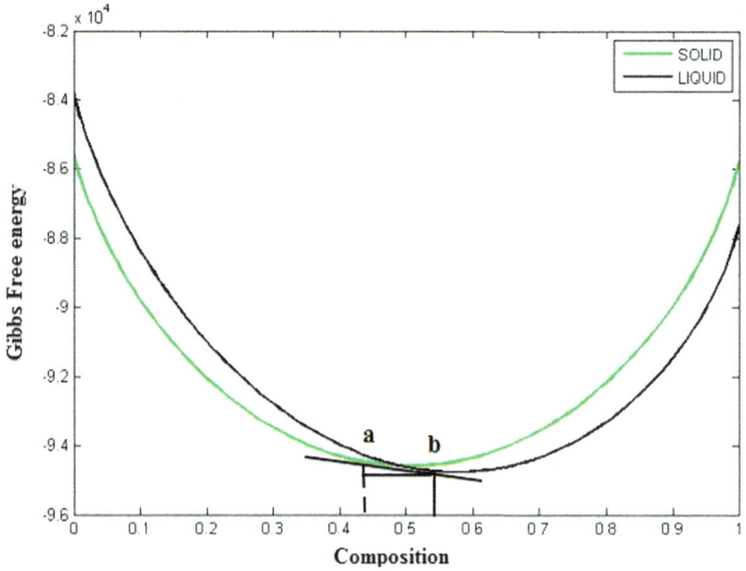

Fig. 1 GFE and composition plot of Ni–Cu binary system

$$G^{\alpha}(c) - G^{\beta}(c) = 0 \quad (3)$$

This equation can be solved by using the MATLAB library function FZERO or with proper algorithms. The roots can be used as the initial points of iteration to solve the nonlinear equation. The flowchart of the algorithm for the analysis of the $P_{\text{diag.}}$ is shown in Fig. 2. At constant temperature, the total GFE of phases is calculated at overall temperature, then equilibrium equations are solved to calculate the composition of stable phases at the temperature, and this method is repeated across various temperature intervals to obtain a complete P_{diag}.

2 Methodology

An algorithm is followed to calculate the P_{diag}. The process map of the algorithm is illustrated in Fig. 2. In this algorithm, the first operation is to provide the starting values of the different variables of the GFE and the GFE for the various phases calculated. Then, the composition of various phases at phase equilibrium is evaluated. This process is carried out over a range of temperatures and the equilibrium phase composition of various phases is stored. Then, the $P_{\text{diag.}}$ is obtained by plotting a graph between the equilibrium compositions of various phases with respect to the temperature.

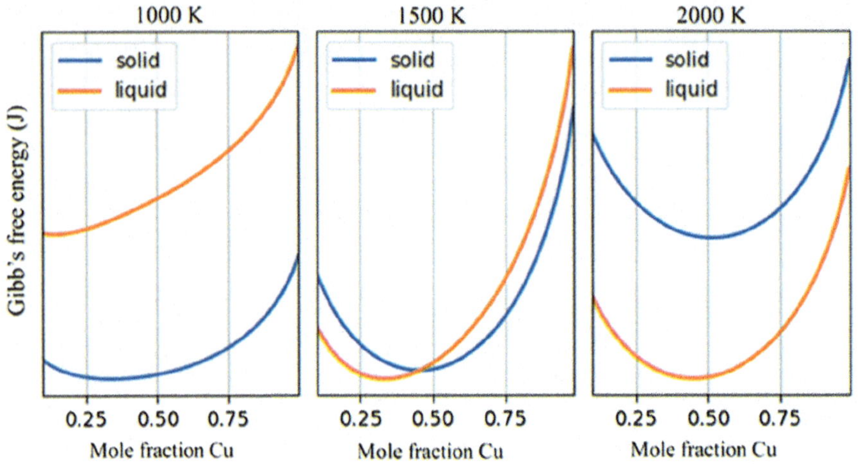

Fig. 2 Illustrates GFE and composition in Ni–Cu alloy system

2.1 Nickel–Copper Binary System

Many researchers have done a thermodynamic evaluation of the Ni–Cu binary system. To calculate the $P_{diag.}$ of the Ni–Cu binary system, the thermodynamic data of pure Ni and Cu were taken from SGTE [7]. Ni–Cu binary alloy was taken as an ideal solid solution. The enthalpy of mixing is taken as zero in this case. The solid elemental substances in their states, stable at 298.15 K, were selected as the reference states of the system (stable element reference $\longrightarrow$ SER). The GFE of liquid, Ni, and Cu solution states dependent on concentration and temperature can be written as the following ideal solid solution model:

$$G^{\varnothing} = C_{Ni}G^{0}_{Ni} + C_{Cu}G^{0}_{Cu} + RT\{\ln C_{Ni} + \ln C_{Cu}\} \tag{4}$$

where C_{Ni} and C_{Cu} is mole fractions, G^{0}_{Ni} and G^{0}_{Cu} are GFE at SER states of respective elements, $R \longrightarrow$ universal gas constant, $T \longrightarrow$ temperature (K) and $\varnothing$ shows the phase.

2.2 Copper Silver Binary System

Unlike the Ni–Cu system, the Cu–Ag system is modeled by a regular solid solution model. In this system, entropy of mixing was added to the total GFE. The thermodynamic data for pure Cu and Ag was taken from the SGTE data and the excess GFE was taken from Redlich-Kister polynomial. The Redlich-Kister parameter is listed in Table 1. Temperature and a concentration-dependent regular solid solution of the

Table 1 Redlich-Kister parameters of Cu–Ag system

Case	Solid	Liquid
L0	$332580.365 - 7.4547 \times \mathrm{T}$	$16914.949 - 14.7721 \times T + 1.54955 \times T \times \ln T$
L1	$-10144.596 + 5.5617 \times T$	$-1963.3 + 0.8623 \times T$

Table 2 Redlich- Kister parameters of Ni–Ag system

Case	Solid	Liquid
L0	$54620.4 + 3.1 \times T$	$54055.6 - 2.2 \times T$
L1	2800	1910

Cu–Ag system are reported below:

$$G^{\varnothing} = C_{\mathrm{Ag}}G^{0}_{\mathrm{Ag}} + C_{\mathrm{Cu}}G^{0}_{\mathrm{Cu}} + \mathrm{RT}\{\ln C_{\mathrm{Ag}} + \ln C_{\mathrm{Cu}}\} + C_{\mathrm{Cu}}C_{\mathrm{Ag}}\sum_{i=0}^{n} L_i(C_{\mathrm{Cu}} - C_{\mathrm{Ag}})^i \tag{5}$$

where L_i is the Redlich-Kister parameters, C and G are the mole fraction, and GFE of respective elements.

2.3 *Nickel Silver Binary System*

The nickel silver binary system was also modeled by regular solid solutions. The thermodynamic data are also taken from SGTE. The Redlich-Kister constants of excess GFE are also taken from the article [8]; the parameters are listed in Table 2.

3 Results and Discussion

The non-linear set of equations is solved and the equilibrium composition of the alpha and phases is obtained using MATLAB, and the Fsolve command. For Cu–Ni alloy, the GFE meets a single point overall ranges of composition and temperature as shown in Fig. 2. The equilibrium composition from Fig. 2 is determined by taking the common tangent points. The $P_{\mathrm{diag.}}$ of the Cu–Ni alloy is presented in Fig. 3. The Ni is entirely soluble in the Cu. Figure 2 indicates that the alloy solid and liquid solution are in a full range of composition. The melting temperature of the Ni and Cu are found to be 1727 and 1359 K, respectively, which equals the reported melting temperature with an error of ±3 K.

The Gibbs variation with temperature of the Cu–Ag arrangement is depicted in Fig. 4, which shows the stability of the various phases at different temperature

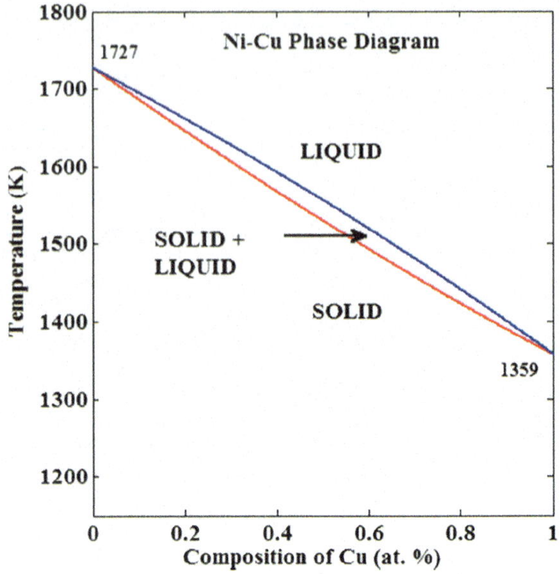

Fig. 3 Ni–Cu binary phase calculations diagram

ranges. The figure indicates that at higher temperatures, the liquid phase is stable, whereas at lower temperatures, the solid phase is stable. Figure 5 shows the GFE variation with the Ag composition in a Cu–Ag alloy system. The figure shows a two-point intersection of the GFE, which indicates that the two phases are stable at the temperature. After obtaining various equilibria compositions, the $P_{\text{diag.}}$ is obtained. The calculated Cu–Ag $P_{\text{diag.}}$ is shown in Fig. 6. The $P_{\text{diag.}}$ is of an eutectic type, and the eutectic temperature was 1056 K, with a composition of 0.6025 at. pct. of Ag, which matches well with the available data [9]. The maximum solubility of copper in Silver and Copper in Silver is found as 14.07 and 4.8 at. %, respectively.

The variation of the GFE with composition in the Ag of Ni-Ag is shown in Fig. 7, wherein it is apparent that the curve does not intersect at 1000 K temperature, however, it intersects at 1500 K. From the equilibria composition of the $P_{\text{diag.}}$ is calculated. The evaluated $P_{\text{diag.}}$ is depicted in Fig. 8. A monotectic and a eutectic reaction was observed at temperatures 1706 K and 1232 K, respectively. Though the eutectic reaction is not visible at this scale, it can be observed at high magnification. In the Ni–Ag system, a monotectic [L $\rightarrow$ FCC (Ni) + L2] reaction at 1708 K and a eutectic [L $\rightarrow$ FCC (Ni) + FCC (Ag)] reaction at 1232 K was found. Which is consistent with Nash and Singleton measured values [8]. The composition of Ag at the respective reaction is also found to be almost equal to Xing Jun Liu et al. thermodynamic examinations of the Ag–Ni system.

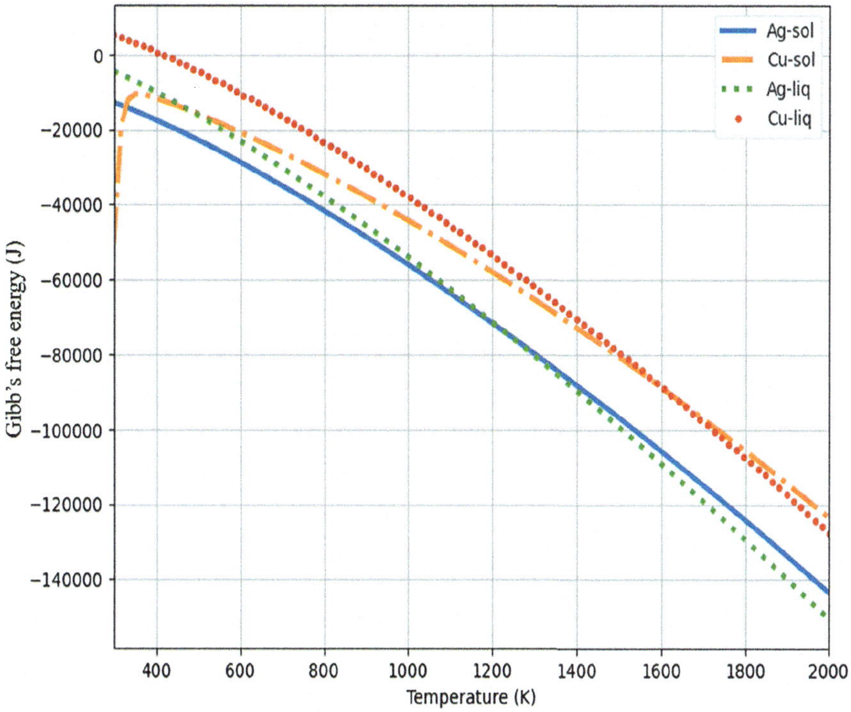

Fig. 4 Variation of GFE of pure copper and Ag with temperature in Cu–Ag alloy system

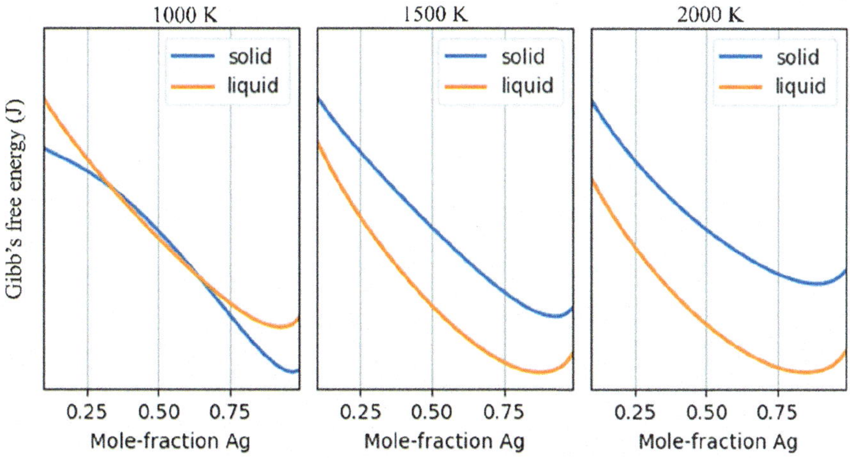

Fig. 5 Schematic diagram showing GFE and composition in Cu–Ag alloy system

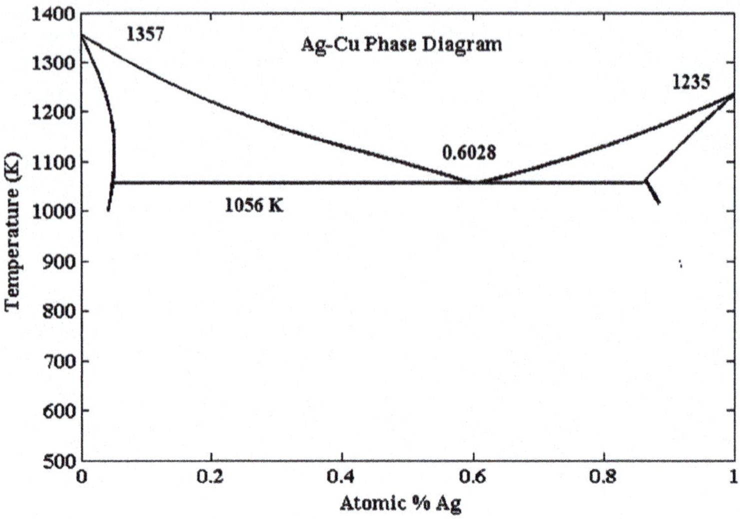

Fig. 6 Illustrates Cu–Ag phase diagram

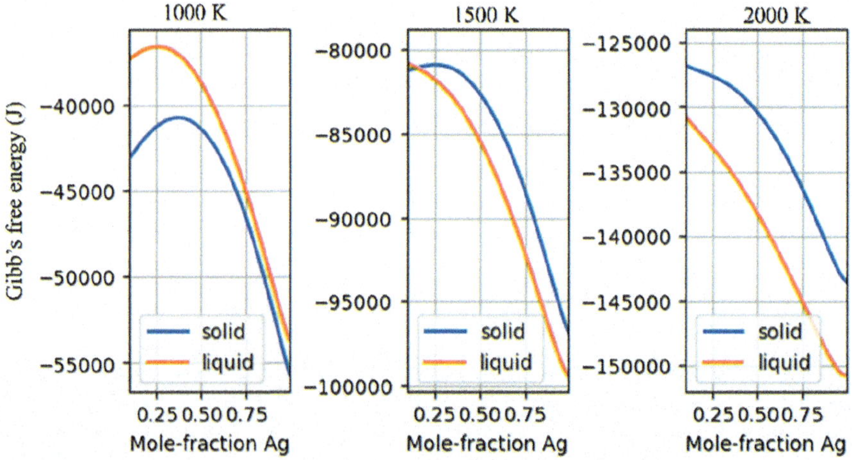

Fig. 7 Schematic diagram showing GFE and composition in Ni–Ag alloy system

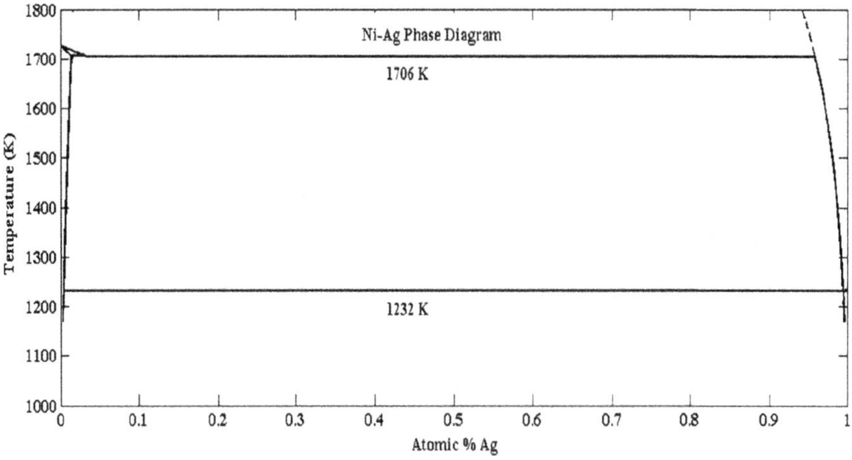

Fig. 8 Ni–Ag binary phase diagram

4 Conclusion

This Ni–Cu, Ni–Ag, Cu–Ag $P_{\mathrm{diag.}}$ was calculated by MATLAB using the CALPHAD method. The melting temperature of the Ni, Cu, and Ag was found to be 1728 K, 1356 K, and 1235 K, respectively, which is consistent with previously reported values. The maximum solubility of copper in Ag is found to be 14.07%. This model can be employed for the mathematical computation of higher component systems P_{diag}.

References

1. Cook GO, Sorensen CD (2011) Overview of transient liquid phase and partial transient liquid phase bonding. J Mater Sci 46:5305–5323. https://doi.org/10.1007/s10853-011-5561-1
2. Tarai UK, Pal S, Robi PS (2023) Kinetics of isothermal solidification during transient liquid phase bonding of Inconel 718 superalloy by differential scanning calorimetry. Proc. Inst. Mech. Eng. Part J. Mater. Des. Appl. 14644207231218113. https://doi.org/10.1177/14644207231218113
3. Tarai UK, Pal S, Robi PS (2023) Reduction of ASZ size and enhancement of mechanical properties of TLP bonded IN718 joint using post bond heat treatment process. Int J Adv Manuf Technol 129:1425–1442. https://doi.org/10.1007/s00170-023-12386-x
4. Li X, Zhang P (2010) Calculation of phase diagram with MATLAB. In: 2010 6th international conference on wireless communications networking and mobile computing (WiCOM). IEEE, pp 1–4
5. Porter DA, Easterling KE (1992) Phase transformations in metals and alloys, Chapman & Hall, London, England
6. Dinsdale AT (1991) SGTE data for pure elements. Calphad 15:317–425
7. Liu XJ, Gao F, Wang CP, Ishida K (2008) Thermodynamic assessments of the Ag–Ni binary and Ag–Cu–Ni ternary systems. J Electron Mater 37:210–217. https://doi.org/10.1007/s11664-007-0315-1

8. Turchanin MA, Agraval PG, Abdulov AR (2007) Phase equilibria and thermodynamics of binary copper systems with 3d-metals. VI. Copper-nickel system. Powder Metall Metal Ceram 46:467–477. https://doi.org/10.1007/s11106-007-0073-x
9. Xie Y, Zhang X (1998) Phase diagram and thermodynamic properties of Ag–Cu alloys. Sci China Ser E Technol Sci 41:348–356. https://doi.org/10.1007/BF02917006

A Comprehensive Review on Wire Arc Additive Manufacturing of Inconel Superalloys

Rupendra Singh Tanwar, Pramod Bhingole, and Suyog Jhavar

Abstract In the realm of Industry 4.0, additive manufacturing (AM) techniques are poised to revolutionize conventional manufacturing processes. Unlike traditional methods where component complexity often escalates costs, AM offers a unique advantage by not adding extra expenses with increasing intricacy. Employing powder or wire feedstock, AM fabricates metal components, with wire-based AM demonstrating a higher deposition rate and reduced fabrication time compared to powder-based methods. This makes Wire Arc Additive Manufacturing (WAAM) particularly suitable for large-scale production. Nickel-based superalloys stand out for their exceptional tensile strength, corrosion resistance, and fatigue endurance at elevated temperatures, rendering them ideal for demanding applications such as aircraft, nuclear, oil and gas, turbocharger rotors, gas turbines, and the chemical industry. This article delineates various AM methodologies employed for fabricating Ni-based alloys, encompassing diverse WAAM approaches utilizing Gas Tungsten Arc Welding (GTAW), Gas Metal Arc Welding (GMAW), Plasma Arc Welding (PAW), and other wire-based AM techniques such as Electron Beam Wire feed AM and Laser Wire feed AM. It delves into the microstructural characteristics, mechanical properties, and the challenges inherent in fabrication and post-processing, crucial for achieving the desired component attributes.

Keywords Additive manufacturing · WAAM · Inconel alloys · GMAW · GTAW · PAW · Ni based alloys

R. S. Tanwar · S. Jhavar (✉)
School of Mechanical Engineering, VIT-AP University, Beside AP Secretariat Amaravati, Inavolu, Andhra Pradesh 522237, India
e-mail: Suyog.jhavar@vitap.ac.in

P. Bhingole
Institute of Infrastructure, Technology, Research and Management, Ahmedabad, India

P. Bhingole et al. (eds.), *Advances in Materials Engineering*, Lecture Notes in Mechanical Engineering, https://doi.org/10.1007/978-981-97-7114-1_10

1 Introduction

Inconel-based alloys are classified as superalloys due to their exceptional mechanical properties, especially when subjected to high temperatures. Their exceptional resistance to extended loading, which approaches their melting point, is an inherent consequence of their creep properties. Moreover, the remarkable resistance of nickel-based alloys to corrosion and oxidation in harsh operational environments has established them as auspicious contenders for a multitude of innovative applications [1]. Among nickel-based superalloys, cobalt-based superalloys, and iron-based superalloys, Inconel-based superalloys emerge as the optimal choice for components requiring exceptional performance at high temperatures to withstand creep, fatigue, and static loading conditions [2].

Additive manufacturing (AM) is an option of conventional manufacturing processes in the era of Industry 4.0 due to high flexibility to fabricate complex part, within the shortest lead time, and minimum material waste. AM is applied in the field of aerospace, medical implants, nuclear, marine, chemical Industry's component. AM opens an opportunity to fabricate plastic, ceramic, metal, multi-material, functionally graded material (FGM), and alloy design [3–5]. There are different ways of classifications of metal AM in this article described in Fig. 1 such as based on heat source it can be categorized laser, electron beam, and arc based. Whereas based on feedstock it can be categorized (1) powder feedstock and (2) wire feedstock. Based on the powder feedstock metal AM subcategories as selective laser sintering (SLS), selective laser melting (SLM), electron beam melting (EBM), laser metal deposition (LMD), laser engineered net shaping (LENS). Whereas wire feedstock AM can be classified into three categories on the basis of heat source electron beam freeform fabrication (EBF^3), wire laser AM (WLAM), and wire arc AM (WAAM). Arc can be generated by different means of welding source such as gas tungsten arc welding (GTAW), gas metal arc welding (GMAW), plasma arc welding (PAW), and hybrid method (combined two AM method).

Today many commercial AM methods are available to fabricate different means of material components. ASTM-F42 committee categorizes AM into seven classes as follows: direct energy deposition (DED), powder bed fusion (PBF), sheet lamination (SL), material extrusion (ME), binder jetting (BJ), material jetting (MJ), and vat photopolymerization (VP). The sequence is arranged in the decreasing order of energy consumption. For example, DED consumed high energy (4–6 kW) to deposit the layers while vat photopolymerization needed (40–60 mW) to cure the layers [6]. This article focuses on various AM methods to fabricate different Inconel alloys, and is mainly concerned about the wire-based AM method and detailed discussion about distinct WAAM methods used to fabricate Inconel alloy components. WAAM fabricated component called WAAMed different Inconel alloys microstructural evolution along the build direction. Further, influence of mechanical properties and microstructure of the Inconel alloys is described briefly. Basically, this review article gives a thorough insight into the different metal AM, WAAM, and wire-based AM methods,

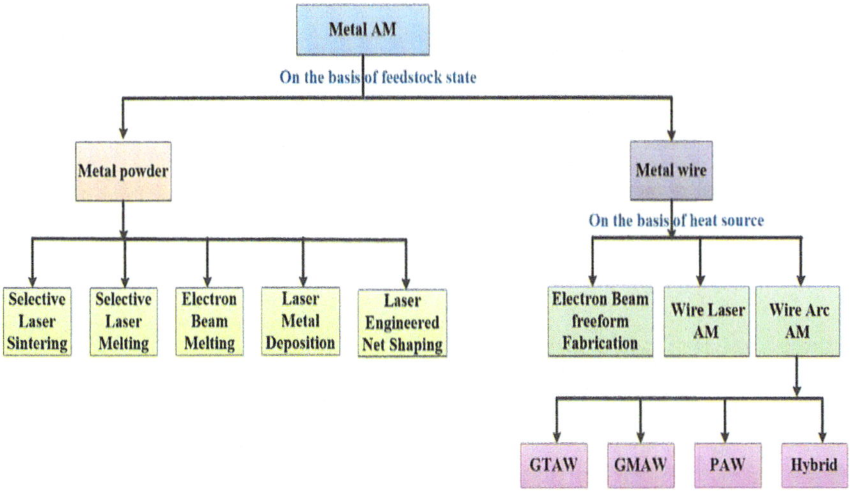

Fig. 1 Classification of metal AM

the characteristics of each method, and suitability of WAAM for the distinct Inconel alloys.

2 AM Methods for Inconel Alloys

2.1 Powder Bed Fusion (PBF)

Metal powder particle melting or sintering by means of heat source to form a layer by joining together. There are many PBF variations available in the market and each one has its traits. The basic traits of PBF are that it has a heat source to melt the metal powder, a build chamber with inert gas atmosphere that protect from oxidation of the molten pool, different mechanisms to control parameters such as heat power, layer thickness, and scanning speed. Usually, laser or electron beam heat source used in PBF. Most PBF system uses laser as a heat source, and based on metal powder particle fusion mechanisms such as sintering or melting, called selective laser sintering (SLS) or selective laser melting (SLM). In SLS, metal powder is sintered not melted to achieve desired surface diffusion of the metal particles, desired porosity, and anisotropic mechanical properties. Whereas in SLM, the metal powder is selectively melted into liquid and solidifies quickly to produce dense and homogenous structure [7].

The PBF process is illustrated in Fig. 2a this whole process performed in an inert (nitrogen and argon gas) chamber, the thickness of powder layer ranges 20–200 μm [8] are spread over the powder bed and distributed evenly using roller or re-coater.

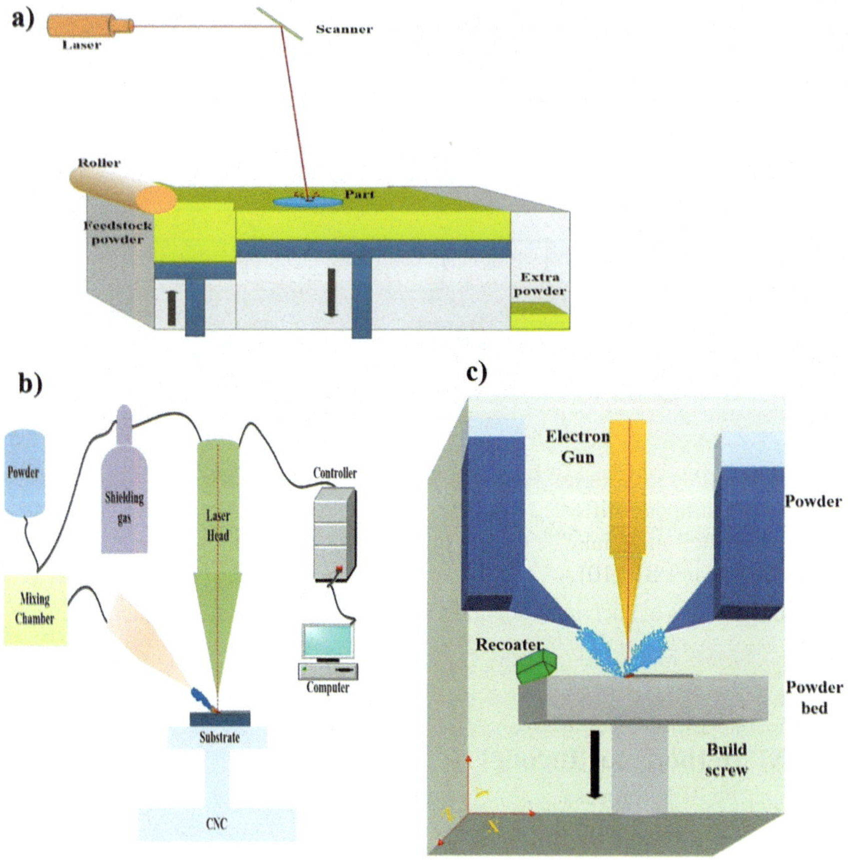

Fig. 2 Powder based AM, **a** powder bed fusion (PBF). **b** Laser metal deposition (LMD), **c** electron beam melting (EBM)

Sometimes roller does not distribute powder uniformly called roller bump; it should be considered during design phase. Whatever thickness is decided, the piston of the feedstock chamber moves up and spreads that powder over the bed. Based on CAD model, scanner changes the position and melt or sintered the deposited powder using laser. At the same time of built, support structure also built using same metal powder to mitigate distortion generated during high temperature interaction. After completion of one layer of scanning, powder bed moves in the downward direction, whereas the feedstock powder piston moves upward direction. This process continues until the desired product is obtained [9].

2.2 *Laser Direct Energy Deposition (LDED)*

Laser metal deposition is a class of direct energy deposition, it is like a welding process. Laser energy source is used to melt the metallic powder as it is deposited by the nozzle. LDED is suitable for creating near net shape component, adding new features in existing product, and repairing damaged components.

The LDED process is illustrated in the Fig. 2b, it is also performed in an inert gas (argon) chamber, which reduces the deviation of laser beam and oxidation. The feedstock (metal powder) is supplied either through blown powder nozzle or coaxial nozzle, which is targeted towards the laser heat source. As ever the metal powder and laser interact each other, forming a melt pool [8]. Laser head follow the path, as instructed by the CAD model and controller. The mixing chamber, mix the appropriate proportion of shielding gas with powder and it is supplied through nozzle. If the nozzle is co-axial the mixed powder supplied along with laser beam. LDED can create a fully dense part with tailored microstructure because of unique capability to accommodate different materials into a single build to optimize solidification rate of deposited materials. Some researchers found that LDED is faster and more economical than PBF [10].

2.3 *Electron Beam Melting*

Electron beam powder-bed fusion also named electron beam melting (EBM). The schematic of EBM process is shown in Fig. 2c, it is an AM method to fabricate Inconel-based alloy to minimizing cracking modalities to the formation of γ' precipitation. This AM uses a high energy electron beam as a heat source, it preheats, sintered, and then melt the powder which is on a bed. After fused of layer, building screw lowered and new layer of powder deposited on the bed, same process continues until final object not achieved [11]. The whole process performed under vacuum and elevated bed temperature (1000 °C for Inconel-based alloys); it provides low thermal gradient throughout the metal powder results part minimizing the residual stresses. In EBM, powder hoppers are filled with metallic powder, and it supplied powder after completion of one layer of fusion. The roller spread the powder uniformly on the bed and unfused powder work as a supporting structure for the component [12].

3 Wire-Based Different AM Methods

WAAM also categorises under DED methods, the first patent was filled in 1925 related to WAAM. In the era of AM technology, WAAM gained a lot of attention since last two decades. It is a revolutionary metal AM technique, which utilizes an electric arc as a heat source. Feedstock material is in the form of wire, which is feeded through wire feeding mechanism, and synchronizing of wire feeding system with the electric arc creates 3D component in a layer-by-layer manner. Nowadays, the whole process either controlled through computer or via automation, which provides more flexibility and controlled metal deposition. As compared to other AM manufactured product, WAAM product has better mechancial and metallurgical properties [6]. GTAW, GMAW, PAW, and CMT welding technology are used for the deposition of metals, among these GTAW, GMAW are commonly used. WAAM offers highest material deposition, complex parts at lower cost, inexpensive tooling, easily configurable, high efficiency. These advantages are achieved at the cost of surface finish which need post processing and machining. The surface quality of WAAM artfacts improves by machining and post processing [13, 14].

Upto now, GTAW, GMAW, PAW, and CMT welding techniques are used in WAAM. Each welding technique has unique feature, efficiency, and characteristics to process different material. Wire feeding system integrated with GMAW, gives more accurate tool path generation and easier programmimg as compared to GTAW and PAW because both have external wire feeding mechanism. For steel and aluminium, GMAW gives higher deposition rate as compared to GTAW and PAW, whereas for Ti alloy GMAW faces the issue of arc wandering, hence GTAW or PAW is suitable for Ti alloy fabrication [13]. On the above of the discussion, Inconel-based alloy can be processed by any of the above-mentioned welding technique [15, 16]. PAW is limited to small size components. Production rate, cost, and metal prcoessed are the criteria to select WAAM method. The next subsections give a detail discussion of different wire-based AM methods such as electron beam wire feed AM, laser wire feed AM.

3.1 GTAW-Based WAAM

Non-consumable electrode arc-welding method named GTAW, is capable of fabricating high quality metallic AM part. The schmatic diagram of GTAW is shown in Fig. 3. The GTAW consists of welding torch, wire, wire feeder, power source, etc. This above-mentioned mechanism combined with CNC and other automation system, GTAW becomes WAAM for the fabrication of complex geometries [17]. The arc formation in GTAW-based WAAM between workpiece and electrode. During fabrication process continue supply of inert gas (argon and helium) to protect component from fusion anomalies, oxidation, and corrosion. To fabricate a component of Inconel-based alloy, aluminium alloy, iron-based alloys, and Titanium alloys, argon gas is the best shielding gas [18]. To fabricate more homogenize component, H_2 (2–5%) mixed with argon reduces shielding atmosphere. On the other hand, increases the temperature of electric arc by addition of helium to argon reduces fabrication time and increase weld penetration. It increases the processing cost and sometime faces the issue to start electric arc. Addition of filler materials separately using wire feeder to achieve 15–35 g/min deposition rate [19]. For improving the production rate two wires are used instead of single wire. A single and two-wire GTAW-based

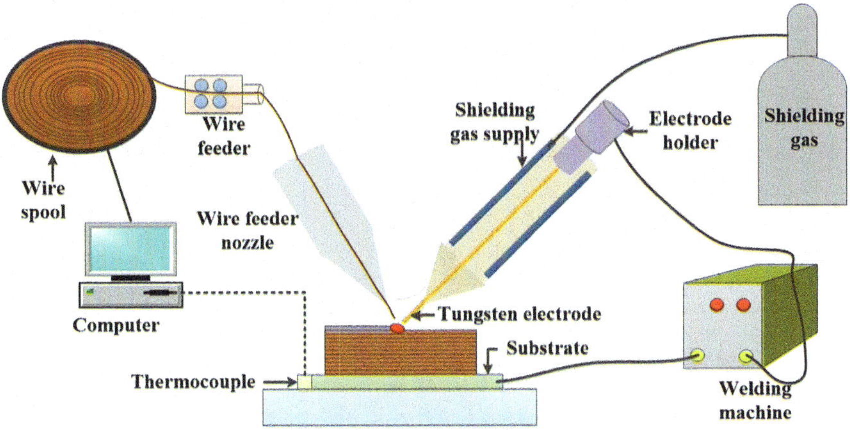

Fig. 3 Gas tungsten arc welding (GTAW) based WAAM

WAAM follows the typical steps of AM. A 3D model is created in computer, then slicing of the 3D model decides tool path of welding torch. The torch movement synchronizing with all other equipemnts of GTAW creates 3D metallic object. Wire feed rate, arc intensity, dwell time, deposition rate are optimized and decided before fabrication [20].

3.2 *GMAW-Based WAAM*

The wire that supplies continously acts as consuable electrode in GMAW-based WAAM. It is extensively used in WAAM. The basic setup of GMAW is shown in Fig. 4 it has welding torch, wire feeding unit, shilding gas supply, power source, etc. These setup combined with computer, CNC, and robot provide more flexibilty to GMAW to create complex part [21]. The arc forms between substrate and electrode. The shielding gas prevents melt pool from oxidation and water vapour. GMAW-based WAAM overcome the limitations of other AM methods such as it has higher deposition rate, affordable, can fabricate medium to large scale component at minium cost for all welded metals. There are many researchers worked towards the WAAM technique to improvise surface finish, microstructure, and mechanical properties [22]. The main goal of the majority of these investigations is to determine whether using a special purpose welding robot or a specific GMAW setup for producing or repairing metal parts using the layered AM principle is process more feasible [23, 24]. To increase deposition rate tandum welding torch is used, tandum torch feeded two wire simultaneously to get thicker section while at the same it needed external cooling to reduce the lead time [25, 26].

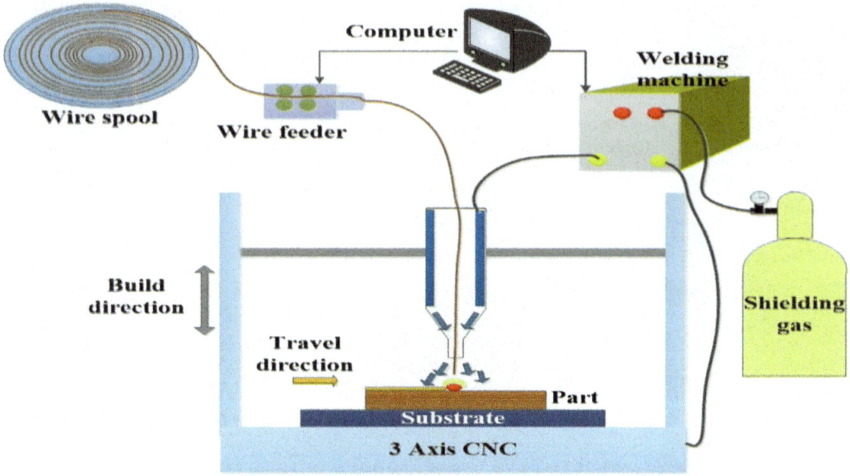

Fig. 4 Gas metal arc welding (GTAW) based WAAM

3.3 PAW-Based WAAM

PAW-based WAAM follows the same principle of GTAW, the high velocity and high temperature plasma is pushed through a nozzle as compared to other WAAM method. It's schematic is shown in Fig. 5. The arc generated between electrode and work-piece [27]. PAW welding technique made a revolutionary step in the development of WAAM. PAW was used to fabricate stainless steel part in 2006 in the area of AM. It consists of PAW torch having a plasma-assisted non-consumable electrode, usually tungsten with inert gas (Argon) as shown in Fig. 5. PAW torch follows the path of wire, which is feeded through wire feeder. Plasma melts the wire and deposited layers on over the other layer. Wire feed rate and Plasma arc (Power) are the two important parameters in case of PAW [28].

3.4 Electron Beam Wire Feed AM

Electron Beam wire feed AM is also a class of wire-based DED AM method, a high vaccum chamber protect the metallic part from oxidation, corrosion and also minimzes the deviation of electron beam [29]. It shows its capability to fabricate dense and oxidation-free components. High intensity of electron gun produces elctron beam, which is melt the wire as shown in Fig. 6. 3D model prepared in a CAD model followed by slicing of these model as per the decided layer thickness. Electron gun follows the tool path of sliced model, electron beam harmonize with the wire feed rate to fabricate 3D component. Electron beam power, three axis working table motion and wire feed rate are important parameter to fabricate defect-free component, these

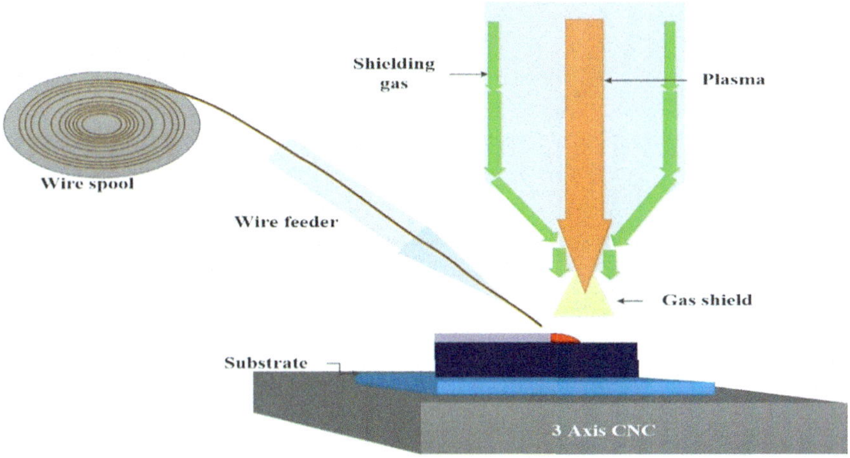

Fig. 5 Plasma arc welding (GTAW) based WAAM

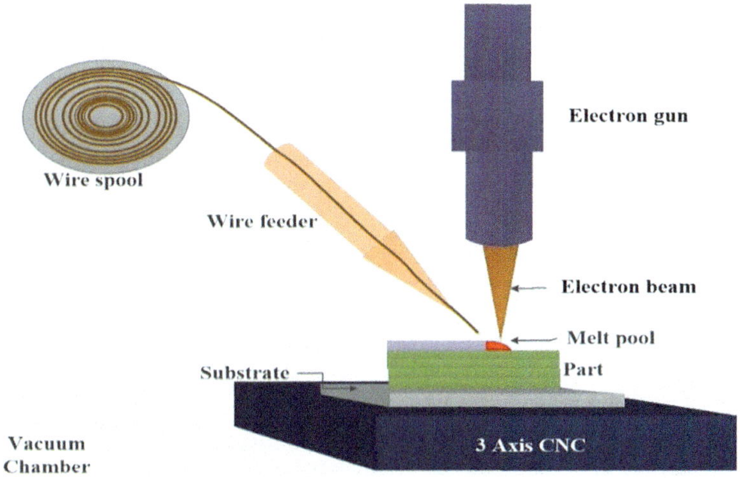

Fig. 6 Electron beam wire feed metal AM

parameters should be optimum for the particular material. The angle between fed wire and electron beam are 40–60 degree, distance of the nozzle to substrate are from 5 to 10 mm. To ensure a constant liquid metal bridge and, thus, a stable metal transfer during the process, a height distance of 0.5–1.5 mm between the substrate and wire tip was selected [30].

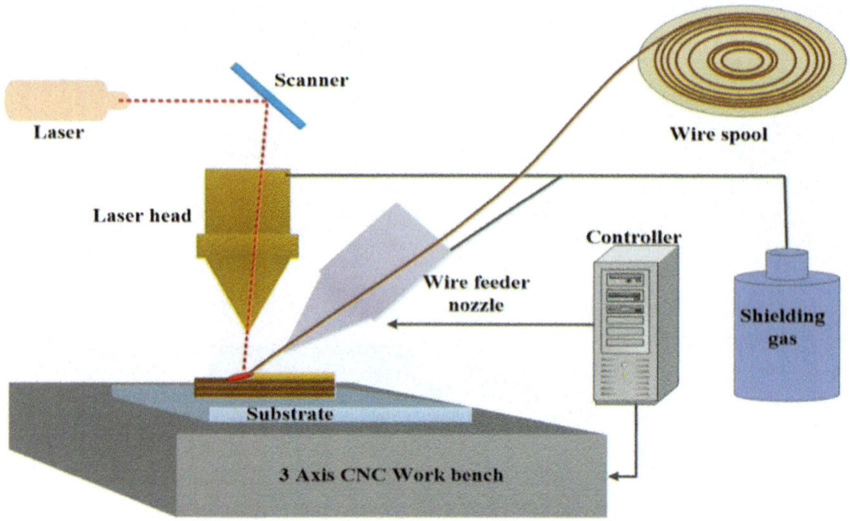

Fig. 7 Laser wire feed metal AM

3.5 Laser Wire Feed Metal AM (LWMAM)

Laser additive manufacturing classified into powder and wire-based laser additive manufacturing, over the last two-decade powder-based laser AM extensively studied by the researcher and industeries, LWMAM as described in Fig. 7. Powder-based laser AM, observed some serious issues such as low material utilization, high cost of metallic powder, imperils the health of operator, and enviornment pollution [31]. Whereas, Laser Wire feed metal AM (LWMAM) shows excellent characteristics such as higher material utilization (approx. 100%), wire available at lower cost, no harm to health operator, and enviornment friendly as compared to powder-based laser AM. LWMAM consist of high power laser, wire feed mechanism, 3Axis CNC or robot system, controller and shielding gas supply etc. [32]. The deposition efficiency of LWMAM system more as compared to powder based system, it reduces the production lead time, and increase production efficiency. The shielding gas, wire feeder, controller, and 3 axis workbench function is same as discussed in the previous methods [33].

4 Microstructure of WAAMed Inconel Alloys

To observe the microstructure and characterize by the various characterization instrument such as optical microscope, SEM, X-ray diffration, energy dispersive spectroscopy analysis, etc. To reveal the excellent microstructure subsequently grinding, polishing, and etching performed.

It is interesting to observe layer by layer deposition, melting, cooling and solidification associated through WAAM leads to noticable microstructure evolution. Grain growth and Change in microstructure depend on heat flux [34]. The grain morphology of Inconel component in terms of shape, size and orientation varied whereas hardness properties vary along the build direction. In macroscopic point of view, build width reduced along the build height [35]. Karmuhilan et al., took the three different region (top, middle, and bottom) of In625 WAAMed component. At the bootom region (near the substrate) dendrite grows perpendicular to the substrate because initially most of heat moved through the substrate, some melting the wire different rate of cooling and heat dissipation are observed from bottom to top, therefore each layer shows unique microstructure [15, 35]. The whole component encomasses columnar dendrites and cellular grains. Near the substrate improved cooling rate are observed and evolved cellular grains without secondary phase. Heat dissipation rate found lower at the middle and top region as compared to bottom region, it reduces cooling rate and significant amount of heat accumulated between the layer hence the middle region has cellular and elongated columnar dendrite structure found. Whreas, at the top region most columnar dendrites are found. Due to accumulated heat between layers residual stresses and distortion introduced within the components. It can be reduced through annealing at various temperature and medium. On the other hand, preheating of substrate is also an effective method to reduce residual stresses. It reduces thermal gradient and homogenize temperature distribution.

During observation of In718 generates different secondary phases observed in the γ matrix, and these secondary phases within a microstructure influence the mechanical properties of the final product. The component observed through SEM, the SEM images have blocky inclusions and lamellar-shaped particles. This feature concludes In718 has carbide and lave phase morphologies respectively in the microstructure. During solidificatin of deposition layers Nb segregates into the interdendritic region, which generates carbides. It can not be dissolved during the heat treatment because it forms compound with other elements. Whereas lave phases can be minimized using heat treatment [36]. WAAMed In825 component's middle portion shows columnar dendrites with different grain size. Due to the different heat transfer mode, columnar dendrite shows almost parallel orientation along the build direction, while this orientation tends toward inside along the build width [35]. This variation at the middle region because of middle region experience heat conduction, whereas side region (build width) experince conduction as well as atomospheric convection. The other microstructure is same as In625. The EDX result reveals that sum of Ni, Fe and Cr are less at near to the substrate as compared to top face of the fabricated component. Microstructure of Inconel825 columnar grains are developed, whereas ductile nature of fracture are observed with uniform distribution of micro-voids and dimples [37]. performed oxidation observation on the WAMMed In625. Ni-base Solid solution of δ-Ni_3Nb phase, lave phases, and (Nb, Ti)C carbide microstructure are formed in the build compnent, at high temperature for the certainamount of time lave phases dissolved, dendrite microstructure disapperared, δ-phase along with grain boundary, metal carbides and dense structured oxidation layer of chromium oxide are formed [38, 39]. To achieve high deposition rate, and improved efficiency achieved through high heat input, it affects adversely on the dimensional accuracy of the product [34]. Surface defects such as surface undulation can be observed due to excessive heat input generated during the deposition process. Researcher fabricated Inconel 82 component and found that the rate of cooling and heat input influences the microstructure, at the beginning of deposition highest cooling rate occurs because of high temperature gradient and high heat conduction hence the cellular microstructure is observed at the bottom of the wall because heat condition though the substrate. Whereas, at the top layer more convention and radiation occur, and the microstructure is transformed into columnar. Reported that the direction of dendritic growth is perpendicular to the substrate [40]. The direction of the heat flux is a significant determinant of the growth direction of grains; in order to follow maximum temperature gradient, grains grow preferentially in the opposite direction to the heat flux direction. Due to the addition of TiC with Nickel 690, sub-grains are replaced with dendrite grains near the substrate whereas at the top deposition of the wall dendrite grains become much finer [41]. Researcher follows the oscillation bead technique to fabricate In625 part through WAAM, the zig-zag way of deposition influences the microstructure, it produces multi-directional columnar grain growth [42]. Using this technique, the direction of the microstructure was changed layer by layer at the same location because the welding order in each layer was in the opposite direction. Researcher uses WAAM with magnetic field to fabricate In625 component, it produces fine dendrite crystals because of magnetic field dendrites are uniformlly oriented along the deposition direction [43].

5 Mechanical Properties of WAAMed Inconel Alloys

This section provides a broad overview about the mechanical properties of WAAMed Inconel alloy, first discusses the deposited different Inconel alloy component through WAAM and then discuss the process modification and material addition that enhances the mechanical properties of WAAMed Inconel alloy component.

The researcher choose two different orientations (horizontal and vertical) to observe the tensile properties of In625, horizontal sectioned component shows better tensile strength and elongation compared to the vertical sectioned component because horizontal section has more grain boundaries [44]. Whereas, the vertical section shows better ductility because it has more elongated columnar grains at differences in orientation along the bead height. The microstructure variations are observed throughout the component, which shows variation in microhardness. The bottom region has cellular dendrite structure, which leads to the hardest section, while middle region has mixed region of cellular dendrite, and elongated columnar grains decrease the hardness value, lastly top region of the build has columnar grains showing least hardness in the whole component [37]. Mechanical properties of the fabricated component improve by preheating the substrate. Compared the In825 component with casting component, In825 component shows higher tensile strength, yield strength, and percentage elongation as compared to the casted sample.

The percentage elongation characteristics same as In625 [37]. Tensile properties are observed along the longitudinal and transverse direction, researcher found that in Inconel 82, transverse specimen showed superior mechanical properties than longitudinal specimen [34]. It is because of grain boundary strengthening mechanism, and exhibits unique anistropy in the mechanical properties of the component. Due to increase in travel speed, solidification time reduces, it does not form precipitates if γ' and γ'' therefore it increases the microhardness due to grain refinement due to variation in microstrcuture, the value of mechanical properties and micro-hardeness value varies along the build direction [40]. The mechanical properties tendency increases from near-substrate to top layer, whereas micro-hardness tendency decreases from near-substrate to top layer observed in Inconel 625. [27] researcher uses pulsed plasma-based WAAM to fabricate In718 alloy, Wang et al., in PAW-based WAAM, during solidifcation process high cooling rates are observed, and it evolved abundant fine columnar dendrites. It grew epitaxially along the build direction. At the bottom region, columnar dendrites with primary dendrite arms, it is replaced with cellular dendrite arms at the top region, and this variation of microstructure from columnar to cellular dendrite changes along the build direction. The Ni–Cr-based -phase makes up the majority of the crystal structure of the Inconel 625 sample that was fabricated through WAAM [45]. Due to the segregation of Nb presence at the interdendritic gaps, undesired delta phases are generated during heat treatment earlier than anticipated. Researcher compared the corrosion behavior of WAAM-HT IN718 and wrought HT alloy, WAAM-HT IN718 exhibits lower corrosion resistance as compared to wrought HT alloy because of Cr_2O_3 and NiO passive film formed and it gives a more porous structure, it reduces corrosion protection [36]. To improve corrosion resistance, x jiang22 et al., added TiC particles in Nickel alloy 690 [41]. Due to addition of TiC, more passive layers of Cr oxide and hydroxide are formed. Fabricated Invar36 component through wire arc AM and observe machinability characterization in end milling [46]. The hardness and force values are more as compared to the wrought one, whereas the surface roughness value is much lower, and most importantly surface roughness value is independent of the cutting direction. The microhardness at the interface is slightly less than that at the passes for the longitudinal section. The inter-pass boundary remelts when the molten weld layer is deposited next to the earlier one [47]. Due to the aligned orientation of microstructure because of magnetic field shows higher yield strength, micro-hardness, and more ductility as compared to without magnetic field In625 component [43].

6 Summary and Conclusion

This paper thoroughly investigates several metal additive manufacturing (AM) technologies, focusing specifically on wire-based AM techniques and providing a detailed analysis of various Wire Arc Additive Manufacturing (WAAM) approaches. The study also examines the microstructure and mechanical characteristics of Inconel alloys fabricated using the Wire Arc Additive Manufacturing (WAAM) technique.

This includes an analysis of tensile properties, corrosion behavior, and hardness across several Inconel variations. This article examines the literature to provide a detailed overview of recent progress in the field of Wire Arc Additive Manufacturing (WAAM), with a specific focus on its use in producing various types of Inconel alloys. Additionally, it explores alterations to WAAM processes with the goal of creating flawless components that have improved microstructural integrity and mechanical performance. The main findings of this review, backed by numerous references, highlight the adaptability of WAAM for Inconel alloy manufacturing and stress the significance of comprehending the relationship between microstructure and properties to enhance component performance. Discussed about different metal AM methods such as PBF, LDED, and EBM, which are used to fabricate Inconel alloy, and detail discussion of wire-based AM methods and review focused on WAAM.

a. WAAM showcases high efficiency in manufacturing Inconel alloy components, resulting in reduced time and material waste and ultimately lowering production expenses. The capacity to adapt to complex designs enables the production of customized components.
b. Ni-based superalloys are preferred for applications that require resistance to corrosion and high temperatures.
c. Microstructural changes in Inconel alloys fabricated by Wire Arc Additive Manufacturing (WAAM) are observed depending on the build direction. At the substrate interface, cellular structures are formed as a result of heat conduction. As we move towards the top layer, columnar patterns are observed due to convective and radiative heat transfer.
d. The mechanical characteristics of Inconel components produced using the WAAM process are affected by these variations in microstructure. Comparative analyses conducted on wrought Inconel alloys demonstrate that WAAMed components have enhanced tensile strength and hardness qualities, which can be related to the development of precipitates. However, this improvement in attributes is accompanied by a reduction in corrosion resistance due to the creation of oxide films.
e. The utilization of magnetic fields and the inclusion of TiC are strategies employed to improve the mechanical properties and corrosion resistance of WAAMed Inconel alloy components.

7 Future Scope

This analysis focuses on the difficulties encountered in attaining uniform microstructures and mechanical qualities in Inconel components produced by Wire Arc Additive Manufacturing (WAAM) and wire-based AM techniques. The analysis indicates that depending just on composition and solidification conditions is inadequate for achieving homogeneity in Inconel components produced using WAAM. Differences in parameters, operating circumstances, and heat transfer modes during deposition lead to distinct microstructures and mechanical properties along the build direction. Future research should prioritize the integration of thermo-mechanical process parameters to achieve consistent microstructures and mechanical properties throughout the whole component. In addition, the investigation of external interventions, such as the application of magnetic fields or other upgrades, has the potential to enhance both the microstructure and mechanical properties of the manufactured components.

References

1. Shahwaz M, Nath P, Sen I (2022) A critical review on the microstructure and mechanical properties correlation of additively manufactured nickel-based superalloys. J Alloys Compd 907:164530. https://doi.org/10.1016/j.jallcom.2022.164530
2. Hosseini E, Popovich VA (2019) A review of mechanical properties of additively manufactured Inconel 718, 30
3. Gisario A, Kazarian M, Martina F, Mehrpouya M (2019) Metal additive manufacturing in the commercial aviation industry: a review 53(June):124–149
4. Culmone C, Smit G, Breedveld P (2019) Additive manufacturing of medical instruments: a state-of-the-art review 27(February):461–473
5. Zhang C (2019) Additive manufacturing of functionally graded materials: a review. Mater Sci Eng A 764. https://doi.org/10.1016/j.msea.2019.138209
6. Srivastava M, Rathee S, Tiwari A, Dongre M (2023) Wire arc additive manufacturing of metals: a review on processes, materials and their behaviour. Mater Chem Phys 294:126988. https://doi.org/10.1016/j.matchemphys.2022.126988
7. Zhong C, Gasser A, Backes G, Fu J, Schleifenbaum JH (2022) Laser additive manufacturing of Inconel 718 at increased deposition rates. Mater Sci Eng A 844:143196. https://doi.org/10.1016/j.msea.2022.143196
8. Armstrong M, Mehrabi H, Naveed N (2022) An overview of modern metal additive manufacturing technology. J Manuf Process 84:1001–1029. https://doi.org/10.1016/j.jmapro.2022.10.060
9. Yang H, Yang J, Huang W, Wang Z, Zeng X (2018) The printability, microstructure, crystallographic features and microhardness of selective laser melted Inconel 718 thin wall. Mater Des 156:407–418. https://doi.org/10.1016/j.matdes.2018.07.007
10. Moradi M (2022) Direct laser metal deposition (DLMD) additive manufacturing (AM) of Inconel 718 superalloy: elemental, microstructural and physical properties evaluation. Optik (Stuttg) 259:169018. https://doi.org/10.1016/j.ijleo.2022.169018
11. Sun S, Koizumi Y, Saito T, Yamanaka K, Li Y, Cui Y (2018) Electron beam additive manufacturing of Inconel 718 alloy rods: impact of build direction on microstructure and high-temperature tensile properties 23(July 2017):457–470
12. Balachandramurthi AR, Moverare J, Mahade S (2019) Additive manufacturing of alloy 718 via electron beam melting: effect of post-treatment on the microstructure and the mechanical properties. https://doi.org/10.3390/ma12010068
13. Vimal KEK, Srinivas MN, Rajak S (2020) Materials today: proceedings wire arc additive manufacturing of aluminium alloys: a review
14. Suyog Jhawar I (2021) Atria Institute of Technology, Banglore.: Wire arc additive manufacturing approaches, pp 183–209
15. Karmuhilan M, Kumanan S (2022) A review on additive manufacturing processes of Inconel 31(April):2583–2592
16. Singh Tanwar R, Jhavar S (2024) Ti based alloys for aerospace and biomedical applications fabricated through wire + arc additive manufacturing (WAAM). Mater Today Proc 98:226–232. https://doi.org/10.1016/j.matpr.2023.11.121

17. Moreira AF, Ribeiro KSB, Coelho RT (2020) An Initial Investigation of tungsten inert gas (TIG) torch as heat source an initial investigation of tungsten inert gas (TIG) torch as heat source for additive manufacturing (AM) process. Addit Manuf 00(2019)
18. Cai X, Dong B, Yin X, Lin S, Fan C, Yang C (2020) Wire arc additive manufacturing of titanium aluminide alloys using two-wire TOP-TIG welding: processing , microstructures, and mechanical properties 35
19. IvánTabernero, Paskual A, Álvarez P, Suárez A (2018) Study on arc welding processes for high deposition rate additive manufacturing. Procedia CIRP 68:358–362. https://doi.org/10.1016/j.procir.2017.12.095
20. Lan B (2021) The influence of microstructural anisotropy on the hot deformation of wire arc additive manufactured (WAAM) Inconel 718. Mater Sci Eng A 823:141733. https://doi.org/10.1016/j.msea.2021.141733
21. Feucht T, Lange J, Erven M, Borg C, Ulrich C, Benedikt K (2020) Additive manufacturing by means of parametric robot programming (0123456789)
22. Ogino Y, Asai S, Hirata Y (2018) Numerical simulation of WAAM process by a GMAW weld pool model, pp 393–401
23. Panchagnula JS, Simhambhatla S (2018) Manufacture of complex thin-walled metallic objects using weld-deposition based additive manufacturing. Robot Comput Integr Manuf 49(May 2017):194–203
24. Jain NK, Sawant M, Nikam SH (2017) Metal deposition: plasma-based processes. Encycl Plasma Technol (2020):722–740. https://doi.org/10.1081/e-eplt-120053919
25. Martina F, Ding J, Williams S, Caballero A, Pardal G, Quintino L (2019) Tandem metal inert gas process for high productivity wire arc additive manufacturing in stainless steel 25(October 2018):545–550
26. Raushan H (2023) Dry sliding and slurry abrasion behaviour of wire arc additive manufacturing—Cold metal transfer (WAAM-CMT) cladded inconel 625 on EN8 steel. Tribol Int 179:108176. https://doi.org/10.1016/j.triboint.2022.108176
27. Wang K, Liu Y, Sun Z, Lin J, Lv Y, Xu B (2020) Microstructural evolution and mechanical properties of Inconel 718 superalloy thin wall fabricated by pulsed plasma arc additive manufacturing. J Alloys Compd 819:30–40. https://doi.org/10.1016/j.jallcom.2019.152936
28. Kumar N (2022) Wire arc additive manufacturing—A revolutionary method in additive manufacturing. Mater Chem Phys 285:126144. https://doi.org/10.1016/j.matchemphys.2022.126144
29. Pu Z (2023) Improvement of tensile superelasticity by aging treatment of NiTi shape memory alloys fabricated by electron beam wire-feed additive manufacturing. J Mater Sci Technol 145:185–196. https://doi.org/10.1016/j.jmst.2022.10.050
30. Pixner F, Buzolin R, Warchomicka F, Pilz A, Enzinger N (2022) Wire-based electron beam additive manufacturing of tungsten. Int J Refract Metals Hard Mater 108:105917. https://doi.org/10.1016/j.ijrmhm.2022.105917
31. Cazic I, Zollinger J, Mathieu S, El Kandaoui M, Plapper P, Appolaire B (2021) New insights into the origin of fine equiaxed microstructures in additively manufactured Inconel 718. Scr Mater 195:113740. https://doi.org/10.1016/j.scriptamat.2021.113740
32. Huang W, Chen S, Xiao J, Jiang X, Jia Y (2021) Laser wire-feed metal additive manufacturing of the Al alloy. Opt Laser Technol 134:106627. https://doi.org/10.1016/j.optlastec.2020.106627
33. Zhang W, Xu Y, Shi Y, Su G, Gu Y, Volodymyr K (2022) Intergranular corrosion characteristics of high-efficiency wire laser additive manufactured Inconel 625 alloys. Corros Sci 205:110422. https://doi.org/10.1016/j.corsci.2022.110422
34. Anand M, Kumar Das A (2022) Grain refinement in wire-arc additive manufactured Inconel 82 alloy through controlled heat input. J Alloys Compd 929:166949. https://doi.org/10.1016/j.jallcom.2022.166949
35. Bhuvanesh Kumar M, Sathiya P, Rajesh Kannan G, Karthikeyan M (2022) Investigation on the microstructure and microhardness of Inconel 825 thick wall fabricated by wire arc additive manufacturing. Mater Lett 317:132115. https://doi.org/10.1016/j.matlet.2022.132115

36. Jang J, Yim J, Lee SH (2022) Phase formation kinetics during heat treatments of Inconel 718 deposits based on cold metal transfer-based wire arc additive manufacturing. Mater Charact 193:112294. https://doi.org/10.1016/j.matchar.2022.112294
37. Senthil TS, Puviyarasan M, Ramesh Babu S, Senthil S (2022) Mechanical characterization of wire arc additive manufactured and cast Inconel 825: a comparative study. Mater Today Proc 62:973–976. https://doi.org/10.1016/j.matpr.2022.04.096
38. Sharifitabar M, Khorshahian S, Shafiee Afarani M, Kumar P, Jain NK (2022) High-temperature oxidation performance of Inconel 625 superalloy fabricated by wire arc additive manufacturing. Corros Sci 197:110087. https://doi.org/10.1016/j.corsci.2022.110087
39. Ravi G, Murugan N, Arulmani R (2020) Microstructure and mechanical properties of Inconel-625 slab component fabricated by wire arc additive manufacturing fabricated by wire arc additive manufacturing. https://doi.org/10.1080/02670836.2020.1836737
40. Yangfan W, Xizhang C, Chuanchu S (2019) Microstructure and mechanical properties of Inconel 625 fabricated by wire-arc additive manufacturing. Surf Coat Technol 374:116–123. https://doi.org/10.1016/j.surfcoat.2019.05.079
41. Jiang X, Di X, Li C, Wang D, Hu W (2022) Improvement of mechanical properties and corrosion resistance for wire arc additive manufactured nickel alloy 690 by adding TiC particles. J Alloys Compd 928:167198. https://doi.org/10.1016/j.jallcom.2022.167198
42. Cheepu M, Ik C, Sang L, Cho M (2020) Microstructural characteristics of wire arc additive manufacturing with Inconel 625 by super-TIG welding 73:1475–1479
43. Wang Y (2021) Effect of magnetic Field on the microstructure and mechanical properties of inconel 625 superalloy fabricated by wire arc additive manufacturing. J Manuf Process 64:10–19. https://doi.org/10.1016/j.jmapro.2021.01.008
44. Karmuhilan M, Kumanan S (2023) Location-dependent microstructure analysis and mechanical behavior of inconel 625 using cold metal transfer (CMT) based wire and arc additive manufacturing. Vacuum 207:111682. https://doi.org/10.1016/j.vacuum.2022.111682
45. Tanvir ANM, Ahsan RU, Ji C, Hawkins W, Bates B, Kim DB (2019) Heat treatment effects on Inconel 625 components fabricated by wire + arc additive manufacturing (WAAM)—Part 1: microstructural characterization, pp 3785–3798
46. Gil Del Val A, Cearsolo X, Suarez A, Veiga F, Altuna I, Ortiz M (2023) Machinability characterization in end milling of Invar 36 fabricated by wire arc additive manufacturing. J Mater Res Technol 23:300–315. https://doi.org/10.1016/j.jmrt.2022.12.182
47. Chintala A, Kumar MT, Sathishkumar M, Arivazhagan N, Manikandan M (2021) Technology development for producing Inconel 625 in aerospace application using wire arc additive manufacturing process 30(July):5333–5341

Additive Manufacturing of Al_2O_3 Ceramic Components: Rheological and Microstructural Characterization

Rupendra Singh Tanwar, Swaraj Sarkar, M. Vignesh, and Suyog Jhavar

Abstract Additive manufacturing has emerged as a promising method for creating intricate structures using a variety of materials, including metals, polymers, and ceramics. Direct Ink Writing (DIW) has proven to be particularly valuable for efficiently and affordably producing ceramic components with complex geometries. This study investigates how printing parameters impact the effectiveness of Al_2O_3 refractory product fabrication via DIW. Specifically, a 95% alumina (Al_2O_3) mixture with low viscosity and high solid loading is employed for DIW 3D printing, utilizing 4% polyvinyl acetate (PVA) and 1% hydroxypropyl methylcellulose (HPMC) water-based solutions as binders. The resulting components are sintered in a furnace based on thermal gravimetric analysis (TGA) and differential scanning calorimetry (DSC) findings. Rheological analysis confirms the printability of the prepared slurry. X-ray diffraction (XRD) identifies the presence of Al_2O_3, PVA, and HPMC elements, while scanning electron microscopy (SEM) images illustrate the bonding between PVA, HPMC, and Al_2O_3.

Keywords Additive manufacturing · Direct ink writing · Ceramic additive manufacturing · Al_2O_3

1 Introduction

The progress in additive manufacturing has revolutionized the creation of complex structures in diverse sectors, including aerospace and biomedical engineering. Direct Ink Writing (DIW) 3D printing is a versatile and efficient method for creating intricate ceramic components. The extraordinary precision of its material deposition control in three dimensions allows for the production of complicated forms and tailored material attributes [1, 2]. This work explores the synthesis and characterization of Al_2O_3 composite ceramic using the DIW-3D printing method and sintering in order to

R. S. Tanwar · S. Sarkar · M. Vignesh · S. Jhavar (✉)
School of Mechanical Engineering, VIT-AP University, Beside AP Secretariat Amaravati, Inavolu, Andhra Pradesh 522237, India
e-mail: Suyog.jhavar@vitap.ac.in

P. Bhingole et al. (eds.), *Advances in Materials Engineering*, Lecture Notes in Mechanical Engineering, https://doi.org/10.1007/978-981-97-7114-1_11

achieve superior mechanical and thermal properties. Alumina, or ceramic aluminum oxide (Al_2O_3), is well known for its exceptional mechanical strength, ability to withstand high temperatures, and chemical inertness. Because of these qualities [3], it is ideal for a variety of engineering applications, including structural parts, biological implants, and thermal barriers [4, 5]. However, the use of conventional production methods like casting and molding frequently results in geometric constraints and limits on design freedom [6]. Conversely, DIW-3D printing offers a viable way around these limitations because it allows ceramic ink to be directly deposited layer by layer, making it easier to create complex structures with unique microstructures. With DIW-3D printing, the primary steps are printing, drying, sintering, and ink formulation. Creating a printable ink with ceramic particles suspended in a binder matrix is the first stage [7, 8]. This ink must have the necessary rheological properties to permit extrusion through a fine nozzle and maintain structural integrity during deposition. The final properties of the produced ceramic and printability are determined by a number of factors, such as particle size distribution, binder composition, and solvent choice. Once the ink is produced, it is placed into a printer equipped with a robotic arm that is precise enough to transform the design blueprint into real structures. During printing, ink is gradually built up to the desired shape by applying layers of ink onto a substrate following a predetermined path. After printing, the deposited layers of the green ceramic structure are dried to remove excess solvent and binder components. Sintering, a thermal treatment process that solidifies the ceramic particles and eliminates any leftover porosity by growing new necks and rearranging old ones, involves heating the dried green body to a high temperature [9, 10]. The final microstructure and mechanical properties of the ceramic composite are significantly influenced by sintering factors such as temperature, time, and environment. Its performance characteristics, like heat conductivity, wear resistance, and toughness, can also be enhanced by incorporating reinforcements or secondary phases into the ceramic matrix. One of the key benefits of DIW-3D printing is its capacity to precisely control the orientation and spatial distribution of reinforcing phases inside the ceramic matrix, which enables the creation of multifunctional composites [2, 11]. Fibers, whiskers, or nanoparticles can be carefully added to the ink formulation to achieve synergistic effects. This will improve the mechanical properties and practicality of the ink [12]. For instance, adding carbon nanotubes (CNTs) to Al_2O_3 composites can increase their electrical conductivity and fracture toughness, opening up possibilities for innovative applications in electronics, energy storage, and aircraft. In addition to mechanical reinforcement, the DIW-3D printing technique enables the incorporation of complex internal architectures, such as lattice frameworks or hierarchical porous structures, that can offer additional capabilities like lightweighting, acoustic damping, or fluid transfer [13, 14]. These intricately planned designs are simple to produce with DIW's layer-by-layer deposition capacity, something that is challenging to accomplish with conventional production procedures [15, 16]. In summary, the combination of DIW-3D printing and sintering provides a workable technique for creating Al_2O_3 composite ceramics with tailored microstructures and enhanced properties. By taking advantage of the design freedom and material flexibility offered by this additive manufacturing technique, researchers and engineers may pave the way for innovation in a range of industries, including electronics, aerospace, automotive, and healthcare. In order to highlight the potential for transformative applications in engineering and other fields, this work will analyze the present advancements, challenges, and prospective future directions in the development of DIW-3D printed Al_2O_3 composite ceramics.

2 Material Selection and Preparation

In Direct Ink Writing (DIW) of alumina, the constituents comprise alumina (Al_2O_3), hydroxypropyl methylcellulose (HPMC), and polyvinyl alcohol (PVA). Al_2O_3 powder serves as the ceramic component, while PVA solvent and HPMC binder are designated as ink constituents.

2.1 Preparation of Slurry

To guarantee ideal printing characteristics and the quality of the finished product, the Al_2O_3 powder is usually evaluated for particle size, distribution, and purity. To create a homogenous ink solution, Al_2O_3 is combined with PVA solvent and HPMC binder at the proper ratios. The extrusion via the printing nozzle is made possible by adjusting the viscosity and rheological characteristics of the ink to preserve shape integrity. To ensure purity and consistency in the ink formulation, distilled water is employed along with PVA and HPMC during the creation of Al_2O_3 composite ceramic via DIW-3D printing [17–19]. Tap water impurities may have an impact on the ink's characteristics, causing unfavorable changes in its viscosity and drying speed, and eventually lowering the quality of the print. Based on the ratios of important components, a precise formulation technique is used to generate ink appropriate for direct ink writing (DIW). Each sample with alumina weighs 5 g, 7 g, and 11 g, respectively. To make each sample, the same proportion of 1 g Hydroxypropyl Methylcellulose (HPMC) and 10 ml Polyvinyl Alcohol (PVA) is used. The ratios of alumina to the mixture are calculated, and the results for the 5 g, 7 g, and 11 g samples are 0.4167, 0.5833, and 0.9167, respectively. After making these calculations, the ink is made by thoroughly combining the distilled water, PVA, and HPMC in the prescribed amounts to get the right consistency. Viscosity is adjusted as needed to satisfy the unique demands of the DIW procedure. The required structures are then created by loading the ink into the DIW printer and extruding it onto the printing substrate, as shown in Fig. 1. This methodical process guarantees the consistency and dependability of the ink composition for accurate research uses in DIW technology [5, 20].

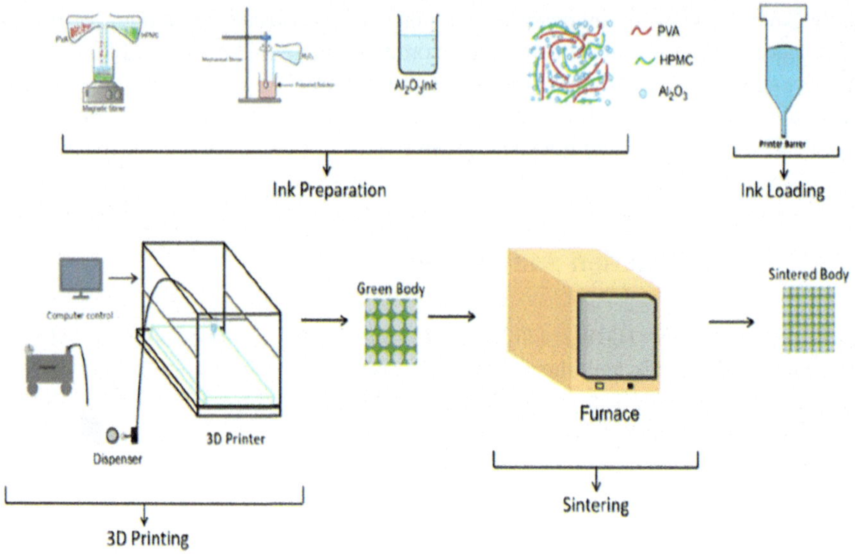

Fig. 1 Steps for the preparation of Al_2O_3 3D printed part

2.1.1 Alumina (Al_2O_3)

In order to ensure optimal printing properties and the caliber of the final product, the Al_2O_3 powder is often assessed in terms of purity, particle size, and distribution. The bulk of alumina, or aluminum oxide, is composed of oxygen and aluminum molecules. Its outstanding mechanical, electrical, thermal, and chemical properties enable a multitude of industrial uses. Alumina is known for its high hardness, stiffness, and thermal stability.

2.1.2 Hydroxypropyl Methylcellulose

Hydroxypropyl methylcellulose, or HPMC, is a multipurpose polymer that finds extensive application in several industries because of its capacity to create films, dissolve in water, and thicken materials. It is used as a surface modification in construction, as a stabilizer and thickening in food and cosmetic products, and as a binder and sustained-release agent in medications. Because of its biocompatibility and ability to work well with other constituents, HPMC is a useful material for a range of production processes and formulations. The rheology, film-forming, and dispersion properties of HPMC, biodegradable cellulose, are essential for DIW and improve the assessment of mechanical properties. Its heat gelation tendency makes it indispensable to the ink-making process as well.

2.1.3 Polyvinyl Alcohol (PVA)

Vinyl acetate monomers are used to create the polymer known as polyvinyl alcohol (PVA). It has several uses because of its unique properties, which include being biodegradable and soluble in water. PVA is a versatile material with a broad range of consumer and industrial applications when all things are taken into account. Biodegradable PVA acts as a binder, controlling the ink's viscosity and enabling proper extrusion during printing. It also enhances the printed items' green strength.

2.2 *Direct Ink Writing*

Direct ink writing (DIW) is a 3D printing technique that uses extrusion and runs at temperatures close to room temperature. It offers one of the largest selections of printable materials among additive manufacturing techniques and uses an ink dispensing device that is computer-controlled to create layer-by-layer three-dimensional patterns, as shown in Fig. 2. When it comes to traditional micromanufacturing methods, DIW has a number of benefits. It usually requires fewer processing steps, doesn't require vacuum or cleanroom procedures, produces less waste, and works well with flexible substrates [21].

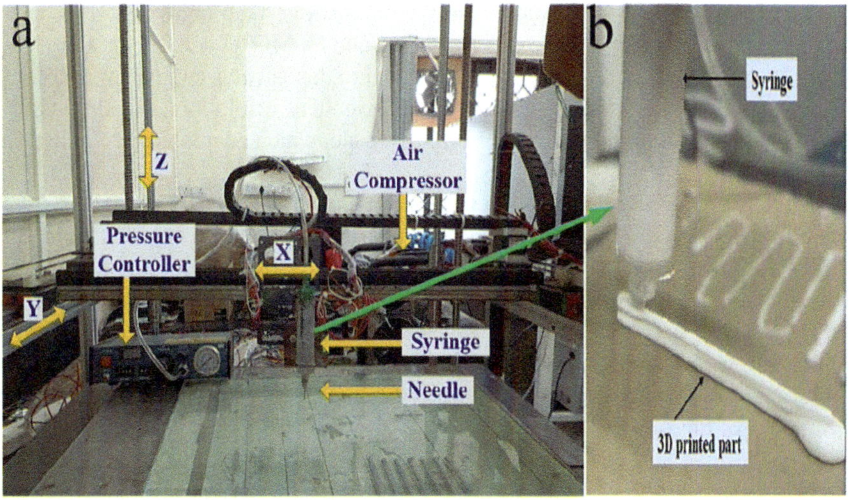

Fig. 2 **a** Modified 3D printer (DIW), **b** Al_2O_3 printing

2.3 Computer-Aided Design (CAD) Modeling

The process of creating Al_2O_3 composite ceramic for Direct Ink Writing (DIW)-3D printing heavily relies on computer-aided design (CAD) modeling. The use of CAD software makes it easier to create complex geometries and exact measurements, which are essential for maximizing the final ceramic components' structural integrity and functionality. When designing the ideal architecture, engineers take mechanical characteristics, porosity, and interconnectivity into account. CAD models are transformed into a format that can be used with the DIW-3D printing program, which prints the model in layers that can be printed. Moreover, CAD enables virtual prototyping and simulation, which gives designers the ability to forecast and optimize the ceramic structure's behavior under diverse circumstances. Support is provided for iterative design procedures, which help to develop and refine the finished product. In the end, computer-aided design (CAD) modeling enables engineers to create intricate ceramic structures with customized characteristics, improving the functionality and adaptability of Al_2O_3 composite ceramic components in a range of applications.

2.4 Printing Process

The first step in creating Al_2O_3 composite ceramic using Direct Ink Writing (DIW)-3D printing is to load the printer's syringe or cartridge with uniformly mixed ink, which consists of Al_2O_3 powder, PVA solvent, and HPMC binder. The printer uses a predefined toolpath created from a CAD model to deposit consecutive layers of ink. The production of intricate geometries with excellent resolution is made possible by controlled extrusion through the printing nozzle, which guarantees accurate layer deposition.

The ink is carefully dried during printing to encourage layer adhesion and avoid distortion. To preserve shape fidelity and microstructural integrity, this technique requires meticulous control of printing parameters such as layer height, extrusion speed, and nozzle temperature. The calibration of the printer guarantees precise placement and motion, which enhances the overall caliber of the ceramic structure that is printed. The printing process is an important phase in the additive manufacturing workflow because it determines the final microstructure and mechanical properties of the Al_2O_3 composite ceramic.

2.5 Sintering

When using Direct Ink Writing (DIW)-3D printing to manufacture Al_2O_3 composite ceramic, sintering is a crucial step. The green ceramic preform is subjected to sintering, a high-temperature procedure that promotes densification and improves

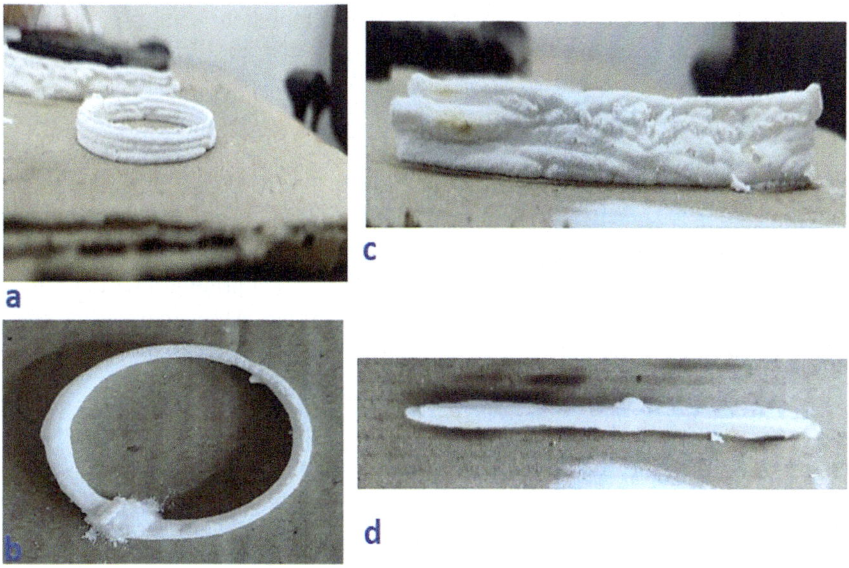

Fig. 3 **a** and **b** Al_2O_3circular part front and top. **c** and **d** Al_2O_3 wall part front and top

mechanical qualities, following printing. First, organic solvents and binders are eliminated by heating, leaving a porous structure in their place. Particle rearrangement and neck development are the results of ceramic particles bonding through surface diffusion as temperature rises, as shown in Fig. 3. Densification results from this process because the substance shrinks and pores are eliminated. Achieving the correct microstructure and mechanical qualities requires careful control of the sintering parameters, including temperature, time, and environment. Prolonged sintering durations can produce grain growth and microstructure coarsening, which can impact material qualities. Rapid heating rates can cause thermal stresses and cracking. In order to reduce residual stresses and avoid cracking, proper cooling is necessary after sintering. All things considered; sintering turns the green ceramic preform into a dense, structurally sound Al_2O_3 composite ceramic that can be used in a variety of ways [22].

3 Results and Discussion

One well-liked additive manufacturing method for creating intricate ceramic structures, such Al_2O_3 pieces, is Direct Ink Writing (DIW). A variety of characterization methods are used to comprehend these parts' mechanical characteristics. In the process of characterizing the mechanical properties of Al_2O_3 parts by DIW. The microstructure analysis was conducted using a ZEISS EV0 10 High Resolution scanning electron microscope (SEM). X-ray diffraction (XRD) patterns were obtained using a Rigaku MiniFlex600 equipped with CuKα radiation, with a scanning rate of 10°/min and diffraction angle (2θ) set between 15 and 20°. Rheology analysis is obtained using Anton Paar MCR102 to find the shear rate, shear stress, and viscosity of liquid. Differential Scanning Calorimetry (DSC) & Thermogravimetric Analysis (TGA) were obtained using Perkin Elmer STA8000 to assess the thermal stability and heat resistance properties of the material. These techniques provide insights into the material's ability to withstand elevated temperatures and its thermal degradation behavior.

3.1 Rheology Analysis

The methodical examination and quantification of a material's flow and deformation properties as a result of applied stresses or forces is known as rheology analysis. It entails examining the behavior of various materials—such as liquids, soft solids, or viscoelastic materials—under various circumstances, including shear, stress, strain, temperature, and time. Understanding a material's mechanical characteristics, such as its viscosity, elasticity, yield stress, and shear modulus, as well as how it flows under different conditions, is the goal of rheology study. Understanding and managing the flow properties of materials is crucial for product development, process optimization, and quality assurance in a variety of industries, including materials science, engineering, food processing, pharmaceuticals, cosmetics, and the petroleum sector. Rheology analysis uses mathematical modeling in conjunction with experimental methods to describe and forecast the behavior of soft materials and complex fluid [3]. The observed pattern, viewed through the prism of rheological analysis, manifests as a general upward tilt in the curve, suggesting that shear stress will correspondingly increase with an increase in shear rate. This empirical relationship suggests that the material under study experiences shear thinning, or a drop in viscosity as shear rates increase. The observed trend in the curve demonstrates a general increase, indicative of a concurrent elevation in viscosity with escalating shear rates. This phenomenon signifies a characteristic behavior of shear thickening within the material under scrutiny, wherein its viscosity exhibits augmentation as the shear rate intensifies. This distinctive trait is emblematic of materials that manifest shear thickening, wherein their viscosity amplifies in response to heightened shear rates, as shown in Fig. 4.

3.2 Diffraction Scanning Calorimetry (DSC) and Thermogravimetric Analysis (TGA)

Thermal deterioration is analyzed by TGA, which analyzes weight variations with temperature. Phase transitions and material properties are characterized by DSC,

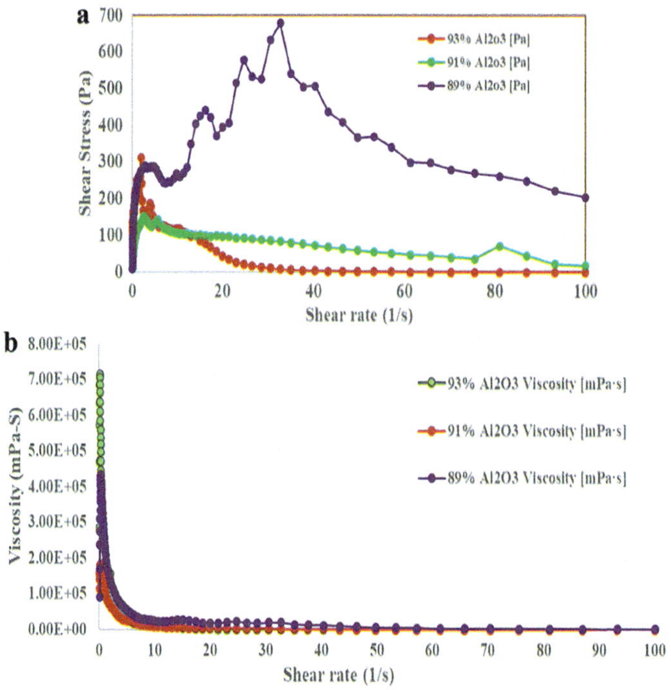

Fig. 4 Rheology results. **a** Shear stress versus Shear rate. **b** Viscosity versus Shear rate

which monitors heat flow in temperature changes. Both support the creation and analysis of materials. The Transition gradient (TG) value observed is 75.32 °C, as shown in Fig. 5, The heat flow that the empty sample holder exhibits when the temperature is gradually raised is represented by the baseline, which is represented by the blue line, as shown in Fig. 6. This baseline is used as a standard by which to evaluate the sample's thermal behavior. The green line, representing the peak, represents the heat flow that is seen during a thermal event in the sample. The thermal event in this case is described as endothermic, meaning that heat was absorbed by the sample from its environment. The amount of heat absorbed by the sample during this process can be inferred from the magnitude of the peak, which is represented by the area under the curve. In addition, the textual annotation located in the graph's lower portion offers more information on the thermal event. It states that 75.32 °C was the highest temperature that was recorded during the procedure. It also reports the sample's enthalpy, or heat absorbed, during the event, which comes out to be 770.3682 J/g. This thorough analysis offers important new information for future research and characterization by quantifying the amount of heat absorption as well as explaining the sample's thermal behavior.

The graphical depiction shows that when temperature rises, the sample's relative humidity noticeably decreases. The aforementioned tendency can be explained by the intrinsic correlation between elevated temperature and the energy state of the water

Fig. 5 Diffraction scanning calorimetry (DSC) of Al_2O_3

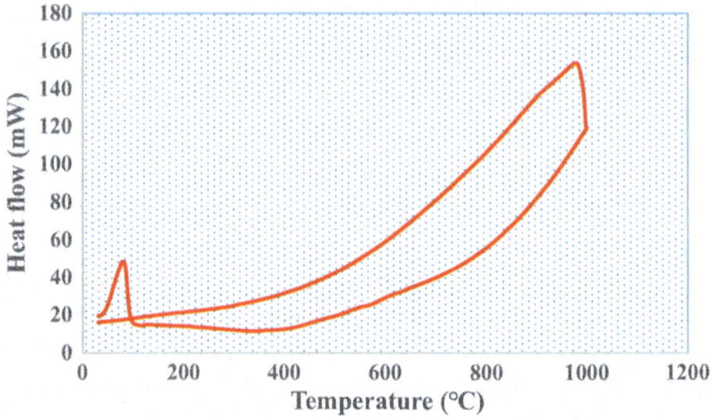

Fig. 6 Thermogravimetric analysis (TGA) of Al_2O_3

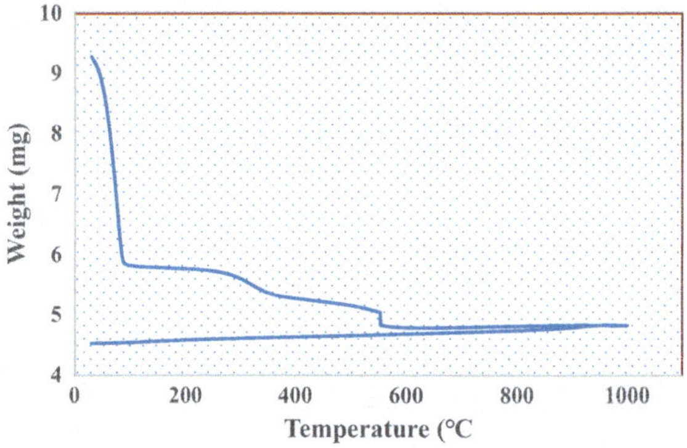

molecules present in the specimen. The kinetic energy of water molecules increases with temperature, increasing their tendency to evaporate. This process results from the molecules' improved ability to overcome intermolecular interactions and change from a liquid to a gaseous state due to their higher thermal energy availability. As a result, the rate of evaporation increases in direct proportion to temperature increase.

3.3 X-Ray Diffraction Analysis (XRD)

The objective of this experiment's XRD investigation is to clarify the Al_2O_3 sample's phase composition and crystal structure, as shown in Fig. 7. Diffraction is produced by X-ray irradiation interacting with the lattice structure after sample preparation and loading. After the data is gathered, it is examined. The main phase of alumina, chemically known as aluminum oxide (Al_2O_3), is called corundum. At normal temperatures, the aluminum oxide crystal known as corundum is the most stable phase of alumina. It occurs naturally in a range of colors, including transparent, brown, blue, and red, depending on the number of impurities. Pure corundum has no color and

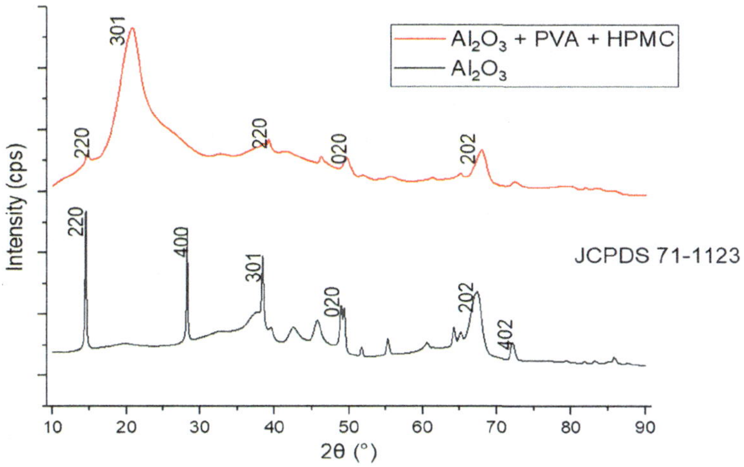

Fig. 7 Phase data & Peak profile of Alumina

is transparent. The powder sample's crystal structure and possible preferred orientations can be inferred from the X-ray diffraction (XRD) spectra. While preliminary examination may identify several features typical of aluminum oxide, more research is required to accurately identify its crystalline shape. Extensive analysis of the XRD spectrum can provide important insights into sample impurities. For example, the sample may not consist entirely of pure aluminum oxide if peaks corresponding to aluminum metal are present [23]. This shows that the sample contains contaminants. In addition, the features of the peaks found in the XRD spectrum provide more information. Sharper peaks usually imply greater grain sizes within the sample, while larger peaks may indicate that the powder is made up of smaller grains. This data facilitates comprehension of the material under study's microstructural characteristics. It is important to take into account a number of parameters that may have an impact on the strength of peaks seen in the XRD spectrum. Peak strength can be influenced by the amount of material used for analysis; more samples may provide more prominent peaks. Peak intensity is also influenced by the sample's capacity to absorb X-rays, which has an impact on the diffraction pattern's overall clarity and resolution. In summary, although XRD analysis offers a useful preliminary characterization of the powder sample, more research and careful evaluation of many impacting elements are necessary to gain a thorough grasp of its crystalline structure and properties.

3.4 Scanning Electron Microscope (SEM)

Scanning Electron Microscopy (SEM), a technique renowned for its high-resolution imaging capabilities, offers a comprehensive analysis of the microstructural features of Al_2O_3, as demonstrated in Fig. 8, in samples subjected to additive manufacturing, provided optimal scanning parameters such as speed and voltage are employed. SEM facilitates meticulous examination of nano-scale attributes of the alloy, encompassing phase distribution, grain morphology, and overall material architecture. Grain structure, a pivotal aspect of microstructural investigation, is directly influenced by the scanning speed during additive manufacturing. Higher scanning speeds yield finer grain structures, indicative of rapid solidification and reduced grain growth. SEM imagery furnishes quantitative insights into grain size and morphology, elucidating the influence of varied processing conditions on grain characteristics. The results of the study reveal a gradient in grain structures from the substrate to the surface due to disparate thermal histories at different depths. Figure 8a, b denote a diminished uniformity in particle dispersion, potentially leading to the formation of voids or suboptimal interfacial bonding between Al_2O_3 particles. Conversely, Fig. 8c, d illustrate a more uniform dispersion of Al_2O_3 particles throughout the matrix material, facilitating augmented interfacial adhesion and densification of the composite structure.

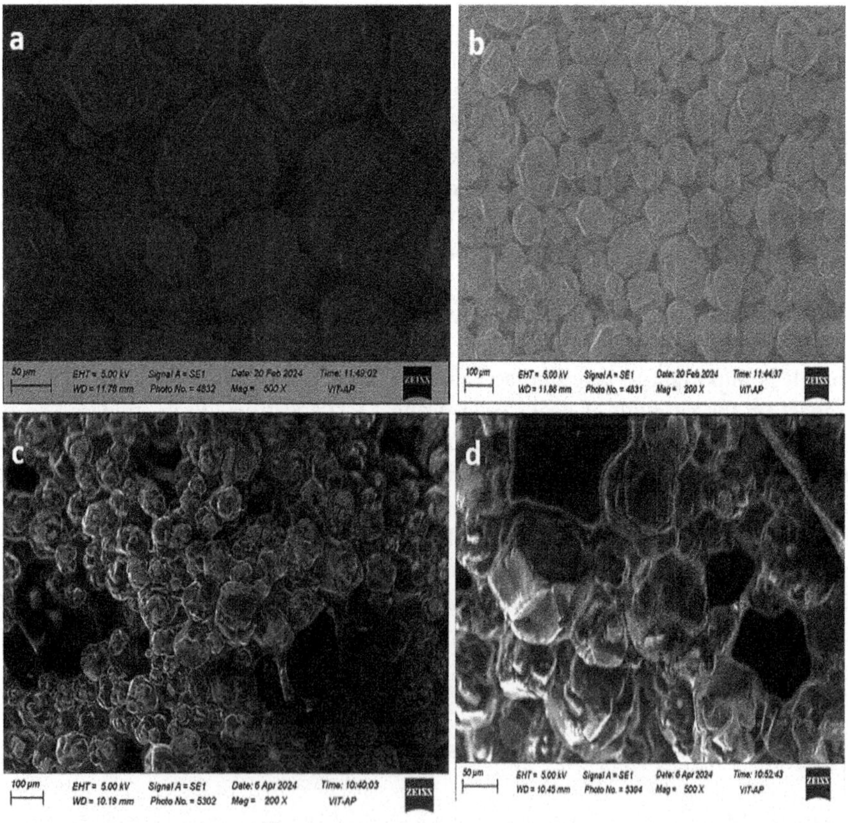

Fig. 8 SEM images. **a** and **b** Al_2O_3 powder. **c** and **d** Al_2O_3 + PVA + HPMC slurry

4 Conclusion

This study focuses on the fabrication of ceramic components using Direct Ink Writing (DIW), preparing Al_2O_3 slurry consisting of 95 wt.% Al_2O_3, 4 wt.% PVA, and 1 wt.% HPMC. The research systematically examines the effects of layer thickness and sintering temperature on the overall properties of these components. The conclusions drawn from the investigation are outlined as follows:

a. A cost-effective, in-house developed 3D printer has been engineered to produce green Al_2O_3 components of varying structures, including walls and cylinders.
b. The ceramic Al_2O_3 slurry, formulated using environmentally friendly and cost-effective materials such as PVA and HPMC, facilitates the fabrication of complex ceramic components tailored for specific applications.
c. Rheological characterization reveals that both shear stress and viscosity of the ceramic slurry decrease in relation to shear rate, indicating its suitability for printing.
d. Differential Scanning Calorimetry (DSC) and Thermogravimetric Analysis (TGA) results provide insights into the sintering temperature range.
e. X-ray Diffraction (XRD) analysis confirms the presence of Al_2O_3, PVA, and HPMC elements, while Scanning Electron Microscopy (SEM) images illustrate the bonding between Al_2O_3, PVA, and HPMC constituents.

Acknowledgements The authors acknowledge the funding supported by VIT-AP University through a project VIT-AP/SPORIC/RGEMS/2023-24/001. Additionally, the authors thank Mr. E. Krushna Murty, Mr. Prasad Loya, and Mr. Ranjan Rana (Technical Assistant, VIT, Amaravati) for their assistance during the testing process.

References

1. Michna S, Wu W, Lewis JA (2005) Concentrated hydroxyapatite inks for direct-write assembly of 3-D periodic scaffolds. Biomaterials 26(28):5632–5639. https://doi.org/10.1016/j.biomaterials.2005.02.040
2. Ling Q, Yang L, Tang S, Fan Z, Liu X, Jiang W (2022) Direct ink writing of hierarchically porous Al2O3 matrix composites with enhanced wettability of Al. J Manuf Process 84:1580–1588. https://doi.org/10.1016/j.jmapro.2022.10.018
3. Xing H (2018) Preparation and characterization of UV curable Al2O3 suspensions applying for stereolithography 3D printing ceramic microcomponent. Powder Technol 338:153–161. https://doi.org/10.1016/j.powtec.2018.07.023
4. Wang HZ, Gao L, Guo JK (2000) Effect of nanoscale SiC particles on the microstructure of Al2O3 ceramics. Ceram Int 26(4):391–396. https://doi.org/10.1016/S0272-8842(99)00069-3
5. Liu X, Zou B, Xing H, Huang C (2020) The preparation of ZrO2–Al2O3 composite ceramic by SLA-3D printing and sintering processing. Ceram Int 46(1):937–944. https://doi.org/10.1016/j.ceramint.2019.09.054
6. Jhavar S, Paul CP, Jain NK (2014) Experimental investigation on geometrical aspects of micro-plasma deposited tool steel for repair applications. Int J Mod Phys Conf Ser 32:1460347. https://doi.org/10.1142/S2010194514603470

7. Luo ZP, Song YG, Zhang SQ (2001) A TEM study of the microstructure of SiCp/Al composite prepared by pressureless infiltration method. Scr Mater 45(10):1183–1189. https://doi.org/10.1016/S1359-6462(01)01148-4
8. Kota N, Charan MS, Laha T, Roy S (2022) Review on development of metal/ceramic interpenetrating phase composites and critical analysis of their properties. Ceram Int 48(2):1451–1483. https://doi.org/10.1016/j.ceramint.2021.09.232
9. Shi Z, Yang JM, Lee JC, Zhang D, Lee HI, Wu R (2001) The interfacial characterization of oxidized SiC(p)/2014 A1 composites. Mater Sci Eng A 303(1–2):46–53. https://doi.org/10.1016/S0921-5093(00)01943-2
10. Lin Q (2022) Kinetic analysis of wetting and spreading at high temperatures: a review. Adv Colloid Interface Sci 305. https://doi.org/10.1016/j.cis.2022.102698
11. Shen P, Fujii H, Matsumoto T, Nogi K (2003) Wetting of (0 0 0 1) α-Al2O3 single crystals by molten Al. Scr Mater 48(6):779–784. https://doi.org/10.1016/S1359-6462(02)00518-3
12. Zhang FZ, Kato T, Fuji M, Takahashi M (2006) Gelcasting fabrication of porous ceramics using a continuous process. J Eur Ceram Soc 26(4–5):667–671. https://doi.org/10.1016/j.jeurceramsoc.2005.07.021
13. Kufelt O, El-Tamer A, Sehring C, Meißner M, Schlie-Wolter S, Chichkov BN (2015) Water-soluble photopolymerizable chitosan hydrogels for biofabrication via two-photon polymerization. Acta Biomater 18:186–195. https://doi.org/10.1016/j.actbio.2015.02.025
14. Li H, Liu Y, Zeng Q, Liang J (2021) 3D printed ceramic slurries with improved solid content through optimization of alumina powder and coupling agent. J Manuf Process 64(February):1206–1213. https://doi.org/10.1016/j.jmapro.2021.02.047
15. Nwogu NC, Kajama M, Gobina E (2015) A study of gas diffusion characteristics on a micro porous composite silica ceramic membrane. Compos Struct 134:1044–1050. https://doi.org/10.1016/j.compstruct.2015.07.121
16. Konegger T, Williams LF, Bordia RK (2015) Planar, polysilazane-derived porous ceramic supports for membrane and catalysis applications. J Am Ceram Soc 98(10):3047–3053. https://doi.org/10.1111/jace.13758
17. Rao PR, Muralidharan K, Momayez M, Runge K (2021) A multiscale microstructural characterization of SiO2–Al2O3 foams. Mater Charact 181. https://doi.org/10.1016/j.matchar.2021.111433
18. Zhang G, Jiang J, Wang H, Qian L, Lan H (2021) Continuous DLP-based ceramic 3D printing using a composite oxygen-rich film. J Manuf Process 64(February):341–348. https://doi.org/10.1016/j.jmapro.2021.01.041
19. Pinargote NWS, Smirnov A, Peretyagin N, Seleznev A, Peretyagin P (2020) Direct ink writing technology (3d printing) of graphene-based ceramic nanocomposites: a review. Nanomaterials 10(7):1–48. https://doi.org/10.3390/nano10071300
20. Zheng T, Wang W, Sun J, Liu J, Bai J (2020) Development and evaluation of Al2O3–ZrO2 composite processed by digital light 3D printing. Ceram Int 46(7):8682–8688. https://doi.org/10.1016/j.ceramint.2019.12.102
21. Su J (2022) Three-dimensional printing of gyroid-structured composite bioceramic scaffolds with tuneable degradability. Biomater Adv 133(November):2021. https://doi.org/10.1016/j.msec.2021.112595
22. Gaudillere C, Serra JM (2016) Freeze-casting: Fabrication of highlyporous and hierarchical ceramic supports for energy applications. Boletín de la Sociedad Española de Cerámica y Vidrio 55(2):45–54. https://doi.org/10.1016/j.bsecv.2016.02.002
23. Wang N, Sun F, Zhou Q, Liao DM, Yan YW (2019) First-principles study on hydrogen diffusion mechanism of α-Al2O3/Ti2O3 interface. Fusion Eng Des 142:40–44. https://doi.org/10.1016/j.fusengdes.2019.04.069

Comprehensive Review of Foaming Agents in Melt Processing for Lightweight Metallic Foam Production

Manan Shah, Dhaval Makwana, and Bhingole Pramod

Abstract Metallic foams are advanced lightweight materials with numerous applications in industries such as aerospace, automotive, and construction. Melt processing is a widely employed technique for the production of metallic foams, wherein foaming agents play a crucial role in controlling the foam formation process. This comprehensive review provides an in-depth analysis of foaming agents used in melt processing for lightweight metallic foam production. Various types of foaming agents, including blowing agents, and foam stabilizers are discussed, along with their mechanisms of action and effects on foam microstructure and pore size. The review covers a range of metallic foam systems, with a focus on aluminium and magnesium foams. Furthermore, optimization strategies for foaming agent selection, concentration, and processing parameters are explored, along with recent advances and future research directions in the field. Overall, this review serves as a valuable resource for researchers, engineers, and practitioners interested in the development and application of lightweight metallic foams produced via melt processing techniques.

Keywords Lightweight metallic foam · Pore size distribution · Viscosity · Blowing agents

1 Introduction

Lightweight metallic foams have garnered significant interest across various industries due to their unique combination of strength, stiffness, and low density [1]. Aluminum and magnesium foams represent a class of advanced materials that have

M. Shah · D. Makwana
Mechanical and Aerospace Engineering, Institute of Infrastructure Technology, Research and Management, Ahmedabad, Gujarat 380026, India

B. Pramod (✉)
Institute of Infrastructure Technology, Research and Management, Ahmedabad, India
e-mail: pramod_bhingole@yahoo.com

P. Bhingole et al. (eds.), *Advances in Materials Engineering*, Lecture Notes in Mechanical Engineering, https://doi.org/10.1007/978-981-97-7114-1_12

gained significant attention across a spectrum of industries due to their exceptional properties and versatility [2, 3]. Characterized by their lightweight nature, high strength-to-weight ratio, and unique porous structure, aluminum and magnesium foams offer a multitude of applications spanning from aerospace and automotive engineering [4] to biomedical and energy sectors [5–7].

In recent years, there has been a growing emphasis on the development and utilization of lightweight materials to enhance efficiency, reduce energy consumption, and minimize environmental impact. Aluminum and magnesium foams emerge as promising solutions to meet these objectives, offering substantial weight savings without compromising on mechanical performance. Their low density, coupled with inherent strength and stiffness, makes them ideal candidates for replacing conventional materials in structural components, resulting in improved fuel economy, extended range, and enhanced payload capacity in aerospace and automotive applications [3, 4, 8]. Moreover, the biocompatibility and biodegradability of magnesium foams have paved the way for their utilization in biomedical applications, ranging from orthopedic implants and tissue engineering scaffolds to drug delivery systems. These foams offer a unique combination of mechanical support, tissue ingrowth, and controlled degradation, facilitating improved patient outcomes and minimizing long-term complications [2, 9–11].

Among the various methods employed for the production of lightweight metallic foams, melt processing techniques stand out for their ability to create intricate structures and tailor foam properties to specific application requirements. Central to the success of melt processing techniques is the use of foaming agents, which play a crucial role in generating the porosity essential for lightweight foam structures [12, 13]. The selection and optimization of foaming agents are pivotal in achieving desired foam characteristics, including pore size, distribution, density, and mechanical properties. Consequently, a comprehensive understanding of foaming agents and their effects on the melt processing of metallic foams is essential for advancing foam production techniques and enhancing foam performance.

This review provides a thorough examination of foaming agents utilized in melt processing for lightweight metallic foam production. It explores the various types of foaming agents, their mechanisms of action, and their influence on foam microstructure and properties. Additionally, the review highlights recent advancements in foaming agent technology, discusses challenges and limitations associated with their use, and identifies potential future directions for research and development in this field.

2 Exploring Foaming Agents in Aluminum and Magnesium Foams

2.1 *Blowing Agents*

Blowing agents play a crucial role in controlling the expansion and density of metallic foams. By adjusting the type and concentration of blowing agents, manufacturers can tailor the foam's properties to meet specific application requirements [14, 15]. Blowing agents such as titanium hydride and calcium carbonate are frequently employed in aluminum and magnesium foam production to achieve desired foam morphology and mechanical properties. Metallic hydrides, such as titanium hydride (TiH_2) and magnesium hydride (MgH_2), are commonly used as foaming agents in the production of metallic foam. These hydrides release hydrogen gas when heated, which serves as the blowing agent, promoting the formation of gas bubbles within the molten metal [12, 16]. The decomposition reactions of metallic hydrides can be represented as follows in Eqs. (1) and (2):

For Titanium Hydride (TiH_2):

$$2TiH_2 \rightarrow 2Ti + 2H_2 \quad (1)$$

For Magnesium Hydride (MgH_2):

$$2MgH_2 \rightarrow 2Mg + 2H_2 \quad (2)$$

In these reactions, the metallic hydride decomposes into its respective metal (titanium or magnesium) and hydrogen gas. The released hydrogen gas forms bubbles within the molten metal, leading to foam expansion and the formation of a porous structure.

Carbonates, such as calcium carbonate ($CaCO_3$) [17], sodium bicarbonate ($NaHCO_3$), Dolomite [$(Ca, Mg)CO_3$] [18, 19]. These compounds decompose upon heating, releasing carbon dioxide gas, which acts as the blowing agent. The decomposition reactions of carbonates can be represented as follows in Eq. (3), (4) and (5):

For Calcium Carbonate ($CaCO_3$) [6]:

$$CaCO_3 \rightarrow CaO + CO_2 \quad (3)$$

For Sodium Bicarbonate ($NaHCO_3$):

$$NaHCO_3 \rightarrow Na_2CO_3 + H_20 + CO_2 \quad (4)$$

For Dolomite [20, 21]:

$$CaMgCO_3 \rightarrow CaO + MgO + CO_2 \quad (5)$$

The choice between metallic hydrides and carbonates as foaming agents depends on various factors, including the specific requirements of the foam, the metal being processed, and the desired processing conditions. Both types of foaming agents undergo controlled decomposition to release gas, which facilitates the formation of metallic foam with tailored properties.

2.2 Foam Stabilizers

Foam stabilizers, such as surfactants and particulate reinforcements, are often incorporated into the melt to enhance foam stability and mechanical properties. Surfactants reduce surface tension, promoting uniform bubble distribution and preventing coalescence. Particulate reinforcements, such as ceramic or metallic particles, act as nucleation sites for gas bubble formation, leading to improved foam structure and mechanical strength. In aluminum and magnesium foam production, Ca or SiC [22] in different amounts are added in the melt to increase the viscosity. Too low viscosity leads to the bubbles growth in upward direction and higher viscosity leads to the suppression of bubbles [23]. The viscosity of the melt depends on the temperature and concentration of the foam stabilizers.

3 Foaming Parameters Influence on Microstructural Characteristics and Pore Size Distribution in Lightweight Alloy Foams

3.1 Effects of Blowing Agents on Microstructure and Pore Size Distribution

Blowing agents are compounds that undergo decomposition or vaporization when subjected to heat, liberating gases that generate bubbles within the molten metal. The following mechanisms occur during the foaming process and significantly affect the pore size distribution in aluminum and magnesium foams.

Gas Evolution: The gas emitted by the blowing agent forms bubbles within the molten metal, subsequently solidifying to establish pores in the foam structure [24]. The volume and speed of gas evolution from the blowing agent directly impact the pore size distribution in the resulting foam. The volume (V) of gas (in m^3) released when the blowing agent is heated to specific temperatures is calculated using Eq. 6 [16]:

$$V = \frac{m}{M} \cdot \frac{R.T}{p} \tag{6}$$

where:

m	represents the quantity of blowing agent (in grams)
M	indicates the molecular weight of the blowing agent (in grams per mole)
p	is the pressure (1.01×10^5 Pascal)
$R = 8.314\ J\ K^{-1}\ mol^{-1}$	(the universal gas constant)
T	is the temperature.

Theoretical calculations were conducted by Iva Nova et al. [16] to determine the amount of various blowing agents in 1 kg of aluminium based on Eq. 6. It was discovered that with 0.5 wt.% of TiH_2, 0.0077 m^3 of H_2 gas evolved, and similar computations were performed for Calcium hydride and Magnesium carbonate. It is deduced that an increase in the quantity of blowing agents results in a corresponding increase in the evolved gas. Furthermore, with the augmentation of TiH_2 from 0.5 to 2 wt.%, there is a nearly 74% reduction in density in aluminium foam. Higher rates of gas evolution typically yield finer pores, whereas slower rates lead to the formation of larger pores. Xia et al. [25] experimented AZ31 foam with $CaCO_3$ as the blowing agent and upon investigation, it was found that in specimens featuring a pore size of 1.2 mm, the cell edges may become excessively narrow to adequately supply molten material, thereby significantly compromising mechanical properties. Consequently, the plateau strength decreases. Conversely, in specimens with pore sizes of 1.8 and 2.0 mm, larger pore sizes induce density inhomogeneity within the foams resulting in reduced plateau strength.

Nucleation and Growth: Blowing agents act as nucleation sites for pore formation during foam expansion. The size and distribution of these nucleation sites influence the final pore structure of the foam. As the decomposition temperature of blowing agents is reached (460 °C for TiH_2 powder [5, 26], the hydrogen gas induces expansion in the melt, whereby the internal gas pressure, P_{H_2}, exceeds external forces. These external forces comprise the pressure within the bubble arising from the bubble-melt interfacial energy, P_C, and the ambient pressure, P_A [27]. Equation (7) describing the continuous growth of the hydrogen bubble in the melt is expressed:

$$P_{H_2} \geq P_A + P_C \tag{7}$$

where $P_C = 2\sigma/r$ (σ represents the surface tension of the melt; r denotes the radius of the bubble). Consequently, pressure is directly proportional to the inverse of the bubble radius, indicating that smaller bubbles experience greater pressure. C. C Yang. et. al suggested that TiH_2 particles have the capacity of wetting with the aluminium melt and cause the reaction very rapidly when in contact with melt aluminium and causes the generation of supersaturated hydrogen.

The growth of pores during solidification depends on gas availability and the cooling rate of the metal matrix. In Al–Si–Mg alloy, within the initial 120 s of cooling with a chill plate at the crucible bottom, foam pores appear smaller and homogenous. However, after 420 s, they enlarge and become spherical. The relationship between foam pores, solidification time, and the radius of spherical pores is described by Eq. 8 [27]:

$$R(t) = \left[\frac{K}{\sqrt{D} \pm \frac{\sqrt{K^2}}{D} + 2(0.9253K)} \right] t^{\frac{1}{2}} \tag{8}$$

where $K = D\,(C - C_w)\rho_g \sqrt{3\pi}$

t pore formation time
D diffusion coefficient of hydrogen (blowing gas)
ρ_g density of the gas
C gas concentration in the liquid
C_w gas concentration in the pore.

Thus, blowing agents act as nucleation sites, significantly influencing the size and distribution of pores within the foam [28, 29]. Overall, these findings underscore the multifaceted interplay of blowing agents, solidification dynamics, and material properties in shaping the microstructure of metal foams, offering insights crucial for optimizing foam production processes and tailoring foam properties for diverse applications.

Foaming Temperature: Foaming temperature plays a critical role in the production of aluminium and magnesium foams via melt processing techniques. The temperature at which the foaming process occurs directly influences the formation, morphology, and properties of the resulting foam. It determines the onset of gas evolution from blowing agents and subsequent foam expansion. At higher temperatures, blowing agents decompose more rapidly, leading to increased gas generation and enhanced foam expansion. Conversely, lower foaming temperatures may result in slower gas evolution and reduced foam expansion. Thus, optimizing foaming temperature is crucial for controlling foam density and pore structure.

Movahedi et al. [13] prepared foam A356 Aluminium foam at 650, 675, and 700 °C with TiH_2 as the blowing agent. At low temperatures, the cell walls were thick and pore size was smaller, and at high temperatures (700 °C) the cell walls coalescence is observed. When the temperatures increases, the pressure inside the bubbles will increase and the overpressure makes the cell wall thin leading to the easy rupture of the wall. Also, the shape of the pores is influenced by density and surface tension of the melt and is governed by Eotovos number. At 650 °C, the pores were round ion shape and with increasing temperature (700 °C), the pores became hexagonal in shape. Excessive temperatures can cause defects such as coalescence and collapse of pores, compromising foam quality. Also, increased foaming temperature decreases the compressive strength by 75% and absorption strength by 50%.

Blowing agents play a crucial role in shaping the microstructure of aluminium and magnesium foams, thereby influencing their mechanical, thermal, and acoustic properties [20, 30, 31]. Through systematic investigation and optimization of blowing agent type, concentration, and processing parameters, researchers can achieve precise control over microstructural characteristics, enabling the development of advanced metal foams with tailored properties for diverse applications.

3.2 Assessing the Influence of Foam Stabilizers on Viscosity in Melt Foaming Production of Aluminium and Magnesium Foam

Foam stabilizers, such as Ca or SiC, are added to the melt to control bubble nucleation, growth, and coalescence. These additives can alter the viscosity of the melt by modifying the interfacial properties between the gas and liquid phases. The viscosity of the melt influences foam expansion dynamics during the foaming process. Lower viscosity promotes faster gas diffusion and bubble coalescence, resulting in more rapid foam expansion. Conversely, higher viscosity may hinder bubble coalescence and lead to slower expansion rates. Several studies have investigated the impact of different types and concentrations of foam stabilizers on melt viscosity in aluminium and magnesium foam production. These studies have shown that foam stabilizers generally reduce the surface tension of the melt, leading to lower viscosity due to decreased resistance to flow. However, the specific influence on viscosity depends on factors such as the chemical composition of the stabilizer, its concentration, and the alloy composition.

Zhen Lun Song [23] added Ca in the pure alumnium melt to produce Al foam with TiH_2 powders. With the addition of the Ca, from 0 to 1.8 wt.% the apparent viscosity (η/mPa s) increased from 1 to 8. The Ca additions from 0 to 2.5 wt.% decrease the pore diameter from 1.8 to 0.5 mm. The more the viscous melt the thickness of the cell walls increases and it prevents the bubble for flowing out of the film. The melt viscosity may not be too high as it may hinder the distribution of blowing agents and thus the non-uniform pores may be generated. Yang et al. [27] suggested that the 0.29 Nm torque of the melt is optimum for creating an Al–Mg–Si foam which is 7 times greater than the original melt. The aluminium powders were added to the melt which reacted to become alumina particles in the presence of air and the alumina particles increased the viscosity leading the high stirring torque of the motor. X. C. Xia used 2 wt.% Ca in AZ31 melt as the thickening agent to form magnesium form with $CaCo_3$ as the blowing agent. The parameters for adding thickening agents like Ca are reviewed from Xu Zhou et al. [26] which depicted addition of 3 wt.% of ca in pure Alumnium melt at 680 °C and stirred for 10 min at 800 Rpm.

The influence of foam stabilizers on viscosity in the melt foaming production of aluminium and magnesium foam is a complex and multifaceted phenomenon. Understanding this influence is crucial for optimizing foam properties and production processes to meet the requirements of various industrial applications. Further research efforts are needed to explore the mechanisms underlying the influence of foam stabilizers on melt viscosity and their subsequent impact on foam properties, paving the way for the development of advanced foam production techniques and materials.

4 Conclusion

In conclusion, this comprehensive review has provided a detailed examination of foaming agents utilized in melt processing for the production of lightweight metallic foams. Through an exploration of various foaming agent types, including blowing agents and foam stabilizers, alongside their mechanisms of action and effects on foam properties, a comprehensive understanding of the role of foaming agents in metallic foam production has been elucidated.

Furthermore, the review has highlighted optimization strategies for foaming agent selection, concentration, and processing parameters, providing valuable insights for researchers, engineers, and practitioners seeking to tailor foam properties to meet the demands of diverse applications in industries such as aerospace, automotive, and construction. Overall, this review serves as a comprehensive resource for those involved in the development and application of lightweight metallic foams, offering a foundation for further research and innovation in the field of foam production and utilization.

Acknowledgements This work was supported by Department of Science and Technology; Science and Engineering Research Board, Government of India (Grant No. ECR/2016/001518). Authors acknowledge the Siemens Centre of Excellence; Centre of Advanced Defence Technology (CADT), IITRAM (Project No: IITRAM/COE/CADT/22-23/IRP05) for support.

References

1. Esmaeelzadeh S, Simchi A, Lehmhus D (2006) Effect of ceramic particle addition on the foaming behavior, cell structure and mechanical properties of P/M AlSi7 foam. Mater Sci Eng A 424:290–299. https://doi.org/10.1016/j.msea.2006.03.013
2. Patel P, Bhingole PP, Makwana D (2018) Manufacturing, characterization and applications of lightweight metallic foams for structural applications: review. Mater Today Proc 5:20391–20402. https://doi.org/10.1016/j.matpr.2018.06.414
3. Robert T, Helmholtz N, Berlin Z, Mukherjee M, García-Moreno F, Berlin H-Z, Banhart J (2011) Magnesium and magnesium alloy foams. In: Porous metals and metallic foams, pp 139–146
4. Sangeetha M, Prakash S, Sridharraja KS, Muralimano J (2022) Development, testing and microstructural study of aluminum foam in automobile application. Int J Ambient Energy 43:2568–2576. https://doi.org/10.1080/01430750.2020.1723690
5. Ashby MF, Evans AG, Fleck NA, Gibson LJ, Hutchinson JW, Wadley HNG (2000) Metal foams: a design guide. Library of Congress Cataloguing-in-Publication Data
6. Soloki A, Esmailian M (2015) Carbonate-foaming agents in Aluminum foams: advantages and perspectives. Metall Mater Trans B 46:1052–1057. https://doi.org/10.1007/s11663-014-0262-1

7. Parveez B, Jamal NA, Anuar H, Ahmad Y, Aabid A, Baig M (2022) Microstructure and mechanical properties of metal foams fabricated via melt foaming and powder metallurgy technique: a review. https://doi.org/10.3390/ma15155302
8. Surace R, De Filippis LAC (2010) Investigation and comparison of aluminium foams manufactured by different techniques. In: Advanced knowledge application in practice, pp 95–118
9. Liu C, Ren Z, Xu Y, Pang S, Zhao X, Zhao Y (2018) Biodegradable magnesium alloys developed as bone repair materials: a review. Scanning 2018:1–15. https://doi.org/10.1155/2018/9216314
10. Bairagi D, Mandal S (2022) A comprehensive review on biocompatible Mg-based alloys as temporary orthopaedic implants: current status, challenges, and future prospects. J Magn Alloys 10:627–669. https://doi.org/10.1016/j.jma.2021.09.005
11. Makwana D, Pramod Bhingole (2020) Fabrication of Bioresorable cellular magnesium alloy sandwich composite coated with hydroxyapatite for bone implant applications
12. Madgule M, CG S, Patel GC M, Singhal P, Pandit D, Malik V (2022) Influence of foaming agents on mechanical and microstructure characterization of AA6061 metal foams. Proc Inst Mech Eng Part E: J Process Mech Eng. https://doi.org/10.1177/09544089221097534
13. Movahedi N, Mirbagheri SMH, Hoseini SR (2014) Effect of foaming temperature on the mechanical properties of produced closed-cell A356Aluminum foams with melting method. Met Mater Int 20:757–763. https://doi.org/10.1007/s12540-014-4021-2
14. Georgy K, Kumar KCH, Mukherjee M (2022) Optimization of Mg blowing agent content for foaming aluminum. Metall Mater Trans B 53:1089–1102. https://doi.org/10.1007/s11663-021-02403-3
15. Barode J, Aravind U, Bhogi S, Muduli B, Mukherjee M (2021) Mg and Mg-based blowing agents for aluminum foam. Metall Mater Trans B 52:292–304. https://doi.org/10.1007/s11663-020-02008-2
16. Nová I, Fraňa K, Machuta J, Nováková I (2019) Theoretical calculations of the foaming properties of powder agents for the production of aluminium foams. Manuf Technol 19
17. Bhosale D, Devikar A, Sasikumar S, Vinod Kumar GS (2021) Foaming Mg alloy and composite using MgCO3 blowing agent. Metall Mater Trans B 52:931–943. https://doi.org/10.1007/s11663-021-02066-0
18. Makwana D, Pramod B (2023) Dry sliding wear and heat flux mapping of closed-cell Mg–2Zn–2Ca foam fabricated by stir casting route. Mater Lett 330:133379. https://doi.org/10.1016/j.matlet.2022.133379
19. Koizumi T, Kido K, Kita K, Mikado K, Gnyloskurenko S, Nakamura T (2011) Method of preventing shrinkage of aluminum foam using carbonates. Metals (Basel) 2. https://doi.org/10.3390/met2010001
20. Papadopoulos DP, Omar H, Stergioudi F, Tsipas SA, Michailidis N (2011) The use of dolomite as foaming agent and its effect on the microstructure of aluminium metal foams-comparison to titanium hydride. Colloids Surf A Physicochem Eng Asp 382:118–123. https://doi.org/10.1016/j.colsurfa.2010.12.005
21. Vinod Pulagara N, Saini S, Sekhar Dondapati R (2015) A study of manufacturing and mechanical properties of Mg-foam using dolomite as the blowing agent: a review. Progr Nanotechnol Nanomater 4:7–10
22. Ravi Kumar NV, Ramachandra Rao N, Gokhale AA (2014) Effect of SiC particle content on foaming and mechanical properties of remelted and diluted A356/SiC composite. Mater Sci Eng A 598:343–349. https://doi.org/10.1016/j.msea.2014.01.050
23. Song Z-L, Ma L-Q, Wu Z-J, He D-P (2000) Effects of viscosity on cellular structure of foamed aluminum in foaming process. J Mater Sci 35:15–20. https://doi.org/10.1023/A:1004715926692
24. Ramirez AMM, Vintila RR, Drew RAL (2019) Morphology of aluminum alloy foams produced with dolomite via partial sintering of precursors. Materials 12. https://doi.org/10.3390/MA12101691
25. Xia X, Feng J, Ding J, Song K, Chen X, Zhao W, Liao B, Hur B (2015) Fabrication and characterization of closed-cell magnesium-based composite foams. Mater Des 74:36–43. https://doi.org/10.1016/j.matdes.2015.02.029

26. Zhou X, Li Y, Chen X (2020) Development of AlMg35–TiH2 composite foaming agent and fabrication of small pore size aluminium foams. J Mater Process Technol 283. https://doi.org/10.1016/j.jmatprotec.2020.116698
27. Yang CC, Nakae H (2003) The effects of viscosity and cooling conditions on the foamability of aluminum alloy. J Mater Process Technol 141:202–206. https://doi.org/10.1016/S0924-0136(02)01048-8
28. Xia XC, Chen XW, Zhang Z, Chen X, Zhao WM, Liao B, Hur B (2013) Effects of porosity and pore size on the compressive properties of closed-cell Mg alloy foam. J Magn Alloys 1:330–335. https://doi.org/10.1016/j.jma.2013.11.006
29. Hangai Y, Utsunomiya T, Kuwazuru O, Kitahara S, Yoshikawa N (2015) Deformation and plateau region of functionally graded aluminum foam by amount combinations of added blowing agent. Materials 8:7161–7168. https://doi.org/10.3390/ma8105366
30. Georgy K, Banerjee P, Mukherjee M (2023) The effect of heat treatment on the compression property of Al–Si–Mg foams produced using Mg blowing agent. Mater Today Commun 36. https://doi.org/10.1016/j.mtcomm.2023.106445
31. Matijasevic-Lux B (2006) Characterisation and optimisation of blowing agent for making improved metal foams. https://doi.org/10.14279/depositonce-1314

Interfacial Characteristics of Ti64–In82 Bimetallic Structure Fabricated Using Wire-Arc Additive Manufacturing

Rupendra Singh Tanwar, Swaraj Sarkar, and Suyog Jhavar

Abstract Additive manufacturing presents a promising avenue for the fabrication of intricate structures comprising metals, polymers, ceramics, and various other materials. Inconel and Ti-based alloys are renowned for their resilience in extreme environmental conditions. Bimetallic structures, representing a subset of multi-material constructions, offer tailored solutions to diverse engineering challenges. This study concentrates on leveraging Wire Arc Additive Manufacturing (WAAM) technology to produce bimetallic structures using Inconel 82 and Ti6Al4V (Ti64) alloys. Challenges such as delamination and other complications stem from the disparate thermal characteristics between Inconel 82 and Ti64 alloys, exacerbated by the formation of brittle intermetallic phases at the interface. The following phase identified: α-Ti and β-Ti transitioning to α-Ti, β-Ti, and Ti2Ni, further progressing to β-Ti, Ti2Ni, TiNi, and γ-Ni. The bimetallic structures underwent comprehensive characterization employing Vickers hardness tests, X-ray diffraction (XRD), scanning electron microscopy (SEM), and scratch testing. The microhardness varies from 319 to 562 HV, during the scratch test 10.11 and 9.854 coefficient of friction fond at 30 N and 40 N load respectively. Evaluation of the interface between the two materials revealed the presence of delamination due to intermetallic compound formed between both metals.

Keywords Additive manufacturing · Wire arc additive manufacturing · Inconel82 · Titanium · Multi-material AM

1 Introduction

The development of multi-material constructions offers numerous advantages compared to conventional materials, catering to a wide array of applications. Bimetallic structures represent a form of multi-material architecture typically

R. S. Tanwar · S. Sarkar · S. Jhavar (✉)
School of Mechanical Engineering, VIT-AP University, Inavolu, Beside AP Secretariat, Amaravati, Andhra Pradesh 522237, India
e-mail: suyog.jhavar@vitap.ac.in

P. Bhingole et al. (eds.), *Advances in Materials Engineering*, Lecture Notes in Mechanical Engineering, https://doi.org/10.1007/978-981-97-7114-1_13

comprising two distinct metals fused together to harness their individual attributes, spanning mechanical, electrical, optical, thermo-physical, and corrosion/oxidation resistance properties [1]. The fundamental idea underlying all additive manufacturing technologies is the progressive layering of materials. In the field of manufacturing, a range of methods are utilized, such as sheet lamination, vat photopolymerization, powder bed fusion (PBF), material extrusion, material jetting, and direct energy deposition (DED) [2]. WAAM is a DED technique where the energy source is an electric arc and the wire serves as the feed, as shown in Fig. 1. Multi-material constructions can be created with this additive manufacturing technique in a single step. The Direct Energy Deposition (DED) method has been effectively used to create both compositionally graded and directly bonded bimetallic structures [3]. Sharp interfaces are a defining feature of directly bonded bimetallic structures, and these interfaces can cause mismatches in the characteristics of constituent materials, which can result in problems including delamination and cracking in areas with poor interfacial bonding. Functionally graded materials (FGMs) are innovative composite materials characterized by a gradual variation in chemical composition or microstructure, resulting in diverse properties across one or more dimensions within the material or its components. FGMs have garnered significant attention across various fields, particularly in demanding environments such as aerospace engineering, nuclear power generation, sensors, biomedical implants, optoelectronic devices, and absorption systems. Traditional fabrication methods for FGMs include vapor deposition, powder metallurgy, and centrifugation. However, these methods face challenges in producing FGMs with intricate shapes or substantial composition changes. In particular, powder metallurgy struggles to achieve high structural integrity in FGM parts and often requires expensive and complex post-treatment processes. These limitations hinder the widespread application and advancement of FGMs [4]. In recent times, there has been a surge in interest among researchers regarding the utilization of additive manufacturing (AM) in the realm of functionally graded materials (FGMs). This concept has arisen to solve the issues. Compared to single-composition materials, FGMs have many advantages due to their gradual change in material composition across all dimensions [5]. However, creating bimetallic structures with immiscible constituent materials because of mismatches in their chemical, metallurgical, or thermophysical properties poses challenges for both FGM and direct deposition techniques. Diffusion barriers or bond layers are two unconventional methods that have been proposed as a solution to these problems in the production of different materials [6]. While surface modifications or coatings to improve features like wear resistance, and selectively increasing mechanical and electrical properties have been the focus of numerous additive manufacturing research efforts, bulk development of bimetallic structures remains a significant issue [7]. A superalloy based on nickel that is well-known for its ability to withstand high temperatures and corrosion, Inconel 82 is widely used in a wide range of sectors. It is especially useful in the aerospace industry because of its remarkable mechanical qualities, which include noteworthy tensile strength, fatigue resistance, creep resistance, and rupture strength, along with excellent oxidation resistance at high temperatures [8]. In general, titanium alloys are renowned for their high strength, low density, excellent corrosion resistance, elevated heat resistance, and favorable biocompatibility. Among titanium alloys, Ti6Al4V stands out for its well-balanced properties and finds widespread applications. On the other hand, nickel-based alloys offer high strength, oxidation resistance, and corrosion resistance at high temperatures. Notably, Inconel 82 possesses impressive strength and corrosion resistance at elevated temperatures while serving as a transitional material between different thermal expansion coefficients (TEC) [9]. Combining the exceptional attributes of both titanium and nickel-based alloys presents promising prospects for applications in harsh environments, such as aerospace or nuclear industries, where high temperatures and corrosive conditions prevail. However, achieving a viable combination of Ti6Al4V and Inconel 82 has posed challenges. The significant disparity in TEC between these alloys can lead to the formation of intermetallic compounds, potentially compromising their properties. To address this issue, functionally graded materials (FGMs) offer a solution by providing a gradient change in composition. This allows for a smooth transition between the properties of different materials, minimizing the risk of failure. Designing an FGM utilizing Ti6Al4V and Inconel 625 enables the amalgamation of these two alloys [10, 11]. The gradient change in composition not only facilitates the combination of their distinct properties but also broadens their utility by offering a range of mechanical and thermal properties across the gradient. Wire arc additive manufacturing (WAAM) stands out among various additive manufacturing technologies as it effectively addresses the challenges encountered in processes like laser metal deposition (LMD). WAAM employs metal wires as its feedstock, which are melted by a heat source and deposited layer by layer. This method offers several advantages, including low cost, straightforward operation, high deposition rates, and the capability to fabricate large-sized components. In this investigation, wire arc additive manufacturing (WAAM) was employed for the deposition of functionally graded materials (FGMs). The formation of intermetallic compounds and the significant difference in thermal expansion coefficients (TEC) between Ti6Al4V and Inconel 82 pose considerable challenges when employing wire arc additive manufacturing (WAAM) for the fabrication of functionally graded materials (FGMs) transitioning from Ti6Al4V to Inconel 82. Nonetheless, in this study, an FGM spanning from Ti6Al4V to Inconel 82 was successfully produced using WAAM, with the utilization of Inconel 82 wire. Experimental characterization of the FGM included analysis through techniques such as scanning electron microscopy (SEM), X-ray diffraction (XRD) for phase evolution, Vickers hardness testing for microhardness assessment, and a scratch test to evaluate its properties.

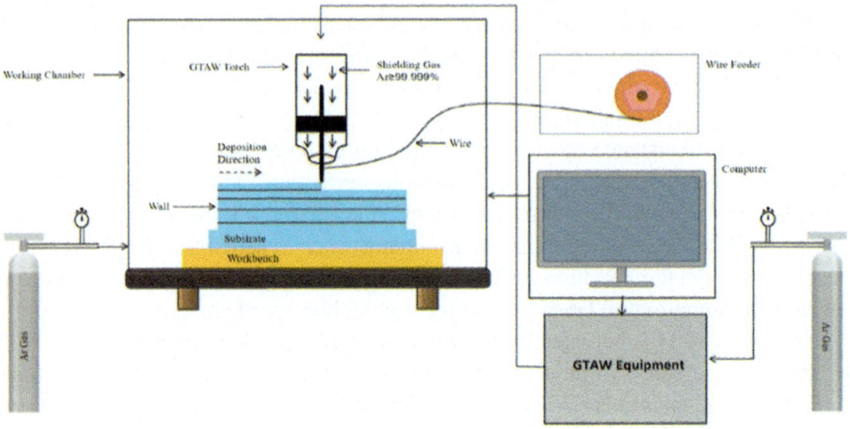

Fig. 1 Schematic diagram of WAAM system

2 Material and Methods

2.1 Wire Arc Additive Manufacturing Setup

The fabrication of the functionally bi-metallic component transitioning from Ti6Al4V to Inconel 82 was achieved using a wire arc additive manufacturing (WAAM) system. This system comprised a power supply, a control cabinet, a control panel, an argon gas shielding system consisting of shielding gas and a trailing shielding cover, a working table, dual wire feeders, a gas tungsten arc welding (GTAW) torch, wire device, and a heating blanket. A wire of Inconel 82 with a diameter of 1.15 mm was utilized to deposit material onto the surface of a Ti6Al4V substrate measuring $100 \times 100 \times 10\ mm^3$.

2.2 Materials Used

Titanium Grade 5, or Ti6Al4V, has a special composition and set of qualities that make it a reliable material for use in aerospace and aviation applications. It features a dual-phase structure with both alpha and beta phases, with 90% titanium, 6% Aluminium, and 4% vanadium, as shown in Table 1. The titanium atoms in the alpha phase are arranged in a hexagonal close-packed (hcp) configuration, which is similar to tightly stacked hexagonal tiles. These phases are balanced when Aluminium and vanadium are present. By applying heat treatment, its mechanical properties are considerably improved.

Inconel is a superalloy based on nickel and chromium that is well known for its remarkable ability to withstand high temperatures, oxidation, and corrosion. These

Table 1 Properties of Ti6Al4V

	Density, g/cm^3	Young's modulus, GPa	Shear modulus, GPa	Bulk modulus, GPa	Poisson's ratio	Yield stress, MPa (Tensile)	Ultimate stress, MPa (Tensile)	Hardnes, rockwell C	Uniform elongation, %
Min	4.43	104	40	96.8	0.31	880	900	36 (common)	5
Max	4.51	113	45	153	0.37	920	950	–	18

Table 2 Composition of Inconel 82

Element	Ni	Cr	Fe	Mo	Cu	Ti	C	Mn	S	Si	Al
IN82	39.4	22	23.6	3.15	2.54	1.13	0.037	0.76	0.02	0.24	0.16

Table 3 Composition of Ti64Al4V

Element	Al	V	Fe	C	O	H	Ti
Ti64	6	4.15	0.04	0.012	0.085	<0.01	Bal

Table 4 Properties of Inconel 82:

Mechanical property	Value
Tensile strength, Rm	690 MPa (100 ksi) minimum
Yield strength, Rp0.2	415 MPa (60 ksi) minimum
Elongation, A5	30% minimum
Hardness, HB	195–240 HB
Density	8.14 g/cm^3
Melting point	1390 °C (2534 °F)

alloys are mostly made of nickel, with notable additions of iron, chromium, and other elements such as cobalt, molybdenum, and niobium, as shown in Table 2. Because Inconel alloys have unique qualities and attributes that are not usually present in the metals that make them up, they are extremely valuable in a variety of industries. Table 3. Shows the chemical composition of Ti64Al4V. One of the most noteworthy characteristics of Inconel alloys is their ability to maintain stable properties and structural strength at high temperatures, as shown in Table 4. Because of this characteristic, they are very suitable for use in corrosive environments and high temperatures, such as gas turbine engines, nuclear reactors, chemical processing, and aerospace. Additionally, Inconel alloys have good mechanical qualities and can be made with ease utilizing common methods like forming, machining, and welding. Their exceptional capacity to withstand oxidation makes them even more suitable for settings with high oxygen content and temperatures. Because of their special combination of qualities, Inconel alloys are widely sought after in sectors where materials that can survive harsh environments are needed. In order to satisfy the changing needs of contemporary technology, ongoing research and development activities are focused on innovating and diversifying Inconel alloys.

2.3 *Material Characterization Techniques*

The samples were initially cut into a 3 mm diameter size, followed by polishing with emery paper. Subsequently, they underwent further polishing until achieving a mirror surface finish. To reveal the microstructure, the samples underwent etching using Glyceregia solution, composed of 15% HCl, 5% HNO_3, and 10% glycerol.

The microstructure analysis was conducted using a ZEISS EV0 10 High Resolution scanning electron microscope (SEM). X-ray diffraction (XRD) patterns were obtained using a Rigaku MiniFlex600 equipped with CuKα radiation, with a scanning rate of 10°/min and diffraction angle (2θ) set between 15° and 20°. Microhardness measurements of the cross-section samples were performed using Fine Spavy Associates and Engineering FMV1-MC-AT Vickers microhardness tester, with loads of 100–1000 g applied along the building direction, each with a dwell time of 10 s starting from the substrate. Scratch test measurements of the cross-section samples were performed using DUCOM TR-101-IAS instrument, with scratches applied along the direction using loads of 30, 40, and 50 N, each with a dwell time of 10 s.

2.4 Fabrication Process

The Wire Arc Additive Manufacturing (WAAM) process offers compatibility by allowing users to define the metallurgical composition of materials. This method involves the step-wise addition of materials in layers to fabricate three-dimensional components, guided by a digital model where chosen materials are functionally graded. Wires are used and interchanged during deposition, replacing metallic powder commonly used in powder-bed processes [12]. By employing an electric arc as the heat source, WAAM overcomes limitations such as low deposition rates, high equipment, and raw material costs associated with powder-bed methods. WAAM's economical and versatile nature allows for the manufacturing of larger components in shorter timeframes with minimal material wastage. Additionally, it is utilized for repairing metallic components to restore them to full density, reducing processing time [13]. However, the dimensional accuracy of WAAM-fabricated parts depends on the precision of the motion system guiding the welding torch. Compared to other additive manufacturing techniques, WAAM boasts lower machine and material costs. Recent advancements include altering the wire feed speed ratio using twin-wire feeding arrangements to control deposition rate and composition effectively. During the fabrication process, a wire feeder feeds Inconel 82 wire into the machine, and the torch is supplied with weld. Applying a high-pulse current causes an arc to form between the wires and the titanium substrate, melting the wire and enabling it to deposit on the substrate. On the work table, the titanium substrate is firmly placed. A 5 mm thick layer is applied to a 10 mm thick Titanium plate during each deposition round, and the process is followed by a 15-s cooling time. Each round's deposition distances from the nozzle are 0.5, 1.5, and 3 mm, in that order. After the deposited layer was cooled down the deposited layer was cut to perform the test, as shown in Fig. 2.

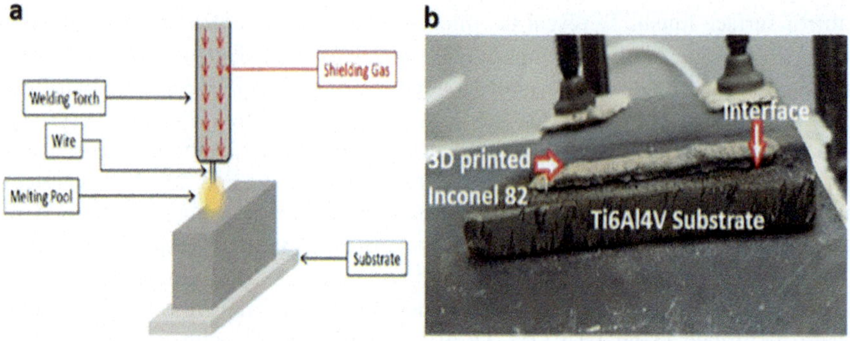

Fig. 2 Fabrication process using WAAM system

3 Results and Discussion

Important insights into parameter optimization for Inconel-82 additive manufacturing were obtained from our work. Nine experiments with various voltage levels and scanning rates were conducted as part of the study. After a thorough investigation, the most advantageous combination was found to be 20.7 V and 250 mm/min scanning speed. This choice was chosen in conjunction with regard to material integrity, overall process efficiency, and deposition efficiency in mind. The application of this novel arrangement has the potential to revolutionize additive manufacturing processes for high-performance alloys like Inconel-82 by yielding enhanced material characteristics and structural stability.

3.1 Scanning Electron Microscope (SEM)

Scanning Electron Microscopy (SEM), renowned for its high-resolution imaging capabilities, is utilized for meticulous examination of the microstructural attributes of In82 specimens subjected to additive manufacturing, contingent upon the precise adjustment of scanning parameters such as speed and voltage. SEM enables comprehensive scrutiny of the intricate nano-scale features of the alloy, encompassing phase distribution, grain morphology, and overall material morphology. Grain structure, a pivotal facet of microstructural analysis, is directly influenced by the scanning speed during additive manufacturing. Accelerated scanning rates yield finer grain morphologies, indicative of accelerated solidification and diminished grain growth. SEM imagery provides quantitative insights into grain dimensions and configurations, elucidating the influence of varied processing conditions on grain characteristics [14, 15]. Experimental findings reveal the presence of varying grain structures at different depths, engendering a gradient from the substrate to the surface. At a depth of 10 μm from the substrate, columnar dendritic growth is evident at a magnification

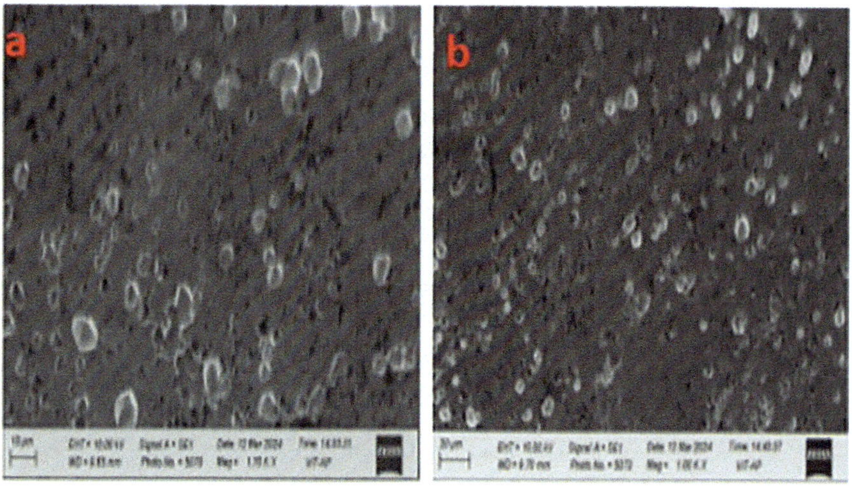

Fig. 3 SEM images (**a** at 10 μm and **b** at 20 μm)

of 1.70KX, an accelerating voltage of 10 kV, and a field width of 9.65 mm. This dendritic configuration, owing to its directional obstruction of heat transfer, may induce anisotropic behaviour and alterations in mechanical properties. Furthermore, at a depth of 20 μm from the substrate, combined dendritic growth of grain structures is observed at a magnification of 1.00KX, an accelerating voltage of 10 kV, and a field width of 9.70 mm. Distinctive grain morphologies contribute to enhanced crack resistance; columnar grains bolster overall material strength, while equiaxed grains impede crack propagation. Various formations within the bimetallic structures, including lamellar, columnar dendritic, equiaxed, and flower-shaped eutectic structures. Phase analysis delineated the evolution of phases in the following sequence: α-Ti and β-Ti transitioning to α-Ti, β-Ti, and Ti2Ni, further progressing to β-Ti, Ti2Ni, TiNi, and γ-Ni. The investigation showcases the presence of columnar grains exhibiting dendritic growth at a depth of 10 μm, as depicted in Fig. 3a. In Fig. 3b, a combination of columnar and mixed formations is evident at a depth of 20 μm. These observations underscore the structural diversity within the analyzed samples, offering an in-depth microstructural insight into the material's properties across various scales.

3.2 *X-Ray Diffraction Analysis (XRD)*

X-ray Diffraction Analysis (XRD), aims to elucidate the phase composition and crystal structure of the Inconel 82 filler material. The diffraction patterns generated result from X-ray irradiation interacting with the lattice structure post-sample preparation and loading. Following data acquisition, a thorough examination is undertaken. Predominantly, the austenitic phase is identified as the primary phase, denoting

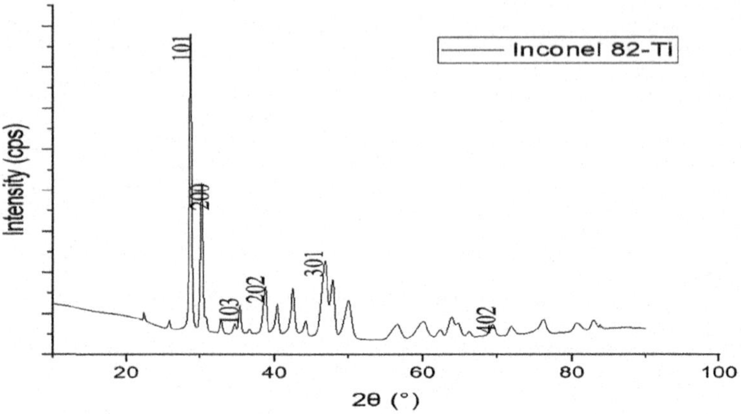

Fig. 4 XRD images (**a** at 10 μm and **b** at 20 μm)

the material's face-centered cubic (FCC) nature, which confers ductility, excellent weldability, and corrosion resistance. The material's malleability, stemming from its ductility and plastic deformation of the crystal structure under stress, allows for shaping without fracture. The secondary phase comprises carbide precipitates, which form robust, wear-resistant particles impeding dislocation movement, thereby augmenting material hardness and strength. Carbides aid in refining grain structure and enhancing mechanical properties such as toughness and fatigue resistance by inhibiting grain growth during solidification or heat treatment [8]. Lattice parameter calculations elucidate unit cell dimensions, revealing alterations in material characteristics due to processing and heat treatment. Graphical representations derived from these computations provide insight into material development patterns. The following phase evolution occurred as: $\alpha + \beta \rightarrow \alpha + \beta + \text{Ti2Ni} \rightarrow \beta + \text{TiNi} \rightarrow \gamma + \text{Laves}$ [16]. XRD peak profiles depict the intensity of diffracted X-rays at various angles, with labelled peaks indicating the presence of different crystalline phases, including nickel chromium oxide (NiCr2O4), in Inconel 82 [16]. Relative peak intensities enable quantification of phase abundance. The phase data view graphically illustrates the relationship between variables x, y, and z, offering a comprehensive visual representation of their interactions, thereby enhancing understanding of Inconel 82's characteristics and behaviour as a material, as shown in Fig. 4. The tangent line and function represented by the blue and red lines, respectively, furnish a nuanced comprehension of the material's phase properties.

3.3 Vickers Hardness Test

The Vickers hardness test, employed to assess the material's surface resistance to deformation or indentation through varied loads and penetration lengths, reveals

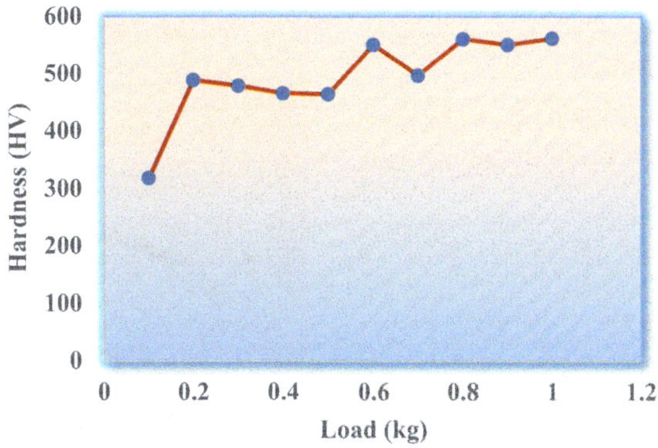

Fig. 5 Graphical visualization of load and HV and related to hardness test

hardness values of 319 HV, 489 HV, 479 HV, 466 HV, 465 HV, 551 HV, 497 HV, 561 HV, 551 HV, and 562 HV corresponding to loads of 0.1 kg, 0.2 kg, 0.3 kg, 0.4 kg, 0.5 kg, 0.6 kg, 0.7 kg, 0.8 kg, 0.9 kg, and 1 kg, respectively, as shown in Fig. 5. The recorded values demonstrate a moderate increase in hardness with increasing applied load, suitable for applications necessitating a balance between toughness and wear resistance. However, the implications of these findings vary depending on the material type. While hardness is a significant parameter, considerations should also encompass additional qualities such as toughness, particularly in scenarios involving impact or shock loading. The observed hardness values underscore the multifactorial nature of hardness determination, indicating a lack of consistent trend across the tested parameters [17]. The graphical representation within the study elucidates the correlation between the Vickers hardness number and associated hardness values. The y-axis delineates the Vickers hardness number, a metric assessing a material's resistance to deformation typically derived from indentation tests. On the x-axis, hardness values, potentially linked to specific loads or material attributes, are illustrated. A continuous curve on the graph depicts the fluctuation of hardness concerning the Vickers hardness number. Each point along the curve corresponds to specific measurements or datasets derived from Vickers Hardness testing, showcasing a seamless transition between these points. This graphical portrayal offers a quantitative understanding of the material's deformation resistance under defined testing conditions, furnishing valuable insights into its hardness characteristics.

3.4 *Scratch Test*

In the investigation of Super alloys, the application of scratch testing is imperative for gaining comprehensive insights into their mechanical and tribological properties. Such knowledge is pivotal for designing components intended for demanding environments like gas turbines and aircraft. The study utilized three distinct loads (30, 40, and 50 N) to systematically evaluate the material's response to varying stress levels,

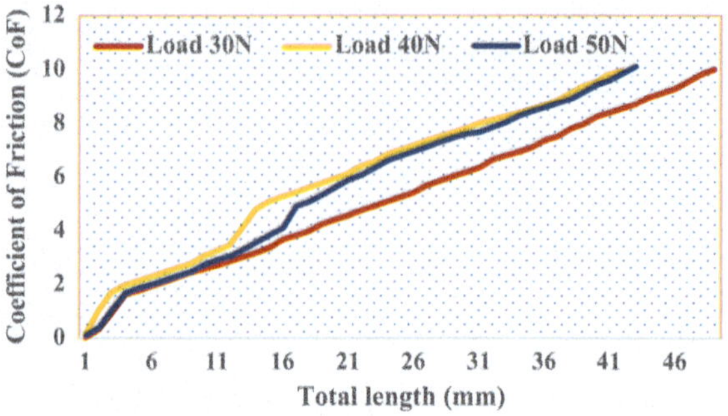

Fig. 6 Coefficient of friction versus total length at interface

the maximum at 30 N whereas at 40 N found lower value. These load conditions facilitate a detailed analysis of the alloy's scratch resistance and its behaviour under different stress levels, as shown in Fig. 6. The findings derived from these assessments are instrumental in enhancing the durability and performance of components exposed to harsh operating conditions in gas turbine and aerospace applications. The graphical representation illustrates the correlation between the Coefficient of friction (CoF) and the total length of the scratch, providing valuable insights into the material's tribological characteristics.

4 Conclusion

This study assesses the feasibility of employing the Wire Arc Additive Manufacturing (WAAM) process to directly deposit In82 onto Ti64, with the aim of guiding the production of dissimilar alloys. Furthermore, the interfacial properties between the IN82 superalloy and Ti64 are explored through microstructural analysis. The key findings from this investigation can be summarized as follows:

a. Microstructural examinations through SEM and XRD unveiled various formations within the bimetallic structures, including lamellar, columnar dendritic, equiaxed, and flower-shaped eutectic structures. Phase analysis delineated the evolution of phases in the following sequence: α-Ti and β-Ti transitioning to α-Ti, β-Ti, and Ti2Ni, further progressing to β-Ti, Ti2Ni, TiNi, and γ-Ni.
b. During the fabrication process of In82 on Ti64, the formation of NiCr intermetallic compound occurs. This formation contributes to issues such as delamination, cracking, and debonding between the two metals, primarily due to their mismatched thermal and mechanical properties.
c. The microhardness values of the Functionally Graded Material (FGM) component varied from 319.0 to 562 HV. The lowest microhardness was observed in the substrate region, while the highest microhardness was recorded at the end of the direction of fabrication. This increased microhardness can be attributed to factors

such as solution strengthening, precipitates, and a high density of dislocations within the material.

d. The coefficient of friction reaches its maximum value of 10.11 at a normal force of 30 N, while its minimum value of 9.854 is observed at a normal force of 40 N.

References

1. Singh Tanwar R, Jhavar S (2023) A review on additive manufacturing of SS-Ni multi-material fabrication. Mater Today Proc. https://doi.org/10.1016/j.matpr.2023.03.044
2. Ji S, Sun Z, Zhang W, Chen X, Xie G, Chang H (2020) Microstructural evolution and high temperature resistance of functionally graded material Ti-6Al-4V/Inconel 718 coated by directed energy deposition-laser. J Alloys Compd 848:156255. https://doi.org/10.1016/j.jallcom.2020.156255
3. Blakey-Milner B (2021) Metal additive manufacturing in aerospace: a review. Mater Des 209. https://doi.org/10.1016/j.matdes.2021.110008
4. Li W (2016) Effect of substrate preheating on the texture, phase and nano hardness of a Ti-45Al-2Cr-5Nb alloy processed by selective laser melting. Scr Mater 118:13–18. https://doi.org/10.1016/j.scriptamat.2016.02.022
5. Meng W (2020) Additive manufacturing of a functionally graded material from Inconel625 to Ti6Al4V by laser synchronous preheating. J Mater Process Technol 275(April):2019. https://doi.org/10.1016/j.jmatprotec.2019.116368
6. Frazier WE (2014) Metal additive manufacturing: a review. J Mater Eng Perform 23(6):1917–1928. https://doi.org/10.1007/s11665-014-0958-z
7. Su Y, Chen B, Tan C, Song X, Feng J (2020) Influence of composition gradient variation on the microstructure and mechanical properties of 316 L/Inconel718 functionally graded material fabricated by laser additive manufacturing. J Mater Process Technol 283:116702. https://doi.org/10.1016/j.jmatprotec.2020.116702
8. Zhao K, Zhang G, Ma G, Shen C, Wu D (2020) Microstructure and mechanical properties of titanium alloy: zirconia functionally graded materials prepared by laser additive manufacturing. J Manuf Process 56(February):616–622. https://doi.org/10.1016/j.jmapro.2020.05.044
9. Lin Z, Song K, Yu X (2021) A review on wire and arc additive manufacturing of titanium alloy. J Manuf Process 70:24–45. https://doi.org/10.1016/j.jmapro.2021.08.018
10. Arsha AG, Jayakumar E, Rajan TPD, Antony V, Pai BC (2015) Design and fabrication of functionally graded in-situ aluminium composites for automotive pistons. Mater Des 88:1201–1209. https://doi.org/10.1016/j.matdes.2015.09.099
11. Xia Y, Teng H, Zhang X, Zheng D, Quan G (2022) A review of the wire arc additive manufacturing of Ti-6Al-4V alloy: properties, defects and quality improvement. Chongqing Daxue Xuebao, J Chongqing Univ 45(4):87–99. https://doi.org/10.11835/j.issn.1000-582X.2020.265
12. Srinivasa Chari V, Jhavar S, Sreenivasa TN, Hanumanthappa H, Durga Prasad C, Shanmugam BK (2023) Impact of post-processing techniques on the wear resistance of plasma beam treatment on SS316L components. J Inst Eng Ser D. https://doi.org/10.1007/s40033-023-00584-6
13. Cunningham CR, Flynn JM, Shokrani A, Dhokia V, Newman ST (2018) Invited review article: Strategies and processes for high quality wire arc additive manufacturing. Addit Manuf 22(June):672–686. https://doi.org/10.1016/j.addma.2018.06.020
14. Toprek D, Belosevic-Cavor J, Koteski V (2015) Ab initio studies of the structural, elastic, electronic and thermal properties of NiTi2 intermetallic. J Phys Chem Solids 85:197–205. https://doi.org/10.1016/j.jpcs.2015.05.019

15. Loh GH, Pei E, Harrison D, Monzón MD (2018) An overview of functionally graded additive manufacturing. Addit Manuf 23(July):34–44. https://doi.org/10.1016/j.addma.2018.06.023
16. Han J (2022) Microstructure and mechanical properties of a novel functionally graded material from Ti6Al4V to Inconel 625 fabricated by dual wire + arc additive manufacturing. J Alloys Compd 903. https://doi.org/10.1016/j.jallcom.2022.163981
17. Stevens EL, Toman J, To AC, Chmielus M (2017) Variation of hardness, microstructure, and laves phase distribution in direct laser deposited alloy 718 cuboids. Mater Des 119:188–198. https://doi.org/10.1016/j.matdes.2017.01.031

Experiment on Modified Bituminous Road with Additive Waste Material

Anant Kumar, Aryansh Chopkar, Khushi Vyas, Kuldeep Suthar, and Raunak Prakash

Abstract In the state of art, transportation is one of the most important aspects of our day-to-day life. To maintain this continuous flow of transportation, the better quality of road network is very important for which Bituminous Road, consisting of Aggregate and Binder, is used. The major issues faced by Bituminous Road are the absorption of water and heavy loads which results in the formation of cracks and potholes. Various research such as the addition of Polymers as additives and the replacement of aggregate with better alternatives is carried out to modify the properties of Aggregate and Binders. In this research, various types of polymers and waste materials are used in different proportions with a conventional mix of Bituminous Road to improve the quality. This paper concluded that conventional Bituminous Road by modifying it with Rubber, LDPE, Coconut Shell, Scrapped Bitumen Road, etc. exhibits better bearing capacity. To overcome the unsustainability and danger of Polymer and Waste products are being utilised in this research work.

Keywords Low Density Polyethylene (LDPE) · Conventional mix (C Mix) · Coconut Shells · Scrap bitumen road · Polymers

1 Introduction

In this world of economic development and industrialisation, demands for better infrastructure have increased abruptly. Hence, utilisation of these waste products for useful outcomes can be immensely beneficial. In this research, Modified Bituminous Road with several waste products as additives to enhance the performance of Flexible Pavement is manufactured. Modified Bituminous Road pavement will help in providing better quality of road infrastructure. Waste materials utilised in this research include Coconut Shell, Crumb Rubber, LDPE, and Scrap Bitumen Road.

A. Kumar · A. Chopkar · K. Vyas · K. Suthar · R. Prakash (✉)
Department of Civil Engineering, Shri Shankaracharya Institute of Professional Management and Technology, Raipur (C.G), India
e-mail: rawnak.prakash@ssipmt.com

P. Bhingole et al. (eds.), *Advances in Materials Engineering*, Lecture Notes in Mechanical Engineering, https://doi.org/10.1007/978-981-97-7114-1_14

Coconut Shells are discarded as waste material after utilisation of coconut flesh. These shells contribute to the waste of about 3.18 million tonnes every year. Coconut Shells have high weather resistance making them an ideal choice for the construction and highway industry because their disposal is uneconomical [1].

Every year around 1.5 billion waste tyres are generated in the world, from automobiles which are either recycled or thrown away. In this research work, Crumb Rubber from waste tyres is utilised instead of disposed of by landfill, burning and recycling as an additive will increase the viscosity, resulting in an increase of the pavement durability over cracks and providing great elastic properties [2].

Disposal of LDPE is either carried out by incineration or landfilling which are not eco-friendly, so utilisation as an additive enhances stiffness, and temperature susceptibility of Pavement. Therefore, these materials are being recycled, maintaining the balance in the ecosystem and improving the performance of pavement against weathering actions [3].

Old and broken Bituminous pavements can be recycled to make new Bituminous pavements which will reduce the total cost of the construction and decrease the waste from the environment. Additional benefits of recycling are the Conservation of Natural Resources, better performance of Bituminous pavements, decrease in the greenhouse gases from the environment and dumping problem, etc. [4].

2 Materials and Methodology

2.1 Coconut Shell

Coconut Shells are solid hard shells of the Coconut Fruit. After consuming the edible flesh, the hard shells are thrown in the garbage and it is no longer considered economical. Coconut Shell waste is produced by households and industries, where coconut is used as the main ingredient [1].

Disposal of these hard Coconut Shells is very toxic and harmful to the environment. Instead of disposing of these Coconut Shells in the environment, they can be utilised in this research work. Coconut Shells, having high resistance towards extreme environmental conditions, are preferably good for utilisation in construction. The Coconut Shells are mainly composed of lignin which is 36.5% of the Coconut Shell. Other than lignin, Coconut Shells are also composed of cellulose about 33.6%, pentosans about 29.3%, and 0.6% ash. Coconut Shells contain 65–75% volatile stuff, but little ash. As compared to conventional Aggregate, Coconut Shells have 24% higher water absorption [5]. When it comes to impact, abrasion, and crushing resistance, Coconut Shells beat other common Crushed Granite Aggregates. Some advantages of Coconut Shells as additives are high modulus of elasticity, high strength, tenacity, and strong bonds with bitumen binders [1].

In this study, tests are conducted for the properties of Coconut Shells, as shown in Table 1.

Table 1 Representation of properties of Coconut Shell

S.no	Sample	Result obtained for the following test			
		Specific gravity	Impact test (%)	Abrasion test (%)	Water absorption (%)
1	A	1.10	8.30	1.80	24.00
2	B	1.53	8.15	1.63	24.00
3	C	1.47	8.05	1.45	15.17

2.2 Crumb Rubber

Crumb Rubber is obtained from small pieces of rubber from waste products like automobile tyres, tubes, etc. Due to the increasing traffic, the demand for rubber tyres and tubes has also increased. This is the reason behind the increase in waste rubber tyres and tubes which are not often recycled. These scrap tyres and rubber tubes pollute the environment. Their disposal by burning is highly toxic and poisonous fumes further pollute the environment.

Crumb Rubber is added into the Bitumen mix to achieve the desired properties that conventional bituminous pavement fails. Recycling Rubber from these scrap tyres helps maintain the balance in the environment and their utilisation in the construction of flexible Bituminous Roads is also very beneficial. The bond between Bitumen and Crumb Rubber is physical where Crumb Rubber diffuses with an aromatic fraction of Bitumen resulting in swelling of Crumb Rubber particles. Crumb Rubber particle swelling is essentially the effect of bitumen oils diffusing into crumb rubber particles [6]. Aromatic oils like naphthenic oils are added while the bonding between Crumb Rubber and Bitumen takes place to increase the workability. Aromatic oils increase the viscosity of the Bitumen and soften it. The properties of Bitumen with Crumb Rubber depend on various aspects including the size of particles, chemical composition (proportion of natural rubber), and specific surface area [7].

When Crumb is mixed in Bitumen mix, its elastic and viscous moduli at low temperature are decreased and binder flexibility is increased [8]. But at high temperature, both moduli increase significantly and loss tangent values are dropped which results in a more elastic binder. The major components of Crumb Rubber are 50% hydrocarbon (natural and synthetic rubber), 32% carbon black, 11% Tetrahydrofuran (THF) extractable, and 4% ash [9]. By analysis, it is observed that the use of Crumb Rubber in bitumen mix gives superior consistency and rheological properties. Crumb Rubber will also save costs due to the high price of pure polymers [7].

Tests conducted on Crumb Rubber for its properties are shown in Table 2.

Table 2 Representation of properties of Crumb Rubber

S.no	Sample	Results obtained for the following test			
		Penetration test (mm)	Softening test (°C)	Flash test (°C)	Ductility test (mm)
1	A	35	62	290	50
2	B	49	61	320	65
3	C	41	59	305	49

2.3 LDPE

Low-Density Polyethylene (LDPE) is a type of polymer usually found in domestic and industrial plastic wastes like shopping bags, beverage bottles, etc. The specific gravity of LDPE is found to be in the range of 0.91 to 0.95 and the density of LDPE lies in the range of 0.912–0.928 g/cm^3. The tensile strength of LDPE is found to be 10 N/mm^2. LDPE has a melting point of 90°C. LDPE is extremely lightweight and moisture-absorbing. It offers good resistance to corrosion. In this research, the LDPE is obtained from waste beverage bottles and shredded them finer than a 4.75 mm sieve. LDPE is also a major concern for the environment. It is non-biodegradable and can last for more than 4500 years. Without proper disposal of LDPE can affect the sewage flow, resulting in Floods and degrading soil fertility [3].

LDPE increases the flexible pavement's resistance to abrasion and slip resistance which improves the overall quality of the bitumen mix [3]. Using LDPE in the construction of modified bituminous roads is the most effective approach for its disposal. In general, the inclusion of polymers makes bitumen more rigid and enhances its temperature sensitivity [10]. Plastic can be divided into two types, Thermosets and Thermoplastics. Thermosets have high strength and durability. When they are heated, they solidify irreversibly. Therefore, can be used in construction-related projects. They reduce the plastic shrinkage and drying shrinkage when used in the construction of bituminous pavement [11].

Tests conducted on LDPE for its properties are shown in Table 3.

Table 3 Representation of properties of LDPE

S.no	Sample	Result obtained for the following test			
		Penetration test (mm)	Softening test (°C)	Flash and fire test	Ductility test (mm)
1	A	586.2	74.30	177°C/181°C	740
2	B	580.6	72.60	172°C/180°C	735
3	C	588.3	75.40	175°C/184°C	755

Table 4 Representation of properties of Scrap Bitumen Road

S.no	Sample	Result obtained for the following test			
		Specific gravity	Impact test (%)	Abrasion test (%)	Water absorption (%)
1	A	1.98	12.36	48.00	75.0
2	B	1.81	15.49	34.50	73.0
3	C	1.70	14.80	35.00	78.0

2.4 *Scrap Bitumen Road*

During the construction of bituminous roads over already constructed roads, old pavement is demolished. This Scrap Bitumen Road can be recycled as a form of Aggregate as well as a Bitumen source. The specific gravity of scrap bitumen road is found to be 2.44. The raw materials like Aggregate and Bitumen can be used again for the construction process. If this pavement is not disposed of correctly, this will pollute the environment and cause serious hazards [4]. The recycling of demolished pavement is not very widely used in India. So, this may become a new alternative for pavement construction which will result in savings of material, money, and energy [4].

In this study, tests are conducted for the properties of Scrap Bitumen Road, as shown in Table 4.

2.5 *Aggregate*

Aggregates are very common construction materials used in almost all types of construction work. Aggregates are commonly categorised into two types, Fine Aggregates and Coarse Aggregate, based on their size. One of the most significant aspects of Aggregate blend is gradation. A specified suggested particle size distribution for the mixture's Aggregates is necessary for any particular material type used in nowadays building construction. Aggregate gradation refers to the size distribution, which is also known as Aggregate distribution [12]. The primary ingredient in asphalt mixture is a blend of mineral Aggregates. As a result, Aggregate gradation greatly affects the properties of asphalt mixtures and also the performance of asphalt pavements. The fineness modulus of Aggregate by performing sieve analysis is 6.93.

Tests are conducted to find the values of Specific Gravity, Impact Value, Abrasion Value, and Water Absorption of Aggregate, which are as shown in Table 5.

Table 5 Representation of properties of aggregate

S.no	Sample	Result obtained for the following test			
		Specific gravity	Impact test (%)	Abrasion test (%)	Water absorption (%)
1	A	2.2	15.36	22.50	0.70
2	B	2.6	21.60	36.56	0.47
3	C	3.5	12.50	26.50	0.68

Table 6 Representation of properties of Bitumen

S.no	Sample	Result obtained for the following test			
		Penetration test (mm)	Softening test (°C)	Flash and fire test	Ductility test (mm)
1	A	64.0	48.0	225°C/246°C	80.0
2	B	72.0	55.0	220°C/232°C	92.1
3	C	64.7	50.6	227°C/235°C	91.1

2.6 Bitumen

Bitumen, an extremely viscous material, can be found in nature or be extracted as a byproduct of petroleum distillation [13]. Bitumen is essentially made up of highly condensed polycyclic aromatic hydrocarbons, with up to 5% sulphur, 1% nitrogen, 1% oxygen, and 2000 parts per million (ppm) of metals. It also contains 95% carbon and hydrogen (±87% carbon and ±8% hydrogen) [14]. With a boiling point of 525 °C, it is the densest component of crude oil [15]. Bitumen is usually used as a binder which makes bonds between Aggregate and other constituents. Modified Bitumen has several benefits, including a longer fatigue life, improved adhesion between Aggregates and binder, reduced temperature sensitivity, improved age resistance, increased resistance to deformation at high temperatures, and a delay in cracking [15].

Tests conducted on Bitumen for its properties are shown in Table 6.

2.7 Proportion of Materials

All the materials, that are to be added into the bituminous mix of a sample of modified bituminous road pavement, are mixed in appropriate proportions. In this research work, 2 samples of modified bituminous road pavements are manufactured. Firstly, 15% fillers (LDPE & Crumb Rubber) and the rest 85% being Aggregates and Coconut Shell (Sample A). Secondly, 30% Fillers with 70% Aggregate and Coconut Shell (Sample B). The recycled dismantled road can provide up to 70% of the Bitumen

Table 7 Mix proportion of materials

S.no	Sample code	Percentage composition (in %)					
		Aggregate			Filler		
		Aggregate	Coconut shell	Scrapped road	Bitumen	LDPE	Crumb rubber
1	A1C7L3	31.28	6.8	46.92	14.18	0.25	0.58
2	A1C5L5	31.28	6.8	46.92	14.18	0.41	0.41
3	A1C3L7	31.28	6.8	46.92	14.18	0.58	0.25
4	A2C7L3	31.28	6.8	46.92	14.10	0.27	0.63
5	A2C5L5	31.28	6.8	46.92	14.10	0.45	0.45
6	A2C3L7	31.28	6.8	46.92	14.10	0.63	0.27
7	A3C7L3	31.28	6.8	46.92	14.03	0.29	0.68
8	A3C5L5	31.28	6.8	46.92	14.03	0.49	0.49
9	A3C3L7	31.28	6.8	46.92	14.03	0.68	0.29
10	B1C7L3	19.32	5.6	45.08	28.35	0.50	1.16
11	B1C5L5	19.32	5.6	45.08	28.35	0.83	0.83
12	B1C3L7	19.32	5.6	45.08	28.35	1.16	0.50
13	B2C7L3	19.32	5.6	45.08	28.20	0.54	1.26
14	B2C5L5	19.32	5.6	45.08	28.20	0.9	0.9
15	B2C3L7	19.32	5.6	45.08	28.20	1.26	0.54
16	B3C7L3	19.32	5.6	45.08	28.05	0.59	1.37
17	B3C5L5	19.32	5.6	45.08	28.05	0.98	0.98
18	B3C3L7	19.32	5.6	45.08	28.05	1.37	0.59

and Aggregate demands that are required for pavement construction activities. The sample with only Pure Aggregate and Bitumen is taken as a Conventional Mix (C Mix).

These samples are further proportioned based on LDPE, Crumb Rubber, and Bitumen which are shown in Table 7.

2.8 *Collection and Preparation of Materials*

For this research work, most of the materials and ingredients are utilised from waste already available in the environment like Crumb Rubber, Coconut Shell, LDPE, and Scrapped Bituminous Road. This will significantly reduce the amount of waste from the environment and reduce pollution.

The main source of Crumb Rubber is the discarded scrap tyres and rubber tubes of automobiles and vehicles. These tyres and Tubes are easily available anywhere and can be used as an additive in the bituminous mix. For this research work, the

Coconut Shells are collected from households, industries as well and religious places like temples. The hard shells were broken into small particles to use in a bituminous mix. It is generally found in daily-use plastic items like shopping bags, beverage bottles, etc. All the LDPE used for this research work was obtained from these sources only, which helped in the reduction of solid waste from the environment. During the construction of new road pavement, the already laid bituminous pavement is first demolished. The recycling of Aggregate and Bitumen from demolished road pavement will significantly decrease the cost of construction. Some portion of Aggregate is being recycled from dismantled road pavement whereas some part of fresh Aggregate is also used in this research work which can easily be found on construction sites.

2.9 Testing of Sample

The Marshall Stability Method is conducted on the sample to deduce the strength and flexibility of Modified Bituminous Road. Marshall Stability Test is executed for the determination of physical properties like plastic deformation of bituminous mixes. The sample used for the Marshall Stability Method should meet the purpose of the design product with proper proportioning of binders and Aggregates. Air void in the specimen should be in the range of 3–5% which is commonly used as design air void. The compacted specimen should also meet the requirement of stability [15].

A sample of 1500g weight is prepared for the testing by Marshall Stability Method. The homogenous mixture of preheated dry Aggregate with preheated binder is prepared. The additives are added to the mixture and Crumb Rubber is spread evenly in the whole mixture. For the following five minutes, the mixture of Bitumen, additive, and Aggregate is mixed at 160 ± 5 °C followed by preheating of mould at a temperature range of 93–149 °C. The specimen is compacted from both the faces by Marshall Compactor with 50 blows being applied at 150°C and weight is taken. This specimen is placed into the Marshall Testing Machine for crushing with a load of a constant rate of 50.8 mm per minute up to the highest load attained. The thickness of the specimen is measured to the nearest 0.1g. Finally, the Marshall Stability of the Bituminous Mixture can be calculated. The ratio between Marshall Stability to Flow is expressed in kN/mm and it represents the stiffness and strength of the mixture.

3 Results

3.1 Marshall Stability Analysis

The Marshall Stability test is used for evaluating the strength of asphalt. The maximum load that an asphalt sample can bear before deforming or crumbling is measured. In this research, parameters like Air Voids (Fig. 1), Marshall Stability value (Fig. 2), Flow Value (Fig. 3), percent Voids in the Total Mix (V_V) (Fig. 4), and percent Voids Filled with Bitumen (VFB) (Fig. 5) of the sample are calculated by experiment.

Void Ratio must be in the range of 3.8–4.8% as per the design criteria, depending on how much LDPE and crumb rubber are used. The highest void ratio was observed when the binder percentage was 5.5% and the lowest void ratio was observed when 6.5% binder was used.

The highest Marshall Stability was observed when the binder percentage was 6% and the lowest Marshall Stability was observed when 7.58% binder was used as in C Mix.

Flow Value must, depending on how much LDPE and Crumb Rubber are used, fall between 2 and 4 according to the design criteria. The highest Flow Value was observed when the binder percentage was 6.5% and the lowest void ratio was observed when 5.5% binder was used.

According to the design criteria, Voids in Total Mix, depending on how much Crumb Rubber and LDPE are used, must be between 16.5 and 19.5%. The highest

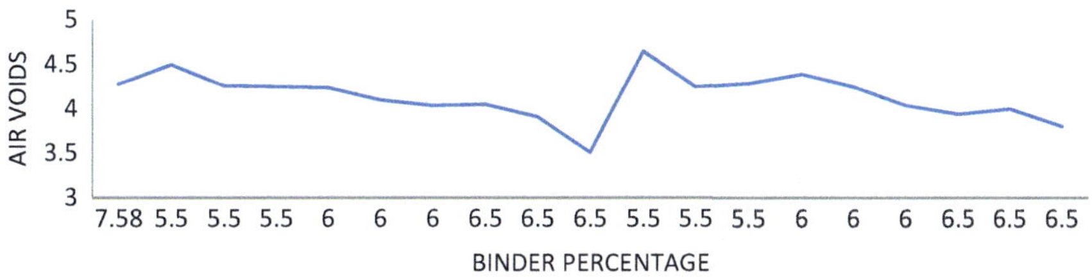

Fig. 1 Relationship between air void and binder percentage

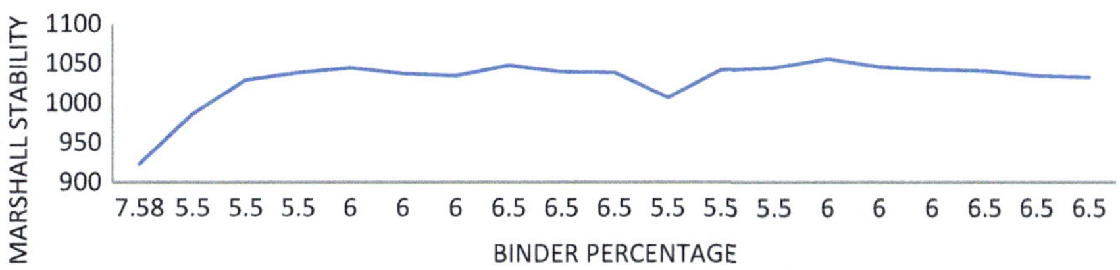

Fig. 2 Relationship between Marshall Stability and binder percentage

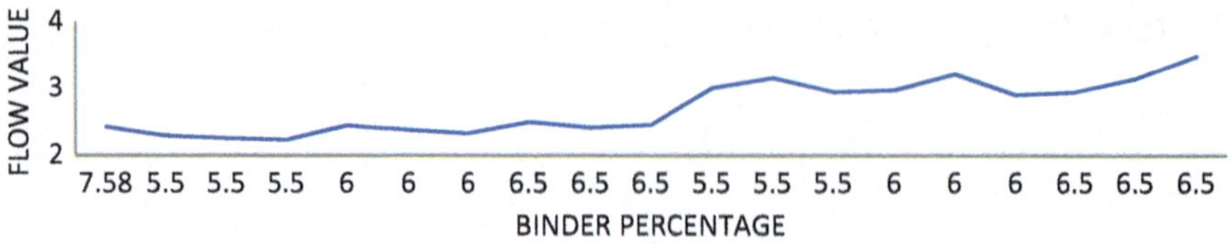

Fig. 3 Relationship between flow value and binder percentage

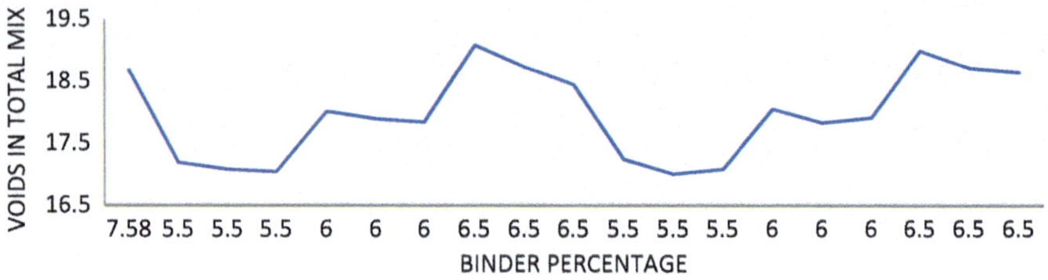

Fig. 4 Relationship between voids in total mix and binder percentage

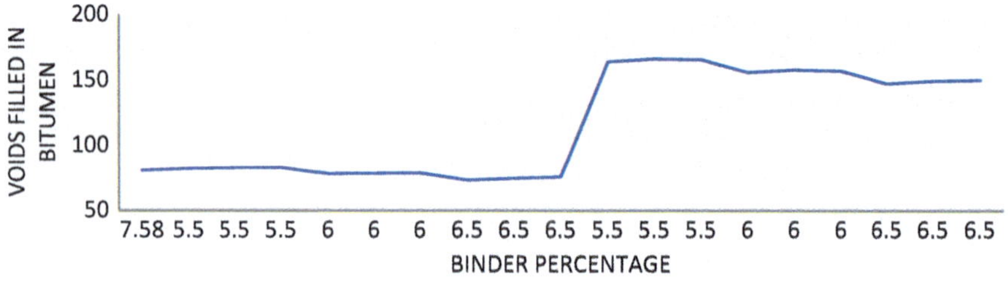

Fig. 5 Relationship between voids filled in bitumen and binder percentage

void in Total Mix was observed when the binder percentage was 6.5% and the lowest void ratio was observed when 5.5% binder was used.

Based on how much LDPE and Crumb Rubber is used, Voids Filled in Bitumen must fall between 50 and 180 according to the design criteria. The highest void filled in Bitumen was observed when the binder percentage was 6% and the lowest void ratio was observed when a 6.5% binder was used (Table 8).

3.2 *Compression Strength*

Compressive strength may be defined as the highest compressive stress a material can withstand before breaking down. It is calculated by dividing the maximum load by the material's cross-sectional area (Fig. 6).

Table 8 Results obtained from Marshall Test

S.no	Sample	Volume of binder, %	Air voids	Marshall stability (kg)	Flow value (mm)	Voids in mineral aggregate	Voids filled in bitumen
1	C MIX	7.58	4.28	923.74	2.43	18.69	81.13
2	A1C7L3	5.5	4.50	986.75	2.30	17.20	82.44
3	A1C5L5	5.5	4.26	1029.56	2.26	17.09	82.97
4	A1C3L7	5.5	4.25	1038.74	2.24	17.05	83.17
5	A2C7L3	6	4.24	1044.85	2.45	18.02	78.25
6	A2C5L5	6	4.10	1037.72	2.38	17.90	78.77
7	A2C3L7	6	4.04	1034.66	2.33	17.85	78.99
8	A3C7L3	6.5	4.05	1047.91	2.50	19.09	73.49
9	A3C5L5	6.5	3.91	1039.76	2.42	18.74	74.87
10	A3C3L7	6.5	3.51	1038.74	2.46	18.45	76.04
11	B1C7L3	5.5	4.65	1007.14	3.01	17.25	164.35
12	B1C5L5	5.5	4.25	1041.79	3.16	17.01	166.67
13	B1C3L7	5.5	4.29	1044.85	2.95	17.09	165.89
14	B2C7L3	6	4.39	1056.07	2.98	18.06	156.15
15	B2C5L5	6	4.25	1045.87	3.22	17.84	158.07
16	B2C3L7	6	4.04	1042.81	2.91	17.92	157.37
17	B3C7L3	6.5	3.94	1040.77	2.95	19.00	147.63
18	B3C5L5	6.5	4.00	1034.66	3.15	18.72	149.84
19	B3C3L7	6.5	3.80	1032.62	3.49	18.65	150.40

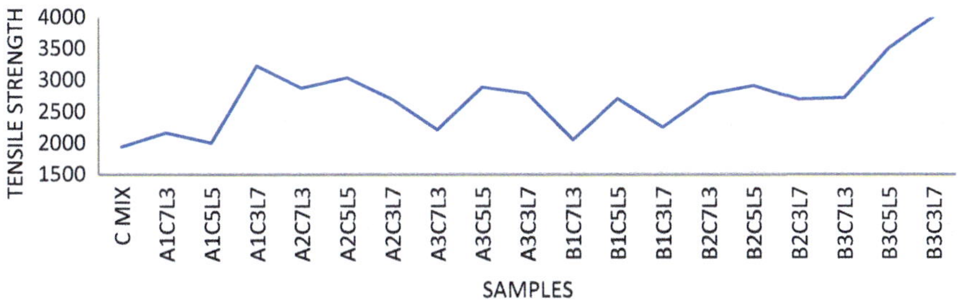

Fig. 6 Representation of compression strength

Depending on the composition, the Compressive Strength of the Mix ranges from 410 to 860 kg/cm^2. The Compressive Strength of all the samples was observed to be greater than C Mix. The highest Compressive Strength was observed in B3C3L7, A1C3L7 and, A2C5L5.

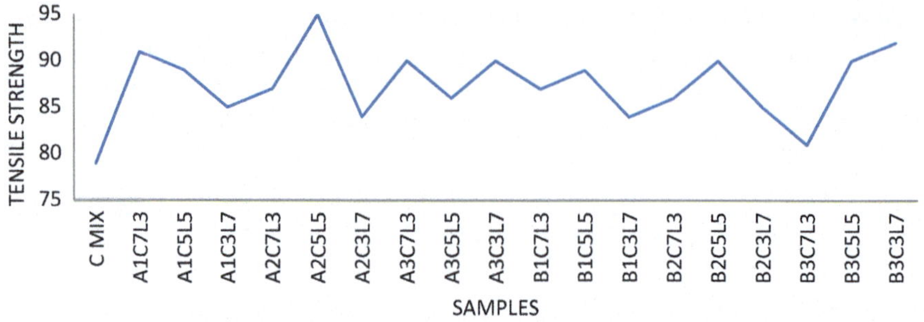

Fig. 7 Representation of Tensile strength

3.3 Tensile Strength

The ratio of the average wet Indirect Tensile strength to the average dry Indirect Tensile strength is then used to calculate the tensile strength ratio or TSR. A material's tensile strength is the highest tension at which it can be loaded without breaking (Fig. 7).

TSRs must be above 85–94% as per the design criteria, depending on how much LDPE and crumb rubber are used. The tensile strength of all the samples was observed to be greater than C Mix. The highest Tensile Strength was observed in A2C5L5, B3C3L7, and A1C7L3.

4 Conclusion

Compared to regular Aggregates, Coconut Shells have 24% higher absorbency. This is due to the fact that Coconut Shells have hard fibrous structure which provides high strength as well as absorption characteristics to the pavement surface. LDPE decreases inflow by 18.2% compared to C Mix. Crumb Rubber helps in increasing compressive and flexural strength by reducing air voids and moisture absorption. Experimentally, it is found that 70% of the asphalt from dismantled roads can be recycled and used for the further construction of a new road. The tensile strength ratio is found to be in the range of 85–94% and compressive strength ranges from 410 to 860 kg/cm^2. The void in mineral aggregate is 18% and air void in the mix ranges from 3 to 5%. The most appropriate proportion of additive is 30% (with 30% Crumb Rubber and 70% LDPE. When 6% binder was used in the sample, the highest Marshall Stability was observed. Research has found that using waste materials with modified asphalt in a fixed ratio helps increase the strength and durability of the flexible pavement.

By improving overall performance, research results show that the addition of waste plastic bottles and rubber tires can increase the durability of bituminous mixtures and extend their life. In addition, it can help prevent environmental pollution from disposing of such waste underground. Polymer coating reduces porosity, minimises moisture absorption, and improves sound quality as it fills the voids present in the pavement. Aggregate asphalt mixtures with polymer coatings have higher Marshall Stability values and adequate Marshall Coefficients, which make them excellent materials for the construction of flexible pavements. The modified asphalt exhibits good results when compared to the standard results. It is observed that a modified road is approximately 5.5% more cost-effective as the aggregate from the scrapped road is recycled reducing the demand for aggregate in road construction.

References

1. Ridwan A et al (2019) Evaluation of the strength of coconut shell aggregate concrete block for parking area. In: IOP conference series: earth and environmental science, vol 277. IOP Publishing, pp 012002
2. Singh B, Kumar L, Gupta M, Chauhan M, Chauhan GS (2013) Effect of activated crumb rubber on the properties of crumb rubber-modified bitumen. J Appl Polym Sci 129:2821–2831
3. Malik EMI (2023) Use of low-density polyethylene in flexible pavement. In: IOP conference series: earth and environmental science, vol 1110. IOP Publishing, pp 012065
4. Dagliya M, Bochare R (2020) Recycling bitumen from dismantled road-a case study. Int J Eng Res Technol 9(06), ISSN: 2278-0181
5. Ting TL, Jaya RP, Hassan NA, Yaacob H, Jayanti DS (2015) A Review of utilization of coconut shell and coconut fiber in road construction. J Teknol 76
6. Jamal M, Giustozzi F (2020) Low-content crumb rubber modified bitumen for improving Australian local roads condition. J Clean Prod 271:122484
7. Nejad FM, Aghajani P, Modarres A, Firoozifar H (2012) Investigating the properties of crumb rubber modified bitumen using classic and SHRP testing methods. Constr Build Mater 26:481–489
8. Alemu GM, Melese DT, Mahdi TW, Negesa AB (2023) Combined performance of polyethylene terephthalate waste plastic polymer and crumb rubber in modifying properties of hot mix asphalt. Adv Mater Sci Eng 2023
9. Navarro FJ, Partal P, Martínez-Boza F, Gallegos C (2005) Influence of crumb rubber concentration on the rheological behavior of a crumb rubber modified bitumen. Energy Fuels 19:1984–1990
10. Awwad MT, Shbeeb L (2007) The use of polyethylene in hot asphalt mixtures. Am J Appl Sci 4:390–396
11. Manju R, Sathya S, Sheema K (2017) Use of plastic waste in bituminous pavement. Int J ChemTech Res 10:804–811
12. Fang M, Park D, Singuranayo JL, Chen H, Li Y (2019) Aggregate gradation theory, design and its impact on asphalt pavement performance: a review. Int J Pavement Eng 20:1408–1424
13. Hasnain Saeed M et al (2019) Sustainable silicon waste material utilization for road-construction: an application of modified binder for Marshall stability analysis. Appl Sci 9:1003
14. Nemade SN, Thorat PV (2013) Utilization of polymer waste for modification of bitumen in road construction. Sci Rev Chem Commun 2:198–213
15. Singhal M, Yadav Y, Mandal R (2016) Use of modified bitumen in highway construction. Int J Innov Res Sci Technol 2:376–382

Material Properties

Investigation of CFRP as a Material for Fabrication of 1U CubeSat Structure for Reduction of Weight with Improved Strength

Anand Belsare, Shruti Kulkarni, Aakanksha Kajale, Deepak Hujare, and Pravin Hujare

Abstract Cube satellites are miniature satellites used for research purposes that have a wide range of applications; technology demonstration, science, educational tools or commercial industry. The cube satellite form factor ranges from one unit '1U' to six units '6U'. A 1U CubeSat typically weighs around 1 kg, the structure is conventionally built using aluminium material. The study presents an alternative material for the cube satellite structure from the conception of the idea to analysis, fabrication and testing of the prototype. The study analyses the viability of composite structures for cube satellites and introduces the potential of High Tg, Prepreg Carbon Fibre Reinforced Polymer (CFRP) as it can drastically reduce the weight of the structure. Before fabrication, analyses were conducted for primary testing, which included structural analyses such as modal tests, static tests, random vibration analyses and impact tests. The fabrication method incorporates modern manufacturing techniques like 3D printing, laser cutting, numerically controlled oscillating knife cutting and other composite fabrication methods as well as other techniques like water-jet cutting, pattern and mould making and vertical machining centre implemented. The designed prototypes are validated by a set of tests as per standardised methods.

Keywords 1U cube satellite · Composites · CubeSat · High Tg Prepreg CFRP · Oscillating knife cutting · Composite fabrication methods

---

A. Belsare · S. Kulkarni · A. Kajale · D. Hujare (✉)
Department of Mechanical Engineering, Dr. Vishwanath Karad, MIT World Peace University, Pune 411038, India
e-mail: deepak.hujare@mitwpu.edu.in

P. Hujare
Vishwakarma Institute of Information Technology, Pune 411048, India

P. Bhingole et al. (eds.), *Advances in Materials Engineering*, Lecture Notes in Mechanical Engineering, https://doi.org/10.1007/978-981-97-7114-1_15

1 Introduction

The 1U CubeSat, is a compact satellite with a cubic configuration, with size 10 × 10x10 cm^3 and weight not exceeding 1.33 kg. Since conception CubeSats have greatly alleviated space launches in the past decade and over 1000 CubeSats have been launched in the twenty-first century [1]. The recent space launches offer affordable and accessible alternatives for low-earth orbit (LEO) missions, benefiting both educational and commercial organizations. CubeSats have gained commercial interest, driven by space and commercial agencies which are typically designed to meet specific educational or technological missions [2].

CubeSat is comprised of various components, including structure, communication, power, attitude determination and control, command and data handling, and the payload [2]. One crucial challenge for CubeSat design is ensuring structural integrity, especially under the harsh conditions of space. Factors such as thermal expansion, density, radiation resistance, as well as mechanical properties like modulus, strength, and toughness must be carefully considered with precise operational loads during the selection of suitable materials for CubeSat structures. Additionally, specialized challenges, such as thermal stress management and payload sensitivity to outgassing and thermal displacements, must also be addressed well in advance [3].

Traditionally, CubeSat structure has been made of aluminum alloy 6061 or 7075, with some limitations in volume efficiency and adaptability. To overcome these challenges, the research investigates the viability of utilizing Composite Fiber Reinforced Polymers (CFRP) as a potential substitute for conventional materials like aluminum alloys. The study aims to propose and validate a novel method of fabrication of a 1U CubeSat structure using CFRP, incorporating industry-standard tests and simulations for comprehensive evaluation.

By exploring the advantages of CFRP, such as its high strength-to-weight ratio, excellent thermal stability, and superior mechanical properties, the potential benefits of using CFRP in CubeSat structure could significantly enhance their performance and reliability. The adoption of CFRP as a structural material promises greater flexibility in designing mission-specific systems and payloads, as well as improved volume efficiency due to the material's ability to perform multiple roles, including thermal management, radiation shielding, and strain actuation.

In the present study, a detailed investigation was conducted for the fabrication of a 1U CubeSat structure by introducing CFRP. The novel method of fabrication of 1U CubeSat is proposed and its outcomes are supported by industry-standard tests and simulation results, which provide valuable insights for the future design and development of CubeSat missions.

2 Modeling and Selection of Materials for CubeSat Structure

In order to ensure proper functionality, a sturdy structure capable of supporting the weight of approximately 2 kg is required for the deployment of CubeSat within the low Earth orbit.

2.1 *Material Properties*

Given the role of the CubeSat and the environmental considerations, a material with a low weight, a wide temperature envelope, and a high strength to weight ratio is preferable. Typically, Aluminium 2024, 6061, 7075-T6, 6082 grades are popularly used for CubeSat structures [4]. A comparison of material properties was conducted to assess the suitability of Aluminium 6082 and prepreg CFRP for the structural design of the CubeSat.

Aluminium alloys possess many challenges and considerations during fabrication of minute geometries for specialized aerospace applications. The geometry of CubeSat structure consists of intricate features with minimum dimensions of 1.2 mm thickness. The machining of intricate geometries of 1.2mm thickness in aluminium alloy, special fixtures are necessary which increases the initial costs. Additionally, the usage of Al 6082 results in significant material waste. In order to prevent cold welding on Al 6082, anodizing is a necessity. Additional coatings are also required to prevent oxidation. These coatings raise the price, take more time and marginally increase the weight.

CFRP has been introduced to the aerospace industry for decades nevertheless it has not yet been commercially used for CubeSat structure due to the complex processes involved while manufacturing CFRP.

Table 1 shows the comparison of Prepreg CFRP and Aluminium 6082 on the basis of its material properties [5].

Other factors to be considered when dealing with manufacturing using Prepreg CFRP are:

Table 1 Comparison of properties between Aluminium 6082 and CFRP

Material	Al 6082	CFRP
Tensile strength (MPa)	295	3450
Modulus of elasticity (MPa)	7e+4	23e+4
Density (kg/m3)	2770	1650
Melting point (°C)	555	1600
Thermal conductivity (W/mK)	180	0.36
Poisson's ratio	0.33	0.77

Table 2 Storage criteria of CFRP

Temperature (°C)	Shelf life
−18	12 months
16	6 months
25	4 weeks

Storage and Handling- Prepreg CFRP should be stored at sub-zero temperatures; −18 °C in a sealed moisture-proof bag, at all times. Shelf life of Prepreg CFRP varies with temperature. CFRP should not reach room temperature before it is cut to shape. The protective covering should not be removed until the lay-up process (Table 2).

CFRP produced by hand lay-up method has low glass transition temperatures of up to 80° to 90 °C while the operating temperature envelope of CubeSats are higher than 90 °C. Thus, hand lay-up CFRP was eliminated from the material consideration. Pre-impregnated carbon fiber was selected due to its characteristic of high glass transition temperature.

A comparative analysis of the material properties of prepreg CFRP and Aluminium 6082 was performed. Prepreg CFRP was selected for the structure of the 1U CubeSat as it fulfils all the necessary criteria such as high strength to weight ratio, high temperature resistance.

2.2 CAD Modeling

The dimensioning and geometry of the 1U Cube-Sat structure were built in accordance with the industry standard guidelines. The 1U CubeSat structure was designed as a cube with the dimensions 10 × 10 × 10 cm with a tolerance of ±0.4 mm.

The design studied was influenced by the Endurosat structure. [6] Fig. 1a shows the CAD Assembly of the proposed CubeSat Structure. A modular design was implemented to simplify the assembly. The geometry was determined, by performing a trade-off material usage and structural strength. The aim was to minimize the material used without compromising the strength.

The geometry was smoothened by rounding the corners to avoid stress concentration. (Fig. 1b) To eliminate stress development in probable regions, notches were provided. (Fig. 1c) The edges of the columns were rounded for stress elimination purposes. Provisions for rods were designed for the assembly of internal components and were constrained by fixtures present on the inside surfaces of the top and base plate. Provisions for kill switch were made at four compatible locations.

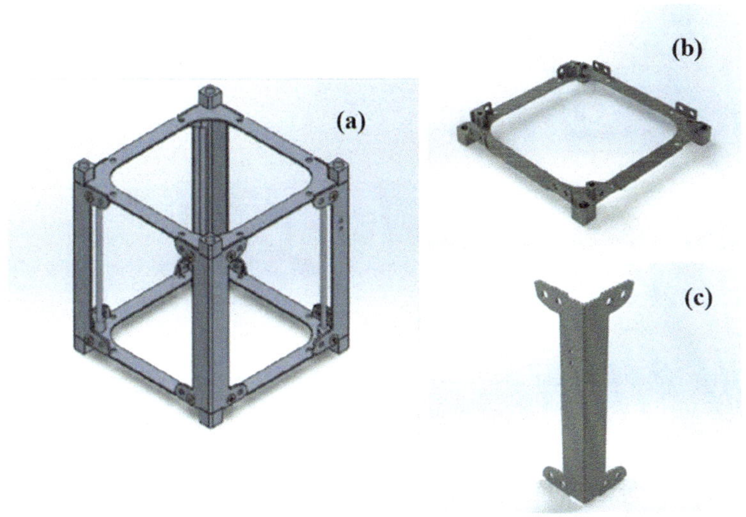

Fig. 1 Components of 1U CubeSat **a** Entire Assembly **b** Top//Base Plate **c** Column of 1U CubeSat

3 Finite Element Modeling and Analysis

FEA stands as the most potent tool for analysing complex working assemblies in real time, providing results for both static and dynamic analyses. It has been widely acknowledged as a cost-effective solution that can significantly reduce design cycle time. Consequently, FEA serves as a precise and reliable method for examining stress levels within the structure of CubeSats [7].

3.1 Loads and Boundary Conditions

For a CubeSat to survive harsh space conditions, it must withstand specific launch conditions. Basic analyses: static structural analysis, random vibration, modal analysis, thermal analysis, and impact testing are performed to predict the behavior of the proposed design in a similar environment. Ansys Workbench 2020 R1 was used for performing the analyses. The specific boundary conditions of each analysis are given hereafter:

Static Structural Analysis. Examines the effects caused by a static load on a particular body. The pre-requisite to perform the following analysis is the magnitude of the static load. The lowermost extruded protrusions represent the fixed support which will be in contact with the lower bound. The load applied is vertically downward with a total force of 2 kg i.e. 19.62 N evenly distributed on the four ends The total deformation and equivalent stress i.e. Von-Misses Stress can be obtained. [8] The total deformation is shown in Fig. 2a which is 0.002961mm.

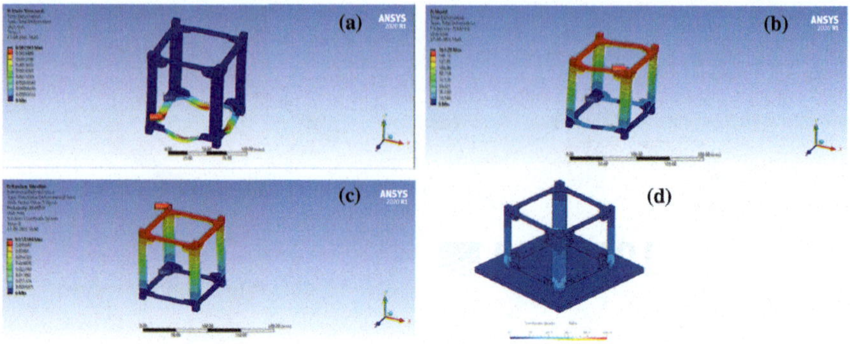

Fig. 2 Various finite element analyses performed on 1U CubeSat **a** Static structural analysis **b** Modal analysis **c** Random vibration **d** Impact test

Modal Analysis. Primarily, a comprehensive dynamic analysis was conducted to ascertain the natural frequencies and mode shapes of the structure under consideration. The analytical process plays a crucial role in understanding how the structure responds to dynamic forces and vibrations, providing essential insights into its dynamic behavior.

To perform this dynamic analysis, the lowermost extruded protrusions were strategically identified and treated as fixed supports. The choice was made based on engineering principles to simulate a realistic representation of the structure's interaction with its foundation. The results obtained from the analysis served as a foundational dataset, characterizing the dynamic response of the structure and revealing how it would behave when subjected to dynamic loads.

The outcomes of the dynamic analysis were not merely confined to theoretical understanding; instead, they were seamlessly integrated into the subsequent phase of our investigation—the random vibration test.

Figure 2b visually encapsulates one of the key findings from our dynamic analysis—the deformation of the structure at a specific frequency, precisely 1941.4 Hz. This frequency is of particular interest as it highlights a mode of vibration that significantly influences the structure's response to dynamic loading conditions.

Random Vibration Analysis. Natural frequencies calculated from the modal analysis are used as inputs to the random vibration test. The CubeSat is subjected to vibrations in the following frequency ranges from 20, 50–800, 2000 Hz, corresponding to ASD values (Acceleration Spectral Density) of 0.026, 0.16, 0.026 g^2/Hz as shown in Table 3. The values were used as boundary conditions to perform the analysis [9].

Thermal Analysis. Thermal analysis is conducted to understand whether the CubeSat can withstand the low Earth orbit space temperatures of −65 to + 125 °C. The basic boundary conditions are solar radiation, albedo, ground IR, and thermal radiation. Solar radiation is the amount of solar energy per unit area of a sphere centred on the sun at a distance d. The output power required corresponds to 3,856 × 1026 watts [10]. The average solar radiation is taken to be 1371.08372 ± 5 W/

Table 3 ASD levels for random vibration

Frequency (Hz)	ASD level (g^2/Hz)-(Acceleration spectral density)
20	0.026
50–800	0.16
2000	0.026

m^2. The albedo, also called the radiation reflected from the Earth, was considered to be 480 W/m^2. A total of 239 W/m^2 was considered as the Earth IR, which is the third radiant heat source on a satellite in low Earth orbit. Finally, thermal radiation was taken into account according to the emission values of the materials used in the CubeSat.

Impact Test. The impact test determines the amount of energy absorbed by a material during fracture. The absorbed energy is a measure of a given material's toughness and acts as a tool to study temperature-dependent brittle-ductile transition. It is to determine whether the material is brittle or ductile in nature. A force of 2 kg was implied at the structural base in contact with the p-pod for a duration of 0.1 s.

4 Experimental Approach

In the CubeSat structure industry, the manufacturing process typically involves four steps that are aided by computer technology. A top-down approach is employed, where a CNC machine is used to carve out the desired geometry from a solid block of aluminum. Tool paths are generated on the basis of a CAD model. The machining of the structure is followed by post-processing [11].

Pre-impregnated CFRP fabrication employs a 'bottom-up' approach, involving mold fabrication, precise fabric cutting and layup, followed by curing and post-processing stages as depicted by Fig. 3. The process ensures high-quality CFRP parts with controlled geometry. The initial phase involves Mold Fabrication, where a mold is designed and created to match the desired geometry. CAD and CAM tools are commonly used to ensure precision. The mold is designed to withstand high temperatures of 120 °C or more for the curing process, limiting material options to fabricate the mold.

The subsequent stage is Fabric Cutting and Layup, where multiple layers of prepreg CFRP fabric are precisely cut and stacked to form a solid part. Accurate cutting facilitates the layup process and influences the accuracy of the final part. A releasing agent is applied to aid demolding, while vacuum bagging is essential to eliminate air bubbles and enhance bonding between the layers [12].

The third stage entails Curing, where the prepreg CFRP fabric layers are heated under vacuum or within an autoclave. The heating cycle duration depends on the resin matrix system used, ranging from 30°C to 180°C based on the resin's glass transition temperature (Tg).

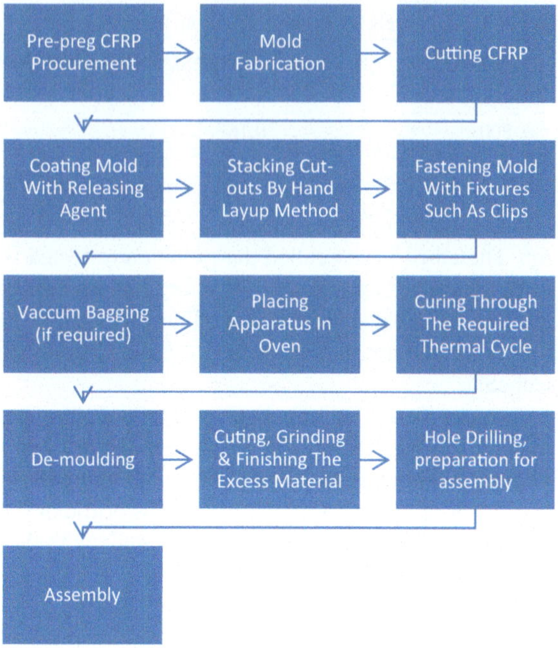

Fig. 3 Pre-Preg CFRP fabrication process flowchart

After the Curing stage, the De-molding and Post-Processing phase takes place. The cured part is carefully removed using a wedge and hammer. Subsequently, it undergoes post-processing to achieve the desired shape within the dimensional tolerances specified. This involves employing a high-speed rotary cutter [12].

4.1 CubeSat Structure in CFRP

The major components of the CubeSat structure are the Structure Columns, Top Component of the Structure and Base Component of the Structure.

With the general process of CFRP manufacturing, Table 4 shows the fabrication prerequisites of these components based on some common parameters.

Structure Columns. The initial prototyping stage the four structure columns were successfully fabricated, which did not contain any intricate geometric elements such as extrusions or flanges.

The fabrication process for the structure column comprised the following steps.

a. Mold Fabrication

The mold for the column component was fabricated using 1.2 mm thick sheet of 404 grade Stainless Steel, which provided excellent heat resistance for sustaining temperatures above 120 °C. Laser cutting was employed to create two development drawings on the Stainless-Steel sheet, followed by bending at a 90° angle with a radius

Table 4 Specifications of components for composite fabrication

Parameters	Structure columns	Top/base plate
Component thickness	1.2 mm	15.2 mm (max)
Layer Thickness	0.2 mm	0.2 mm
No. of layers	6	76
Mold complexity	Simple design, easy to fabricate	Complex design
Intricate features	Absent	Present
Demolding	Easy	Hard
Accuracy of fiber cut required	Low	Highly accurate

of 1.2 mm to achieve the required shape. Notches and holes were incorporated into the mold to ensure the desired accuracy.

b. Fabric Cutting and Layup

Molds precisely matching the component's shape were used to guide the cutting and trimming of excess fabric. For the layup process, six layers of 200 GSM fabric were utilized to achieve the desired 1.2 mm total thickness. A two-part mold was employed, creating a sandwich structure of SS-CFRP-SS that eliminated the need for vacuum bagging or autoclave. Metal clips were used to securely hold the mold together during closure, ensuring precise alignment and accuracy.

c. Curing

The curing process followed the specified guidelines in section 4.3 for the prepreg, ensuring optimal results.

d. De-Molding and Post-Processing

Prior to de-molding, post-processing was performed on the column component. The mold served as a guide for cutting and drilling, allowing for meticulous de-molding using a metal wedge and knife to minimize the risk of damage.

Top and Base Component of the Structure. The upper and lower parts of the assembly are identical components, with feature dimensions as small as 3.5 mm. These components include intricate curves and sharp edges in the geometry. To achieve accuracy and precision, number of iterations with successive modifications in geometry and fabrication techniques were incorporated. In relation to it, three prototyping stages are explained in detail as follows:

a. Prototype 1

Additive manufacturing was used to accurately fabricate the two-part mold. The mold design incorporated necessary features such as protrusions, flanges, and alignment holes right from the beginning. To construct the mold, Onyx was utilized as the primary material, with carbon fibre reinforcement comprising 20% of the wall composition. A 30% infill was implemented to ensure the necessary strength. The

(a) (b)

Fig. 4 **a** Prototype 1 (Top and Base Component of the Structure). **b** Prototype 2-Vertical Flange Feature and Corner Square Cut-out of a 1U CubeSat

mold was 3D printed using the Markforged mark-I printer. Figure 4a shows the 3D printed mold. The reinforced mold can withstand temperatures up to 150 °C and hence was perfectly suitable for the required application considering the maximum curing temperature of 120 °C. Fabric was cut into exact shape with accurate dimensions to ensure the overall dimensional accuracy of the cured component. Silicon based high temperature releasing agent was used for easy removal of the component.

Laser cutting was not feasible to cut the fabric as the laser directly cured the resin, damaging the fabric. Water Jet cutting was not feasible as the prepreg carbon fibre cloth needed to be kept free from any moisture. The small feature parts with dimensions less than 5 mm had issues with fabric integrity while removing the protective film just before the layup process. Furthermore, there was a possibility of air bubble or gap formation due to stacking of multiple layers.

A CNC oscillating knife cutting machine was used for complex parts, while the rest of the simpler geometries were manually cut using laser-cut acrylic stencils. This manual cutting method reduced accuracy but still yielded acceptable results. To minimize damage to the fabric's weave pattern, forceps were carefully used to remove the protective film. During the demolding process, the component suffered significant damage due to weaker bonding between layers. The layers separated due to uneven pressure distribution during the curing process, leading to a drastic reduction in component strength. In Prototype 2, clips were employed to enhance bonding between layers, and clearances in the mold cavity were increased to facilitate easy removal.

b. Prototype 2

The second iteration involved a layer-by-layer stacking process, utilizing a stainless-steel mold with tighter tolerances. Instead of fabricating small geometries, metal inserts were employed. Aluminum 6082 inserts were cut to the required thickness of 6.3 and 7.3 mm using water-jet cutting as shown in Fig. 4b. The desired thickness was achieved through milling. The stainless-steel mold was created using a 1.2 mm thick sheet of 404 grade. CNC turret punch was utilized to create the necessary cut-out shapes, and the sheet was then bent into the required shape. The corner block and vertical flange features of the component incorporated four corner square cut-outs and four vertical extrusions on the inner walls. With fewer cut parts in the second iteration, the fabric was cut by hand to match the desired shape, as the mold had higher tolerances. Prior to the layup process, the mold and clamps were sprayed with a releasing agent.

The difficulty in welding the extrusions arose from the minor thickness of the sheet, which also caused warping. The hand-cut fabric parts and stacking of all the layers outside the mold led to the formation of large pleats when the layup was placed inside the mold. Additionally, the weight of the corner inserts resulted in the corners dipping during the curing process inside the oven, causing deformation of the component.

To minimize warping, the job was securely held in specific fixtures during welding, and any small, warped geometries were promptly straightened out. The issue of corner deformation was addressed and resolved in prototype 3, with further details provided in the study. The layup was pressed onto the main mold surface using a flat, cured carbon fiber piece, effectively eliminating pleats and creases and resulting in a uniform surface.

c. Prototype 3

Prototype 3 was manufactured following a similar methodology as Prototype 2, with a notable difference in the usage of inserts for the corner block features. Instead of employing aluminum inserts, thirty-eight layers of cut prepreg carbon fiber pieces were stacked to form the block feature. This was done to enhance the bonding between the block feature and the main body, as well as to eliminate any deformation caused by the weight of the aluminum inserts. Additionally, to prevent any corner dipping and subsequent component deformation, the main mold was placed on a base plate.

4.2 *Assembly*

The entire CubeSat assembly is shown below in Fig. 5.

The complete CubeSat assembly consists of four structural columns, a base, and a top, assembled using stainless steel nut bolts with a size of M3*12.

5 Results and Discussions

Based on the selected materials, a conventional aluminium structure was compared with a composite structure with CFRP based on the masses of the structures. The total weight of the composite structure assembly including nuts and bolts was determined to be 72 g, optimizing structural weight by a 40% reduction compared to aluminium structure.

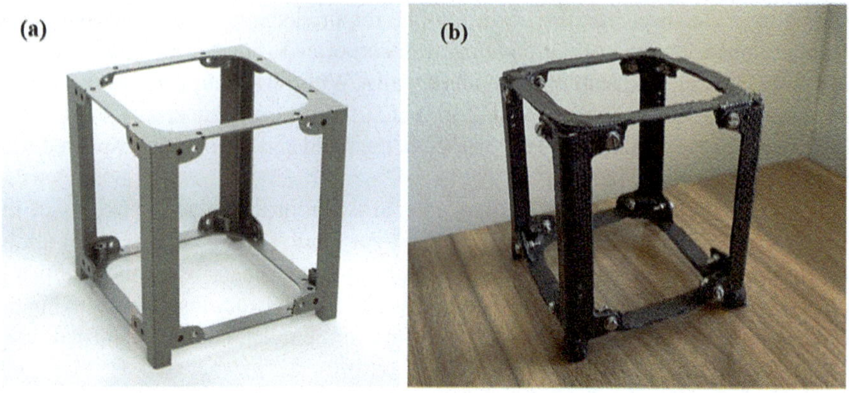

Fig. 5 **a** Proposed structure **b** Final assembly of the 1U CubeSat

5.1 FEA Simulations

Static Structural–he protruding structures at the bottom signify the fixed supports that make contact with the lower boundary. A vertical load of 2 kg, equivalent to 19.62 N, was uniformly applied across the four ends. The deformation was negligible in both the Aluminium 6061 and CFRP materials. The Von Mises Stress for Aluminium 6061 and CFRP, were observed to be 290 and 345 MPa, respectively and both these values fall well within the ultimate Yield Stress limits.

Modal Analysis and Random Vibration–An effective modal analysis was conducted, in which the lowermost extruded protrusions represented fixed supports, and the outcomes were employed as the foundation for the Random Vibration analysis. The resultant total deformation measured 0.051184 mm, and the peak Von Mises stress remained below the ultimate tensile strength of Carbon fiber. There is approximately a 68% probability of attaining this stress value.

Thermal Analysis–In thermal analysis, the Stefan-Boltmann law and the fundamental heat rate law were used to calculate the operating temperature range of the CubeSat. The established standard operating temperature for CubeSat in the low earth orbit ranges from −30 to 120 °C. The resultant calculated temperatures from the analyses indicate a range of 85 °C for aluminum and −30 °C for CFRP, which were falling comfortably within the acceptable limits. Additionally, the linear thermal expansion for aluminum was observed to be 0.2 mm which is considered to be well within acceptable limit.

5.2 Standard Testing Procedures

CubeSat has to undergo the standard necessary testing procedures to withstand the extreme launch and space conditions. These procedures encompass a series of tests including visual inspection, thermal-vacuum bakeout, thermal cycling, and vibration. These tests are established as minimum requirements in the CubeSat design specification document. The tests strictly adhere to the specifications outlined in the 'ESA EECSS-E-ST-10-03C' and 'NASA GEVS: GSFC-STD-7000A' standards [13].

Following these standards, a range of testing procedures are listed which include Physical Properties, Thermal Cycling, Pyroshock Test, Thermal Vacuum, Sinusoidal, and Random Vibration [14]. The subsequent section provides an elucidation of the outcomes obtained from these tests.

Physical Properties. Three parameters are considered: the weight of the entire CubeSat platform, including subsystems, the Centre of Gravity, and the Moment of Inertia of the fully assembled CubeSat platform. The assembly's total weight, including the nuts and bolts, is 72 g, which optimised the structural weight by 40% as compared to the aluminium structure.

Thermal Cycling. Assessment to withstand alternating high and low temperature extremes was conducted, with temperature cycles ranging from -30°C to 60°C. During the test, a dwell time of 1 h was maintained at the extreme temperature values.

To monitor the thermal behaviour of the assembly during the thermal cycling test, a Through Process Data Logging system provided by 'Grant Instruments' was utilized. The instrument used underwent calibration and certification according to the stand ardized procedure established by NCL, ensuring the highest level of result accuracy.

Figure 6 illustrates the thermal behavior of the assembly during the 2-h thermal cycling test: Channel 1: Base Structure, Channel 2: Structure Column, 3: Environment.

It is evident that the surface behavior of each component exhibits a linear variation in response to the environmental temperature, without any unexpected or abrupt changes within the temperature range of −24 to 60 °C. The base structure experienced a minimum temperature of −22 °C and a maximum temperature of 61.3 °C, while the

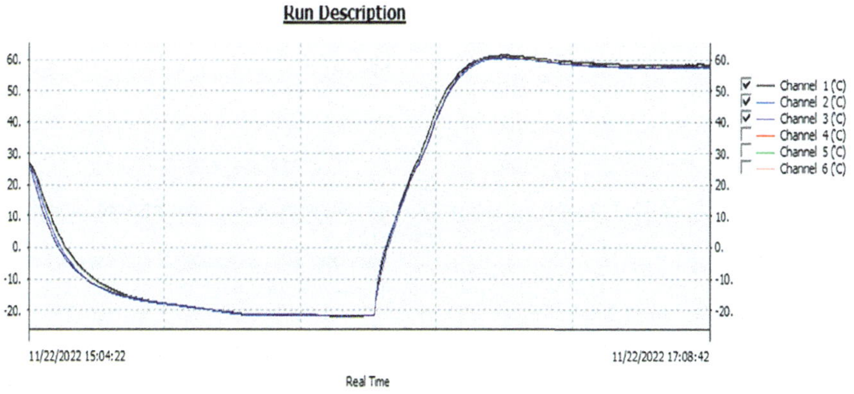

Fig. 6 Thermal cycling test for 1U CubeSat

structure column had a minimum temperature of −21.8 °C and a maximum temperature of 60.5 °C. The dimensions of the assembly remained unchanged before and after the test, confirming the absence of any direct deformation. Furthermore, even after considering the safety factor, the carbon fiber exhibits temperature tolerances well beyond the required limits. Therefore, the thermal cycling test was successful, demonstrating that the structure can withstand the anticipated environment in the Low Earth Orbit (LEO).

6 Conclusion

In the present study, a novel method of the entire fabrication process of a CubeSat structure using CFRP has been explained in detail. It entails analysing opportunities for improvement in the current CubeSat scenario and recommending appropriate solutions. It explains the entire process, from designing the cube satellite to testing and validating the findings and design.

Carbon fiber reinforced polymer has, officially, not yet been used as a material to produce a CubeSat structure. The feasibility of fabrication of a 1U cube satellite structure in pre-impregnated CFRP. It proposes a solution which leads to the optimization of the conventional aluminum CubeSat structure by reducing the weight of the structure by 40% while achieving the required structural strength. The process is thoroughly detailed, where the primary design is simulated to imitate the real-time environment. The validated design is then executed by using modern manufacturing techniques like 3d printing, oscillating knife cutting, etc.

The proposed method of fabrication of 1U CubeSat effectively decreases the weight of the structure in a way that is substantially economically beneficial. All the standard tests were performed on the product to verify the suitability of the 1U cube satellite structure in actual application.

References

1. Nieto-Peroy C, Emami M (2019) CubeSat mission: From design to operation. Appl Sci 9(5):1–24
2. Kang S, Oh H (2016) On-orbit thermal design and validation of 1 U standardized CubeSat of STEP cube lab. Int J Aerosp Eng Article ID 4213189
3. NASA CubeSat Launch Initiative, Basic Concepts and Processes for First-Time CubeSat Developers. https://www.nasa.gov/wp-content/uploads/2017/03/nasa_csli_cubesat_101_508.pdf. Accessed 15 Apr 2024
4. El-Hameed A, Abdel-Aziz Y (2021) Aluminium alloys in space applications: a short report. J Adv Res Appl Sci Eng Technol 22(1):1–7
5. Mrówka-Nowotnik G, Sieniawski J (2005) Influence of heat treatment on the microstructure and mechanical properties of 6005 and 6082 aluminium alloys. J Mater Process Technol 162:367–372

6. Sol B, Promoteur A, Facult G (2019) Mission analysis of the OUFTI-Next nanosatellite. Master Thesis, Faculty of Applied Sciences, University of Liège
7. Hujare D, Girase N, Karnik M (2016) Crack detection in shaft by finite element analysis and experimental modal analysis. Eur J Adv Eng Technol 3(10):18–24
8. Athirah N, Afendi M, Hafizan K, Amin N, Majid M (2014) Stress and thermal analysis of CubeSat structure. Appl Mech Mater 554:426–430
9. Eriksson C (2021) Finite element analysis of stresses in the MIST CubeSat due to dynamic loads during launch. In: Degree project in vehicle engineering, KTH Royal Institute of Technology, Stockholm, Sweden
10. Corpino S, Caldera M, Nichele F, Masoero M, Viola N (2015) Thermal design and analysis of a nanosatellite in low earth orbit. Acta Astronaut 115:247–261
11. Ampatzoglou A, Baltopoulos A, Kotzakolios A, Kostopoulos V (2014) Qualification of composite structure for CubeSat picosatellites as a demonstration for small satellite elements. Int J Aeronaut Sci Aerosp Res 1(1):1–10
12. Auti S, Sheth J, Tulasyan P, Gaikwad A, Bagwe P (2020) Design of shredder machine for ELV tyres. In: Proceedings of international conference on intelligent manufacturing and automation, pp 635–646
13. Tiseo B, Quaranta V, Bruno G, Sisinni G (2019) Tailoring of ECSS standard for space qualification test of CubeSat Nano-satellite. Int J Mech Ind Aerosp Sci 13(4):295–302
14. Asundi S, Fitz-Coy N (2013) CubeSat mission design based on a systems engineering approach. IEEE Aerosp Conf

Physical Phenomena Behind the Enhancement of Joint Properties in Solution-Treated Nickel-Based Superalloy

Swagat Dwibedi, Nilakantha Sahu, and Suyog Jhavar

Abstract The current study investigates the physics governing the improvement in joint strength in solution-treated (ST) Hastelloy X (HX) weld specimens fabricated using the resistance spot welding (RSW) process. Due to its unique composition, HX, prominent for its high-temperature strength and oxidation resistance, poses challenges in achieving robust weld joints. This study investigates the influence of solution treatment on the joint properties of these weldments through a comprehensive analysis of the underlying physical mechanisms. The HX weldments are solution-treated at 1177 °C and soaked for 2 h as post-weld heat treatment (PWHT). The outcome of PWHT has led to enhancement in the tensile properties of the joints, including load-bearing capacity (LBC) and tensile-shear strength, by 43.48 and 45.91%, respectively. This improvement is associated with the elimination of the σ phase and dissolution of M_6C carbides. Further, the growth of Cr-rich precipitates at the grain boundary resists dislocation motion and strengthens the material.

Keywords Welding metallurgy · Solution treatment · Tensile-shear strength · Mode of failure · Hastelloy X

S. Dwibedi
Department of Mechanical Engineering, Indian Institute of Technology Guwahati, Guwahati, Assam 781039, India
e-mail: swagatsudhadwibedi@iitg.ac.in

N. Sahu
Department of Metallurgical and Materials Engineering, Veer Surendra Sai University of Technology, Burla, Odisha 768018, India
e-mail: nksahu_mme@vssut.ac.in

S. Jhavar (✉)
School of Mechanical Engineering, VIT-AP University, Amaravati, Andhra Pradesh 522237, India
e-mail: suyog.jhavar@vitap.ac.in

P. Bhingole et al. (eds.), *Advances in Materials Engineering*, Lecture Notes in Mechanical Engineering, https://doi.org/10.1007/978-981-97-7114-1_16

1 Introduction

Nickel-based superalloys are popularly recognized for their exceptional properties, such as high-temperature strength, outstanding corrosion resistance, and microstructural stability [1–3]. Hastelloy X (HX) is a superalloy (nickel-based) strengthened by a solid-solution process [4]. The existence of solutes like Fe, Mo, and Cr in HX makes it an exciting alloy that provides creep durability, excellent strength, and remarkable oxidation resistance up to 1200 °C [5–9]. This attribute makes HX an indispensable material for use in various demanding applications in gas turbine engines, nuclear reactor components, and furnaces [10–12]. Nevertheless, the high cost associated with these alloys makes it challenging to replace the components frequently, and hence, quite often economical to rectify or join the degraded parts than opt for complete replacement [13, 14]. Interestingly, the welding of HX can be done through various conventional fusion welding processes [15, 16]. However, the welding becomes very complicated due to the formation of various secondary phases like M_6C, $M_{23}C_6$, P, and σ precipitates in the end stage of solidification [17]. The secondary phases offer a convenient site for nucleation, facilitating the initiation of hot cracks. Additionally, the higher heat input in conventional welding processes causes the softening of the heat-affected zone (HAZ), which further lowers strength of the joints [18–20]. Subramani and Manikandan [21] utilized the current pulsation technique to lower the $Cr_{23}C_6$ precipitation micro-segregation and found that the faster cooling rate in pulsed current-gas tungsten arc (PC-GTA) welding leads to the formation of fine equiaxed grain, whereas GTA welding leads to coarser grain in the weld zone. Hence, providing a technique to employ minimal heat and a faster cooling rate to preserve nearly identical properties to those of the parent material is crucial to obtaining a high-quality weld joint.

RSW is a widely acknowledged welding process for joining thin sheets that utilizes a passage of current through the electrodes for a short period of time (typically less than a second) and causes a localized melting and solidification of weld nugget and formation of a weld joint. Hence, it is an extensively used welding technique in automobiles, electronics, etc., due to its rapid processing, low cost, and ease of automation [22–24]. In our previous study, a thin sheet of HX (0.75 mm thick) was successfully welded using the RSW process parameter optimization. The study revealed that with the increase in weld duration, the tensile-shear properties of the weldments were enhanced owing to the rise in nugget diameter. In addition, it was also observed that the failure occurred through the discontinuity in the weld nugget, and an interfacial mode of failure was observed [25]. Further, it has been reported that the presence of M_6C carbides in the joints transforms into an unwanted eutectic phase, which is correlated with the poor mechanical properties of the weld joint [9]. Numerous premature component failures, such as micro-fissures and residual stress corrosion cracking, can be attributed to the characteristics of the weld's microstructure and its strength. Even though the HX is weldable, the micro-segregation due to the non-equilibrium solidification influences the precipitation of the topologically closed packed phases (TCP) in the inter-dendritic zones [26]. These inter-dendritic zones and solidification cracks are proven to deteriorate properties like oxidation resistance, creep performance, and stress corrosion [26, 27].

Hence, to alleviate such issues and recover the strength of the weld joint, PWHT is a widely acknowledged technique utilized in the research community [17, 28–30]. Several researchers have utilized PWHT to alter the weld microstructure and lower the residual stress to obtain its optimum properties. Ghasemi et al. [31] investigated the metallurgical aspects of welding HX using the gas tungsten arc welding (GTAW) process, and it was observed that the solution treatment of the weldments helped to lower the microstructural heterogeneity and eliminate the σ phase. Xie et al. [32] employed the GTAW technique to join the K4750 superalloy with HX. The authors stated that the residual welding stress in the weld joint was substantially relieved in the PWHT samples. Further, it has been found that the PWHT has a significant role in enhancing the mechanical properties of the weld joints, which is attributed to the collapsing of the eutectic phase during the PWHT [30].

With the aforementioned literature, it can be presumed that the PWHT has a noteworthy effect in altering the microstructure and lowering the residual stress of the weld joint. Currently, the literature on the studies concerning the role of PWHT over the RSW weldments appears to be lacking. With this motivation, in this current study, a thin sheet of 0.75 mm thick HX is joined using the RSW process. The PWHT samples are investigated to study the effect of PWHT on the tensile-shear properties of the weldments. Further, the strength of the PWHT samples is compared with the as-welded samples. The insights through this study will not only allow the production of high-quality weld joints using RSW but also provide a detailed aspect of PWHT over the HX weldments joined using the RSW process.

2 Experimentation Details

2.1 Materials and Welding Process

Thin sheets of HX ($82.4 \times 25.3 \times 0.75$ mm^3) are employed in the ongoing investigation. The primary elements (by wt.%) present in the HX are Ni (47.62%), Cr (21.77%), Fe (18.86%), and Mo (8.42%). Before the welding, the samples were thoroughly cleaned with acetone to remove any presence of contamination. Further, the samples are joined using the RSW technique as depicted schematically in Fig. 1. The process parameters and the details of the welding processes are already discussed in our previous work [25].

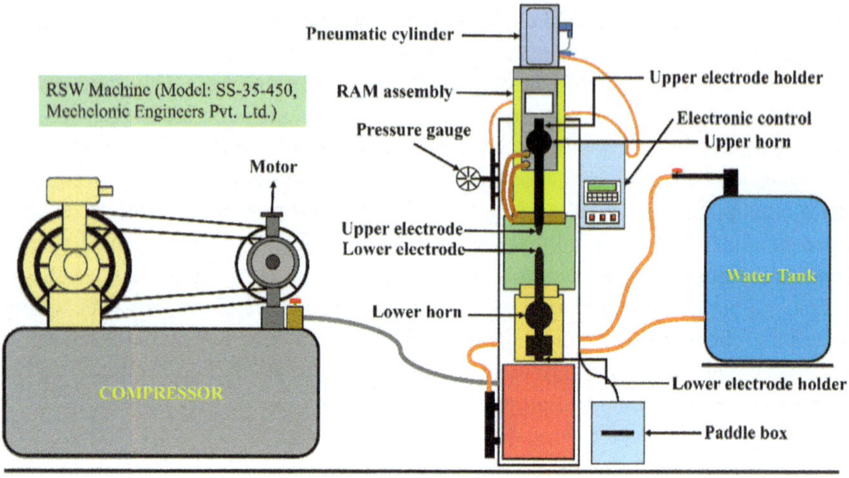

Fig. 1 Schematic representation of the RSW process

2.2 *Post Weld Heat Treatment (PWHT)*

Further, a comprehensive evaluation is conducted to assess the influence of PWHT on the joint properties of the spot-welded specimens. Specifically, a subset of the as-welded (AW) samples are subjected to the heat treatment process. The heat treatment process involved annealing the AW samples, where the samples were heated to the temperature up to 1177 °C [31] in a muffle furnace under a controlled inert gas atmosphere, ensuring precision and consistency in the thermal environment. Subsequently, the samples were allowed to soak at this temperature for 2 h, facilitating the necessary transformations in the microstructure. Further, the samples are water-quenched to carry out the heat treatment process. Figure 2 schematically shows the PWHT cycle of HX weldments. For brevity, the nomenclature for the samples is presented in Table 1.

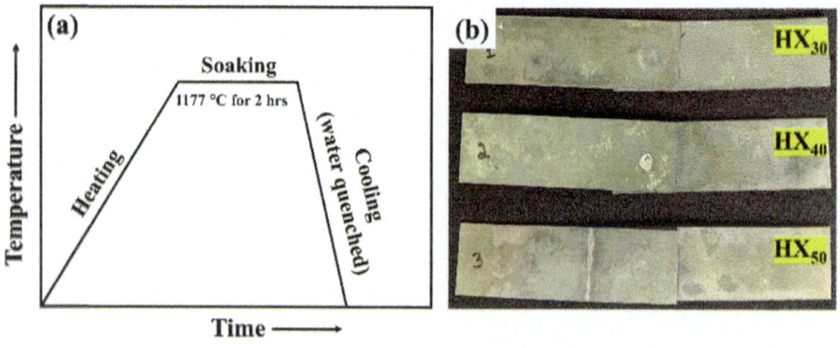

Fig. 2 **a** Schematic representation of ST cycle, **b** image of ST RSW weldments

Table 1 Process parameters employed for the current experimentation

Sample	Pressure (kg/cm^2)	Current (kA)	Squeeze	Weld	Hold	Off
			Time (cycles)			
HX_{30}	4.0	6.4	20	30	20	20
HX_{40}	4.0	6.4	20	40	20	20
HX_{50}	4.0	6.4	20	50	20	20

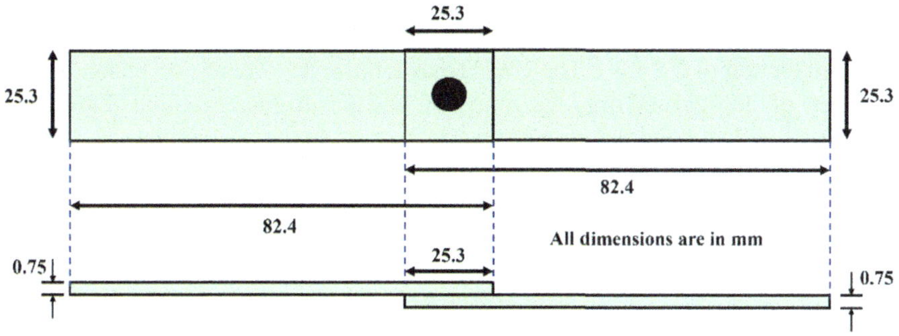

Fig. 3 Schematic representation of tensile-shear test coupon

2.3 Mechanical Properties Evaluation

Tensile-shear tests were performed to analyze the mechanical characteristics of both AW specimens and those subjected to heat treatment (ST). The test is executed using a Universal Testing Machine (Model: INSTRON-3382) at room temperature. The test specimens, as illustrated in Fig. 3, are subjected to a crosshead speed of 2 mm/min during the tensile-shear tests. This investigation assesses the load-bearing capacity (LBC) between the as-welded and heat-treated samples. Furthermore, this also seeks to discern the failure mode of the welded joint.

3 Results and Discussion

3.1 Joint Properties

Figure 4 illustrates the comparative tensile properties of welds before and after the PWHT is carried out on HX weldments. As can be seen from Fig. 4a, the tensile-shear properties of the AW joint increase with the increase in weld time (from HX_{30} to HX_{50}), which is discussed in our earlier study [25]. However, it is worth noting that the PWHT cycle of the AW samples led to a significant enhancement in the LBC and strength of the weld joints. Figure 4b shows the maximum load, tensile-shear, and yield strength (YS) of AW and ST samples. The maximum LBC and tensile-shear strength of the AW joints obtained is 4.07 kN and 213.46 MPa, respectively for the HX_{30} sample. Whereas the heat treatment of the sample led to an increase in maximum load and tensile-shear strength to 5.84 kN and 311.46 MPa, around 43.48% and 45.91% enhancement in the joint properties. A similar trend has been noticed for the other samples also, in which the maximum load and tensile strength have been enhanced from 4.81 to 6.27 kN and 252.43 to 334.39 MPa, respectively, for the HX_{50} sample. The peak load and tensile-shear strength show improvement after PWHT, while a decrement is observed in yield stress. This can be attributed to the relief of residual stresses during PWHT through the initiation of plastic flow, which results in the reduction of YS [33]. The PWHT has a positive influence on enhancing the plasticity of the weld and improving the LBC and tensile-shear strength of the joints. However, a reduction in ductility has been observed in the ST samples. This can be associated with the development of chromium precipitates, which adversely affect the material ductile properties. A similar observation was reported by several researchers, where the presence of chromium carbide precipitates was found to lower the ductility and yield strength [17, 34, 35]. Further, a detailed analysis of the improvement in mechanical properties is discussed. Table 2 summarizes the tensile-shear properties of the joints before and after the ST.

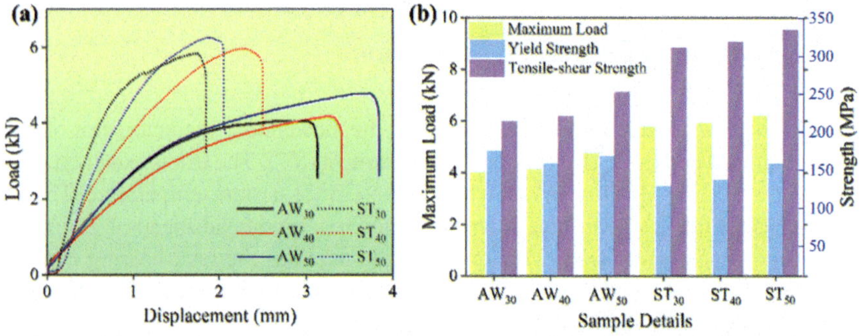

Fig. 4 **a** Load and displacement curve and **b** maximum load, tensile-shear strength and YS, for the AW and ST samples

Table 2 Tensile-shear properties of AW and ST samples

Sample	Maximum load (kN)	Tensile shear strength (MPa)	YS (MPa)
AW_{30}	4.07	213.46	175.28
AW_{40}	4.20	220.68	158.49
AW_{50}	4.81	252.43	168.32
ST_{30}	5.84	311.46	128.65
ST_{40}	5.98	319.05	136.78
ST_{50}	6.27	334.39	157.92

3.2 Weld Nugget Analysis and Mode of Failure Study

Figure 5(a–c) depicts the digital image of the weld nugget of the heat-treated samples. It can be seen that the nugget is free from cracks and hot cracking even after heat treatment. The weld nugget also shows the absence of expulsion of molten metal, which indicates the proper selection of process parameters, contributing to a more efficient and reliable welding process. Further, Fig. 5(d–f) describes the failure mode of the ST samples. Interestingly, all the samples before the heat treatment are shown to fail in pullout mode. Generally, the spot weld fails in three different ways, which include interfacial failure (IF), partial interfacial failure (PIF), and pullout failure (PF). The pullout mode of failure is associated with the withdrawal of weld nuggets from one sheet, and it is associated by greater permanent deformation and energy absorption. Optimal parameter settings promote proper nugget formation and bonding, leading to the desired pull-out failure mode, ensuring high reliability and performance of spot welds.

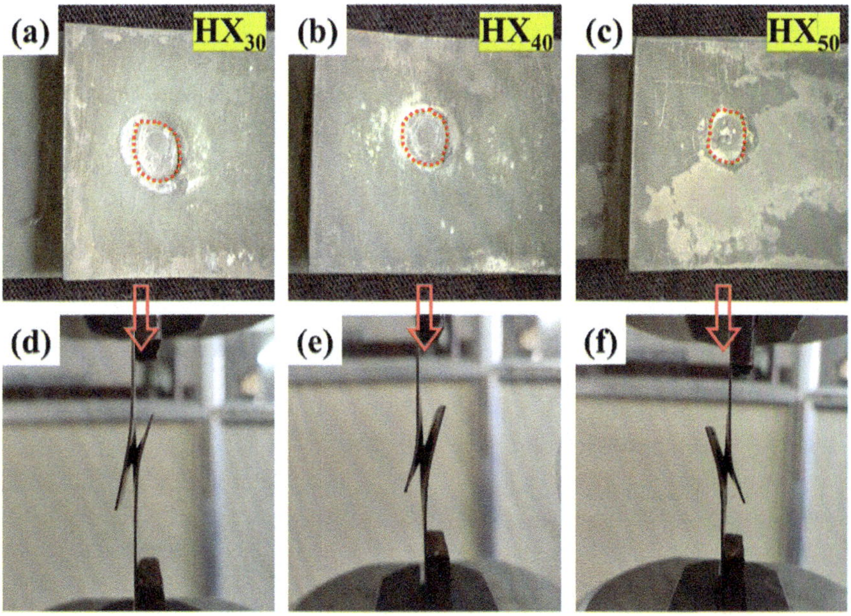

Fig. 5 **a–c** Nugget formed in the ST RSW weldments, **d–f** tensile-shear tests of ST samples showing the mode of failure

3.3 Physics Behind Enhancement in Joint Properties

The PWHT has a significant role in enhancing the mechanical properties of weld joints. Within PWHT, the solution treatment cycle is a crucial step, impacting both the base and weld metal microstructures. Notably, the AW condition often exhibits non-uniformity and the presence of non-equilibrium σ phases, which can compromise weld integrity [31]. Additionally, during welding, the thermal gradient induces tensile residual stresses, which is highly undesirable. Ghasemi et al. [31] studied the role of solution treatment over the weldments and found that ST demonstrates promising results. It effectively eradicates the σ phases and facilitates the dissolution of M_6C carbides, helping in achieving a more homogenous and sounder weld metal. The advancement in the mechanical characteristics of the weld joint can be linked to the grain size refinement and presence of higher volume of γ strengthening precipitates. Bal et al. [36] studied the PWHT of nickel-based superalloy and their findings revealed that PWHT led to a notable increase and growth of chromium-rich precipitates. The formation of such precipitates is due to the depletion of Cr and Mo in the neighbouring regions. The decrease in ductility can also be attributed to these carbides' precipitates. A similar observation has been made by various researchers, who found that the existence of M_3C and M_6C precipitated reduced the ductility of the weldments [9, 37]. These carbide formations demonstrated a significant role in impeding the dislocation movement and providing a strengthening effect to the material under tensile load.

4 Conclusion

In this research work, the PWHT is carried out over the HX weldments fabricated using the RSW process. Subsequently, the joint characteristics of the specimens are investigated carefully, and the influence of PWHT is reported. The primary conclusions that can be drawn from the current study are as follows:

a. The joint strength is found to be increasing with the incorporation of PWHT. The integration of PWHT has led to enhancement of the LBC and tensile-shear strength by 43.48% and 45.91%, respectively, for the case of ST1, samples and a similar enhancement is observed for other samples.
b. The PWHT exerts a notable influence on the failure mode of RSW specimens. In all samples, the consistent pullout mode was observed, which underscored the proficient selection of weld parameters and signified sound joint strength.
c. It can be inferred that the PWHT has a significant role in enhancing the mechanical properties of the weld joint, which is attributed to the eradication of the σ phase and dissolution of M_6C carbides, helping in forming homogenous weld metal. In addition, the Cr-rich carbide precipitates after PWHT in the grain boundary help to impede the dislocation movement and improve the tensile properties.

Acknowledgements The first author (SD) would like to thank Mr. Suraj Prasad (PhD Scholar, MEMS, IIT Bombay) for the initial draft preparation of the manuscript.

References

1. Shahwaz M, Nath P, Sen I (2022) A critical review on the microstructure and mechanical properties correlation of additively manufactured nickel-based superalloys. J Alloys Compd 907:164530. https://doi.org/10.1016/j.jallcom.2022.164530
2. Dwibedi S, Bag S (2024) Critical assessment on the influence of process parameters in micro-plasma arc welding process: A review. Weld Int 38:583–617. https://doi.org/10.1080/09507116.2024.2396917
3. Dwibedi S, Haldar V (2024) Impact of weld time and solution treatment on the performance of IN625 spot joints: An experimental study. J Inst Eng India Ser D. https://doi.org/10.1007/s40033-024-00817-2
4. Sanchez-Mata O, Wang X, Muñiz-Lerma JA, Atabay SE, Attarian Shandiz M, Brochu M (2021) Dependence of mechanical properties on crystallographic orientation in nickel-based superalloy Hastelloy X fabricated by laser powder bed fusion. J Alloys Compd 865:158868. https://doi.org/10.1016/j.jallcom.2021.158868
5. Aghaie-Khafri M, Golarzi N (2008) Forming behavior and workability of Hastelloy X superalloy during hot deformation. Mater Sci Eng A 486:641–647. https://doi.org/10.1016/j.msea.2007.11.059
6. Wang X, Dallemagne A, Hou Y, Yang S (2016) Effect of thermomechanical processing on grain boundary character distribution of Hastelloy X alloy. Mater Sci Eng A 669:95–102. https://doi.org/10.1016/j.msea.2016.05.031
7. Bemani M, Pouranvari M (2020) Microstructure and mechanical properties of dissimilar nickel-based superalloys resistance spot welds. Mater Sci Eng A 773:138825. https://doi.org/10.1016/j.msea.2019.138825
8. Lu YL, Liaw PK, Chen LJ, Wang GY, Benson ML, Thompson SA, Blust JW, Browning PF, Bhattacharya AK, Aurrecoechea JM, Klarstrom DL (2006) Tensile-hold effects on high-temperature fatigue-crack growth in nickel-based HASTELLOY® X alloy. Mater Sci Eng A 433:114–120. https://doi.org/10.1016/j.msea.2006.06.069
9. Sihotang R, Sung-Sang P, Eung-Ryul B (2014) Effects of heat input on microstructure of tungsten inert gas welding used hastelloy X. Mater Res Innov 18:S2-1074–S2-1080. https://doi.org/10.1179/1432891714Z.000000000559
10. Sah I, Kim E-S (2022) High-temperature tensile behavior of diffusion-welded hastelloy X. J Mech Sci Technol 36:3419–3428. https://doi.org/10.1007/s12206-022-0620-x
11. Mevissen F, Meo M (2022) Creep detection of Hastelloy X material for gas turbine components with nonlinear ultrasonic frequency modulation. Mater Charact 191:112099. https://doi.org/10.1016/j.matchar.2022.112099
12. Jinoop AN, Paul CP, Bindra KS (2019) Laser assisted direct energy deposition of Hastelloy-X. Opt Laser Technol 109:14–19. https://doi.org/10.1016/j.optlastec.2018.07.037
13. Li Z, Gobbi SL, Richter KH (1997) Autogenous welding of Hastelloy X to Mar-M 247 by laser. J Mater Process Technol 70:285–292. https://doi.org/10.1016/S0924-0136(97)02939-7
14. Ghasemi A, Pouranvari M (2018) Microstructural evolution mechanism during brazing of Hastelloy X superalloy using Ni–Si–B filler metal. Sci Technol Weld Join 23:441–448. https://doi.org/10.1080/13621718.2017.1410341
15. Pakniat M, Ghaini FM, Torkamany MJ (2016) Hot cracking in laser welding of Hastelloy X with pulsed Nd:YAG and continuous wave fiber lasers. Mater Des 106:177–183. https://doi.org/10.1016/j.matdes.2016.05.124

16. Fu H, Nagasaka T, Tanaka T, Sagara A, Serizawa H, Satou Y (2017) Dissimilar-metals bonding between NIFS-HEAT-2 vanadium alloy and Hastelloy X nickel alloy by controlling intermetallics. Fusion Sci Technol 72:680–685. https://doi.org/10.1080/15361055.2017.1347469
17. Sathishkumar M, Manikandan M (2019) Preclusion of carbide precipitates in the Hastelloy X weldment using the current pulsing technique. J Manuf Process 45:9–21. https://doi.org/10.1016/j.jmapro.2019.06.027
18. Bayley CJ, Mantei A (2009) Influence of weld heat input on the fracture and metallurgy of HSLA-65. Can Metall Q 48:311–316. https://doi.org/10.1179/cmq.2009.48.3.311
19. Viano DM, Ahmed NU, Schumann GO (2000) Influence of heat input and travel speed on microstructure and mechanical properties of double tandem submerged arc high strength low alloy steel weldments. Sci Technol Weld Join 5:26–34. https://doi.org/10.1179/stw.2000.5.1.26
20. Zhang C, Lu P, Hu X, Song X (2012) Effect of buffer layer and notch location on fatigue behavior in welded high-strength low-alloy. J Mater Process Technol 212:2091–2101. https://doi.org/10.1016/j.jmatprotec.2012.05.014
21. Development of welding technique to suppress the microsegregation in the aerospace grade alloy 80A by conventional current pulsing technique-ScienceDirect. https://www.sciencedirect.com/science/article/pii/S1526612518303311
22. Kumar B, Dwibedi S, Bag S, Mahapatra M, Gupta R (2021) A comparative study on microstructural and mechanical behavior of spot welded Ti-6Al-4V alloy in as-welded and solution treated condition. Adv Mater Process Technol. https://doi.org/10.1080/2374068X.2021.1934648
23. Kumar B, Maharana BR, Sahu AK, Dwibedi S, Jena A, Das S, Prasad S (2022) Influence of squeeze time on fracture mode of SS304 spot weldments. In: Recent advances in manufacturing, automation, design and energy technologies. Springer, pp 305–313. https://doi.org/10.1007/978-981-16-4222-7_36
24. Prasad S, Dwibedi S, Purohit A (2024) An experimental study to probe defect formation and failure mode in dissimilar spot joints. In: Tambe P, Huang P, Jhavar S (eds) Advances in mechanical engineering and material science. Springer Nature Singapore, Singapore, pp 307–315. https://doi.org/10.1007/978-981-99-5613-5_23
25. Dwibedi S, Kumar B, Bhoi SR, Tripathy SR, Pattanaik S, Prasad S, Behera R (2020) To investigate the influence of weld time on joint characteristics of Hastelloy X weldments fabricated by RSW process. Mater Today Proc 26:2763–2769. https://doi.org/10.1016/j.matpr.2020.02.576
26. Ogborn JS, Olson DL, Cieslak MJ (1995) Influence of solidification on the microstructural evolution of nickel base weld metal. Mater Sci Eng A 203:134–139. https://doi.org/10.1016/0921-5093(95)09832-1
27. DuPont JN, Lippold JC, Kiser SD (2009) Welding metallurgy and weldability of nickel-base alloys. Wiley
28. Jiang S, Shen J, Nagasaka T, Muroga T, Sagara A, Ohnuki S, Hokamoto K, Rao W, Tanaka S, Inao D, Morizono Y, Kasada R, Zheng P (2023) Formation of intermetallic and its effect on the hardening of welding joint between vanadium alloy and Hastelloy X alloy after heat treatment. Nucl Mater Energy 34:101389. https://doi.org/10.1016/j.nme.2023.101389
29. Wang W, Jiang L, Li C, Leng B, Ye X-X, Liu R, Chen S, Yu K, Li Z, Zhou X (2019) Effects of post-weld heat treatment on microstructure and mechanical properties of Hastelloy N superalloy welds. Mater Today Commun 19:230–237. https://doi.org/10.1016/j.mtcomm.2019.02.004
30. Wang W, Li C, Jiang L, Ye X-X, Yu K, Chen S, Li Z, Zhou X (2018) Evolution of carbide precipitates in Hastelloy N joints during welding and post weld heat treatment. Mater Charact 135:311–316. https://doi.org/10.1016/j.matchar.2017.11.045
31. Ghasemi A, Kolagar AM, Pouranvari M (2020) Microstructure-performance relationships in gas tungsten arc welded Hastelloy X nickel-based superalloy. Mater Sci Eng A 793:139861. https://doi.org/10.1016/j.msea.2020.139861
32. Xie J, Ma Y, Wang M, Li H, Fang N, Liu K (2021) Interfacial characteristics and residual welding stress distribution of K4750 and Hastelloy X dissimilar superalloys joints. J Manuf Process 67:253–261. https://doi.org/10.1016/j.jmapro.2021.05.001

33. Sterjovski Z, Dunne DP, Ambrose S (2004) Evaluation of cross-weld properties of quenched and tempered pressure vessel steel before and after PWHT. Int J Press Vessels Pip 81:465–470. https://doi.org/10.1016/j.ijpvp.2003.12.007
34. Nam T-H, An E, Kim BJ, Shin S, Ko W-S, Park N, Kang N, Jeon JB (2018) Effect of post weld heat treatment on the microstructure and mechanical properties of a submerged-arc-welded 304 stainless steel. Metals 8:26. https://doi.org/10.3390/met8010026
35. Sanchez-Mata O, Muñiz-Lerma JA, Wang X, Atabay SE, Attarian Shandiz M, Brochu M (2020) Microstructure and mechanical properties at room and elevated temperature of crack-free Hastelloy X fabricated by laser powder bed fusion. Mater Sci Eng A 780:139177. https://doi.org/10.1016/j.msea.2020.139177
36. Bal KS, Dutta Majumdar J, Roy Choudhury A (2019) Effect of post-weld heat treatment on the tensile strength of laser beam welded Hastelloy C-276 sheets at different heat inputs. J Manuf Process 37:578–594. https://doi.org/10.1016/j.jmapro.2018.12.019
37. Sathishkumar M, Bhakat YJ, Kumar KG, Giribaskar S, Oyyaravelu R, Arivazhagan N, Manikandan M (2021) Investigation of double-pulsed gas metal arc welding technique to preclude carbide precipitates in aerospace grade Hastelloy X. J Mater Eng Perform 30:661–684. https://doi.org/10.1007/s11665-020-05360-1

Influence of Weld Time on Joint Behavior of IN625-SS316L Weldments

Vivekananda Haldar, Swagat Dwibedi, and Suyog Jhavar

Abstract Dissimilar welding is always a tedious task to achieve due to the variation in thermophysical properties of different parent materials. The fusion welding of Inconel 625 superalloy with AISI 316L is a very challenging task due to the formation of different intermetallic phases during solidification. The current work attempts to understand the impact of welding time during resistance spot welding (RSW) of Inconel 625 (IN625) with AISI 316L stainless steel (SS316L). The welding time is varied between 40–80 s with an interval of 20 s each to obtain sound weld nuggets by keeping another parameter constant. It is observed that with an increase in welding time, the tensile-shear load-bearing capacity (LBC) of the weld nugget is showing a significant improvement along with three-point bending strength. The fracture surface analysis reveals that the specimen with the highest welding time failed in tearing fracture mode, and the other two failed in a partial interfacial mode of failure. The three-point bending test reveals that the energy absorption and maximum flexural stress are higher when the welding time is maximum.

Keywords Dissimilar joining · Weldability · Formability · Fracture mode · Inconel 625 · Austenitic 316L stainless steel

V. Haldar · S. Dwibedi
Department of Mechanical Engineering, Indian Institute of Technology Guwahati, Guwahati, Assam 781039, India
e-mail: vivek176103115@iitg.ac.in

S. Dwibedi
e-mail: swagatsudhadwibedi@iitg.ac.in

S. Jhavar (✉)
School of Mechanical Engineering, VIT-AP University, Amaravati, Andhra Pradesh 522237, India
e-mail: suyog.jhavar@vitap.ac.in

P. Bhingole et al. (eds.), *Advances in Materials Engineering*, Lecture Notes in Mechanical Engineering, https://doi.org/10.1007/978-981-97-7114-1_17

1 Introduction

Nickel-based superalloy Inconel 625 (IN625) is popular for its high corrosion resistivity from room temperature to elevated temperature. Nickel can dissolve higher percentage of alloying elements, especially chromium percentage, which forms a protective layer of Cr_2O_3 and provides corrosion resistivity [1]. Besides, addition of elements like Al, Ti, Nb, and Mo formed Ni_3(Nb, Ti, Al, and Mo) or γ' and γ" [2], which act as a solid-solution strengthener. Thus, it is widely used in industries like petrochemicals, nuclear power generation, and aerospace [3, 4]. On the other hand, austenitic stainless steel 316L (SS316L) is well known for its good corrosion resistivity [5] due to its higher percentage of Cr, workability, and formability [6]. SS316L is one of the unique types of austenitic, which contains mainly two major phases ferrite and austenite phases [7–10]. Moreover, it contains less carbon percentage, reducing the chance of formation of carbides at the grain boundaries during the welding operation [11]. Due to such unique properties, this steel gains huge demand in medical implants, food processing industries, and bellow manufacturing [12, 13].

Similar and dissimilar joints of these alloys are used in different industries to reduce the cost and flexibility of the system [14, 15]. Many researchers have implemented different types of joining techniques for the joining of IN625 and SS316L. Dokme et al. [15] investigated the dissimilar joining of these alloys using continuous and pulse tungsten inert gas welding (TIG). A multi-directional growth was observed in the fusion zone (FZ), located near the SS316L, and less grain growth was observed near the IN625 side. Ajabshiri et al. [16] studied the bonding time effect in dissimilar transient liquid phase bonding of IN625 and SS 316L with Cu interlayer. It was found that the failure of the bonded specimen was ductile in nature. Sasikumar et al. studied the wire arc additive manufacturing of SS316L and IN625 as functionally graded materials by using MIG welding. A sharp microstructural transition was observed at the interphase of these two alloys. It was also observed that all the specimens failed from the SS316L side due to the lower ultimate tensile strength (UTS) value with the comparison of IN625. However, among all welding processes RSW gains its demand due to its characteristics like a high-speed welding process, minimal setup time and equipment cost, and minimal heat distortion due to localized heating [17–21]. Thus, lots of research work has been performed to join different alloys by using RSW. Bemani and Pouranvari evaluated [22] the relationship between mechanical and microstructural properties of dissimilar RSW of Hastelloy X and Nimonic 263 nickel-based superalloy. It was observed that the FZ hardness increased sharply in comparison to other zones due to the formation of Mo-rich carbides. However, the formation of carbides did not affect the brittleness of the FZ. Hussein et al. conducted an experiment on the dissimilar RSW joint of SPFC780Y high-strength steel and Q&P980 steel by varying different parameters like electrode pressure, welding current, holding time, and welding time. The results indicated that the maximum peak load was maximum when the electrode pressure and holding time were minimum.

Despite the considerable amount of research carried out on different types of sheet materials, there has been a scarcity of studies focusing on the RSW process for dissimilar materials such as IN625. As per the literature, there has been limited literature available on the dissimilar joining of IN625 and SS316L. Hence, the present work aims to join IN625 and SS316L by RSW process to make a correlation between the effect of RSW process parameters and the mechanical behavior of the welded joint.

2 Experimentation

In this investigation, 0.5 mm IN625 and SS316L are joined together with the help of RSW. The samples are prepared with a dimension of 82.4×25.3 mm^2. The chemical composition of the base materials (BMs) was measured by energy dispersive X-ray (EDX) analysis and represented in Table 1. Before the welding process the BMs to be joined together are cleaned using acetone. Thereafter, both the sheets are kept one above the other between the electrodes. The entire welding process is described by a schematic of the RSW process (Fig. 1). During the welding process, the electrode force, electrode pressure, and the welding current are kept constant at 1 kN, 6 kgf/cm^2, and 8.1 kA. Besides, weld time is varied from 40 to 80 cycles with a 20-cycle interval. A total of three types of samples were prepared: S_{40}, S_{60}, and S_{80} (Table 2).

The mechanical properties of the RSW joint are characterized by tensile-shear test (TST) and three-point bending test. Both the tests are conducted on a universal tensile testing machine (UTM) with different fixtures and Fig. 2 represents the schematic of the test specimen. The TST is carried with a rate of 1 mm/min.

The obtained load–displacement data is plotted together to make a comparison among them. On the other hand, the flexural load and extension measured from the three-point bend test were also plotted to evaluate the formability of the welded specimen. The fracture surfaces of the TST specimens are studied in detail to evaluate mode failure, whether the failure modes are interfacial failure (IF), pull-out failure (PF), or a combination of these two modes.

Table 1 Chemical composition of the base metals (wt%.)

Material	C	S	Al	Si	Cr	Mo	Ti	Fe	Ni
IN625	0.1	0.2	0.3	0.2	22.8	6.8	0.3	3.8	Bal
SS316L	0.04	0.1	0.1	0.4	17.9	2.2	–	Bal	8.9

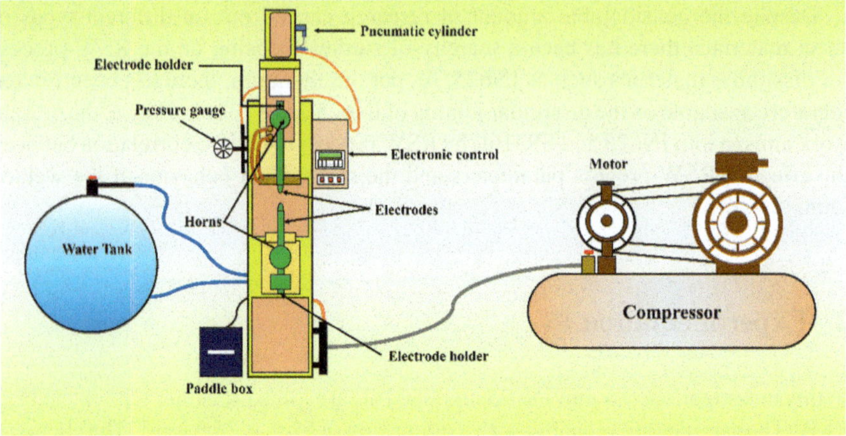

Fig. 1 Schematic representation of the RSW process

Table 2 Process parameters used to join dissimilar metals IN625-SS316L

Sample	Electrode force (kN)	Electrode pressure (kgf/cm^2)	Current (kA)	Weld time (cycles)
S_{40}	1	6	8.1	40
S_{60}	1	6	8.1	60
S_{80}	1	6	8.1	80

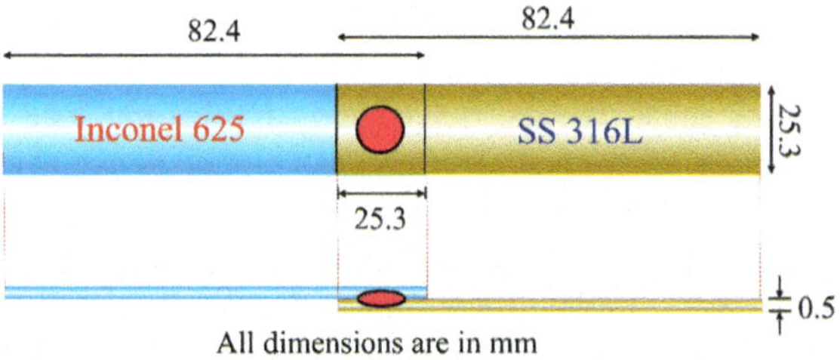

Fig. 2 Schematic of the test specimen

3 Results and Discussion

Figure 3 shows the microstructure of as-received SS316L and IN625 BM. It is observed that both the BMs consist of annealing twins. These annealing twins are formed due to the rolling operation during their production from billet to sheets.

The SS 316L BM has mainly two major phases, i.e., austenite γ-matrix and δ-ferrites at the grain boundaries. As the γ has an FCC structure, it increases the ductility of this steel. Besides, the IN625 BM has a nickel γ-matrix upon which different types of precipitates and irregular shapes of carbides are found. The presence of these precipitates makes this alloy harder than SS 316L. Figure 4 represents the as

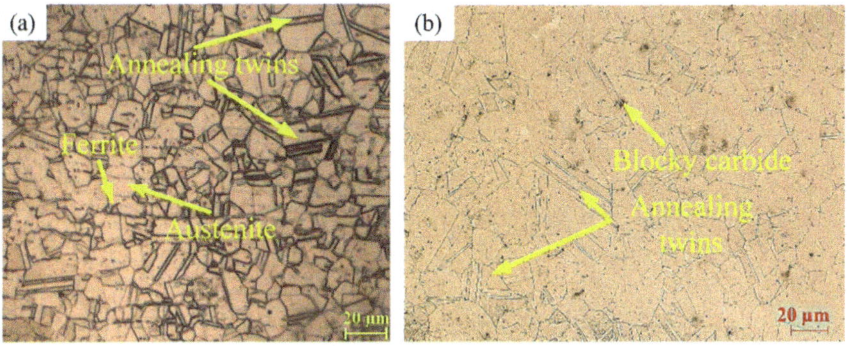

Fig. 3 Microstructure of **a** SS316L and **b** IN625

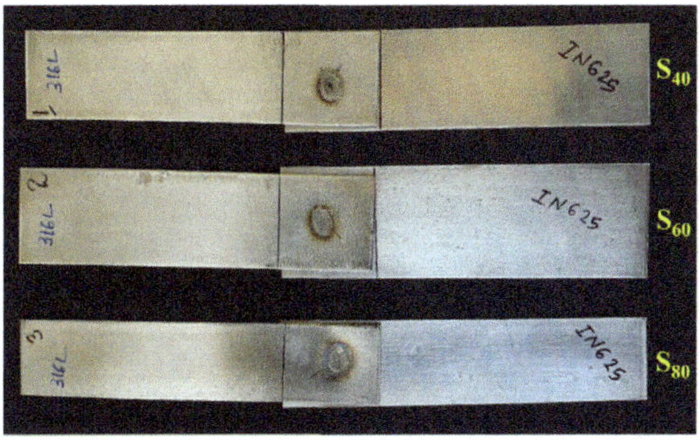

Fig. 4 As welded RSW specimens in different weld time

welded dissimilar RSW joint of IN625 and the SS 316L. It is observed that welded specimens exhibit good weld geometry without any visual defects.

3.1 *Tensile-Shear Performance*

Assessing the quality of RSW is crucial for evaluating weld performance, as it directly reflects the strength of the joints. A reliable method for estimating the mechanical behavior of the welded components is using the TST. This approach helps to implement effective measures for assessing weld performance. Peak load denotes the maximum load RSW can withstand, while failure energy indicates the amount of weld energy absorbed until reaching peak load. Figure 5 represents the load–displacement diagram of S_{40}, S_{60}, and S_{80} obtained from TST. The load–displacement (LD) graph analysis indicates that specimen S_{80} exhibited a maximum peak load (3.80 kN), and S_{40} was found with a minimum peak load (3.24 kN). Besides, S_{60} exhibits maximum displacement. All details of the obtained results are tabulated in Table 3.

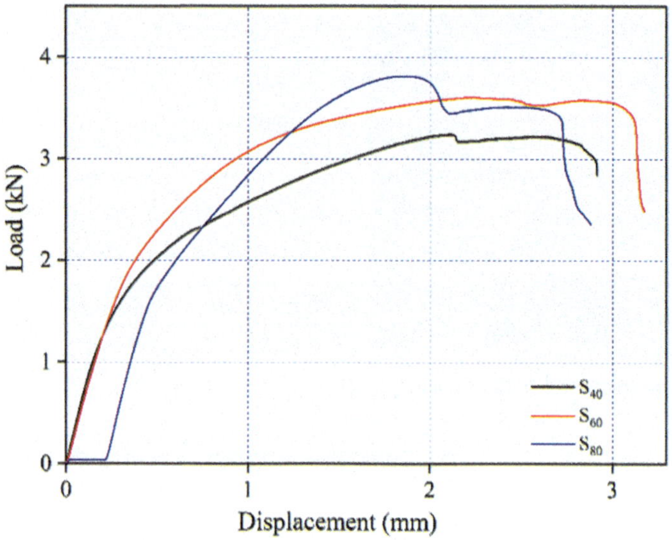

Fig. 5 Load and penetration curve of spot welded IN625-SS316L joints

Table 3 Tensile-shear strength of the joints obtained through tensile-shear tests

Sample	Maximum load (kN)	Ultimate tensile strength (MPa)
S_{40}	3.24	258.96
S_{60}	3.61	288.76
S_{80}	3.81	305.05

The geometry of the nugget zone (NZ) has a great influence on the mechanical properties of RSW specimens, and the NZ fully depends on the heat input. Augmenting heat input has a beneficial impact on enlarging the nugget size, a factor regulated by welding current and welding time by Joule's law [23]. However, it has been demonstrated that electrode force and holding time also notably influence nugget size expansion. In this current study, all the welding parameters are fixed except the weld time. Thus, it can be mentioned that increasing weld time also increases the maximum peak load (Fig. 5).

3.2 *Mode of Failure Study*

Examination of joint morphology can be conducted following the execution of a TST or peel test until fracture ensues. Failure modes are considered a significant qualitative parameter for assessing weld quality. Figure 6 represents the macrograph of the fracture surface of the TST specimens. However, S_{40} and S_{60} specimens failed

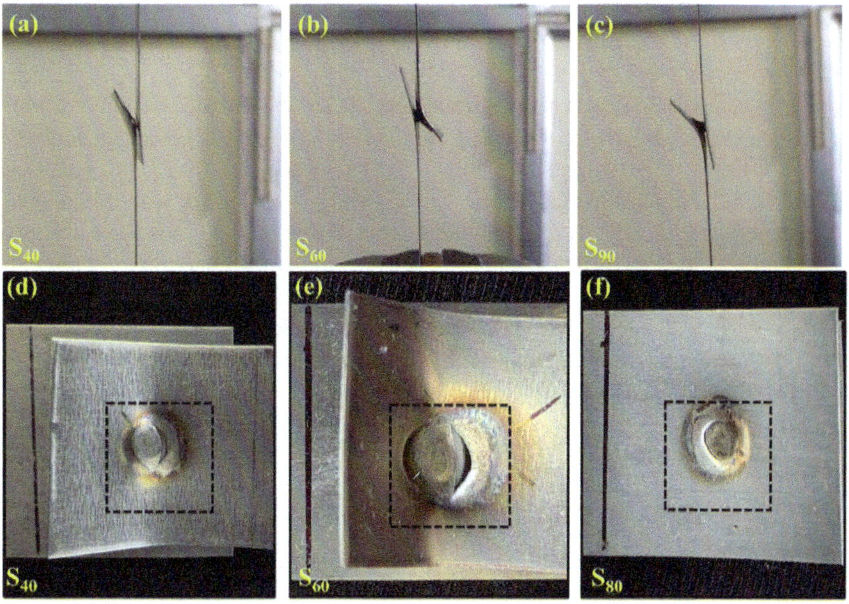

Fig. 6 **a–c** Tensile-shear specimens illustrating the mode of failure of all samples, and **d–f** macrograph of the fracture surface after the tensile shear tests

by following a Partial-IF mode. In which the SS316L side failed by leaving the vast part of the nugget zone on the IN625. On the other hand, S_{80} failed in tearing fracture mode. Few failures took place from the SS316L side by leaving a few parts of the SS316L and the entire NZ on the IN625. The mode of failure depends on the welding current, electrode pressure, welding time, and material properties. As the welding time for S_{40} and S_{60} is lower than the S_{80}, sufficient time is not available to form the desire NZ area. As a result, the nugget zones of S_{40} and S_{60} are weaker with the comparison S_{80}. Moreover, the UTS of the SS316L is lower than the IN625 and the nugget zone of S_{80}. Thus, S_{80} failed in tearing fracture mode.

3.3 *Evaluation of Flexural Strength of the Samples*

The three-point bending test is a mechanical test used to evaluate the bending properties of materials. The test measures critical parameters such as the bending moment, deflection, and ultimate strength of the material. It is commonly used to assess the flexural stress, ductility, and fracture behavior of materials, particularly in sheet metal forming and welding applications. The flexural stress is the highest stress σ_f, which can be measured by following the Eq. (1).

$$\sigma_f = \frac{3FL}{2bd^2} \tag{1}$$

where F is maximum load (kN), L is span length (mm), b (mm) and d (mm) are width and thickness. Comparison of the flexural load-extension plots for all specimens is shown in Fig. 7. It is found that specimen S_{80} has maximum σ_f as 83.17 MPa with 207.22 MPa flexural modulus. Besides, specimen S_{40} is exhibited with minimum σ_f (63.62) and 223.45 MPa flexural modulus. This implies that S_{80} has more ability to resist bending than the other specimens. The area under the curve up to failure, i.e., maximum energy, is presented in Table 4. Figure 8 and the flexural load-extension plot correlate with each other in energy absorption during the three-point bending test. S_{40}, S_{60} and S_{80} are found with 0.273 J, 0.373 J, and 0.428 J energy absorption, respectively.

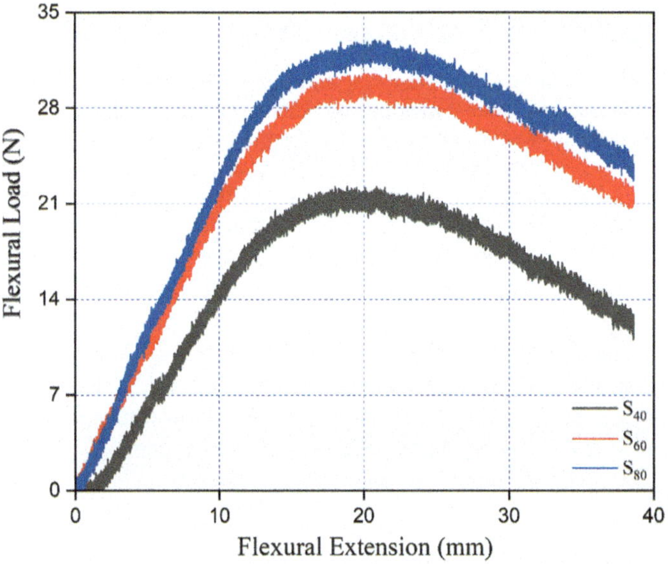

Fig. 7 The load vs. extension curve obtained through a three-point bend test depicting the bending strength of all the spot-welded samples

Table 4 Flexural strength of the samples obtained through a three-point bend test

Sample	Maximum flexural stress (MPa)	Flexural modulus (MPa)	Maximum energy (J)
S_{40}	63.62	223.45	0.273
S_{60}	76.93	201.41	0.373
S_{80}	83.17	207.22	0.428

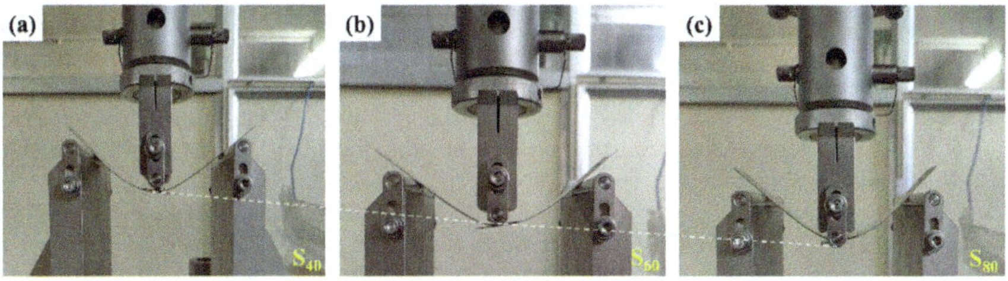

Fig. 8 Illustrates images of the samples undergoing the three-point bend test

4 Conclusion

In this investigation, the effect of welding time is studied on the dissimilar RSW joint of nickel-based superalloy IN625 and austenitic SS316L. The welding time is varied with different values by keeping the other parameters like welding current, electrode pressure constant. The joint quality is measured by the tensile-shear test and three-point bending test to determine the effect of welding time on the mechanical properties of the joint. The major findings from this study can be summarized as.

(a) A defect-free metallurgical joint is achieved using RSW. The as-received SS316L consists of mainly two phases i.e., γ-austenite and δ-ferrite. Besides, the IN625 has mainly a Ni-γ matrix, with different types of secondary phases and blocky carbides. In the presence of such precipitates, IN625 behaves harder than the SS316L.
(b) Long welding time means more heat input during the welding process. As a result of this, specimen S_{80} and S_{40} is found with maximum and minimum tensile shear strength, respectively.
(c) A significant impact is also observed on the mode of fracture of tensile shear specimen. Due to the higher heat input, the S_{80} specimen failed in tearing fracture mode. On the other hand, the S_{40} and S_{60} specimens failed by following the interfacial mode of failure. However, for all cases, failure mainly occurred from the SS316L side.
(d) The three-point bending test revealed that the S_{80} specimen has maximum flexural stress and energy consumption during the deformation. Besides, S_{40} is exhibited with minimum flexural stress and energy due to weak nugget zone.

References

1. Lippold JC, Kiser SD, DuPont JN (2011) Welding metallurgy and weldability of nickel-base alloys. John Wiley & Sons, New Jersey, USA
2. Liu X, Fan J, Zhang P, Cao K, Wang Z, Chen F, Liu D, Tang B, Kou H, Li J (2022) Influence of heat treatment on inconel 625 superalloy sheet: carbides, Γ'', Δ phase precipitation and tensile deformation behavior. SSRN Electron J 930:167522. https://doi.org/10.2139/ssrn.4161698

3. Dwibedi S, Bag S (2024) Critical assessment on the influence of process parameters in micro-plasma arc welding process: A review. Weld Int 38:583–617. https://doi.org/10.1080/09507116.2024.2396917
4. Dwibedi S, Haldar V (2024) Impact of weld time and solution treatment on the performance of IN625 spot joints: An experimental study. J Inst Eng India Ser D. https://doi.org/10.1007/s40033-024-00817-2
5. Haldar V, Pal S (2022) Effect of the rolling direction on the mechanical and microstructural properties of AISI 316L stainless steel welded joint. In: Dave HK, Dixit US, Nedelcu D (eds) Recent advances in manufacturing processes and systems. Springer Nature Singapore, Singapore, pp 381–389. https://doi.org/10.1007/978-981-16-7787-8_29
6. Haldar V, Biswal SK, Pal S (2022) Formability study of micro-plasma arc-welded AISI 316L stainless steel thin sheet joint. J Braz Soc Mech Sci Eng 44:1–16. https://doi.org/10.1007/s40430-022-03871-7
7. Mansor MSM, Yusof F, Ariga T, Miyashita Y (2018) Microstructure and mechanical properties of micro-resistance spot welding between stainless steel 316L and Ti-6Al-4V. Int J Adv Manuf Technol 96:2567–2581. https://doi.org/10.1007/s00170-018-1688-4
8. Dwibedi S (2016) Investigation on autogenous joining of stainless steel sheets of different thickness using micro Plasma Transferred Arc (μ-PTA) process
9. Dwibedi S, Bag S (2021) Development of micro-plasma arc welding system for different thickness dissimilar austenitic stainless steels. J Inst Eng India Ser C 102:657–671. https://doi.org/10.1007/s40032-021-00696-6
10. Dwibedi S, Bag S (2022) Influence of process parameters on microstructural evolution, solidification mode, and impact strength in joining of stainless steel thin sheets. Adv Mater Process Technol 8:1089–1104. https://doi.org/10.1080/2374068X.2021.1927645
11. Trigwell S, Selvaduray G (2005) Effects of welding on the passive oxide film of electropolished 316L stainless steel. J Mater Process Technol 166:30–43. https://doi.org/10.1016/j.jmatprotec.2004.07.091
12. Panda B, Sujata M, Madan M, Bhaumik SK (2014) Stress corrosion cracking in 316L stainless steel bellows of a pressure safety valve. Eng Fail Anal 36:379–389. https://doi.org/10.1016/j.engfailanal.2013.11.007
13. Dadfar MR, Fathi MH, Karimzadeh F, Dadfar MR, Saatchi A (2007) Effect of TIG welding on corrosion behavior of 316L stainless steel. Mater Lett 61:2343–2346. https://doi.org/10.1016/j.matlet.2006.09.008
14. Haldar V, Pal S (2023) Influence of fusion zone metallurgy on the mechanical behavior of Ni-based superalloy and austenitic stainless steel dissimilar joint. J Mater Eng Perform. https://doi.org/10.1007/s11665-023-08335-0
15. Dokme F, Kulekci MK, Esme U (2018) Microstructural and mechanical characterization of dissimilar metal welding of Inconel 625 and AISI 316l. Metals. 8. https://doi.org/10.3390/met8100797
16. Ajabshiri M, Shamanian M, Ashrafi A, Karimi MA (2022) The effect of bonding time on dissimilar joint properties between Inconel 625 and AISI 316L using transient liquid phase bonding method with Cu interlayer. J Mater Eng Perform 31:3311–3325. https://doi.org/10.1007/s11665-021-06411-x
17. Ribeiro RRC, de Souza LGP, de Souza Caliari JC, Teodoro CL, de Freitas Gomes JH (2024) Bibliometric and systematic analysis on electric resistance spot welding of 22MnB5 steel. Int J Adv Manuf Technol. https://doi.org/10.1007/s00170-024-13487-x
18. Dwibedi S, Kumar B, Bhoi SR, Tripathy SR, Pattanaik S, Prasad S, Behera R (2020) To investigate the influence of weld time on joint characteristics of Hastelloy X weldments fabricated by RSW process. Mater Today Proc 26:2763–2769. https://doi.org/10.1016/j.matpr.2020.02.576
19. Kumar B, Dwibedi S, Bag S, Mahapatra M, Gupta R (2022) A comparative study on microstructural and mechanical behaviour of spot welded Ti-6Al-4V alloy in as-welded and solution treated condition. Adv Mater Process Technol 8:2637–2651. https://doi.org/10.1080/2374068X.2021.1934648

20. Kumar B, Maharana BR, Sahu AK, Dwibedi S, Jena A, Das S, Prasad S (2022) Influence of squeeze time on fracture mode of SS304 spot weldments. In: Recent advances in manufacturing, automation, design and energy technologies. Springer, pp 305–313. https://doi.org/10.1007/978-981-16-4222-7_36
21. Prasad S, Dwibedi S, Purohit A (2024) An experimental study to probe defect formation and failure mode in dissimilar spot joints. In: Tambe P, Huang P, Jhavar S (eds) Advances in mechanical engineering and material science. Springer Nature Singapore, Singapore, pp 307–315. https://doi.org/10.1007/978-981-99-5613-5_23
22. Bemani M, Pouranvari M (2020) Microstructure and mechanical properties of dissimilar nickel-based superalloys resistance spot welds. Mater Sci Eng A 773. https://doi.org/10.1016/j.msea.2019.138825
23. Manladan SM, Abdullahi I, Hamza MF (2015) A review on the application of resistance spot welding of automotive sheets. JET J Eng Technol 10:20–37

Study on Dissimilar SS316L-Incoloy 800 Joints with Inconel 82 as Filler Metal

Nilakantha Sahu, Swagat Dwibedi, Sushant Kumar Badjena, and Suyog Jhavar

Abstract This study investigates the joining of SS316L and Incoloy 800 (Alloy 800) using filler (Inconel 82) metal, focusing on the joint characterization of the welded joints. The research aims to elucidate the impact of parameters on joint strength, failure behavior, and microstructural evolution. Plates of Alloy 800 and SS316L with thicknesses of 6 mm are selected for dissimilar specimens, welded in butt configuration using the GTAW process. The experimental procedure includes detailed material preparation, welding process setup, and characterization techniques such as microstructural analysis, hardness testing, and hot impact tests. Results reveal distinct microstructural differences between the base metals and welded joints, with the weld zone exhibiting finer grain size and higher hardness. Charpy impact tests were conducted at room temperature and 700 °C demonstrate a decrease in impact energy at elevated temperatures due to oxide inclusions.

Keywords SS316L · Alloy 800 · GTAW · Impact test · Metallography

N. Sahu · S. K. Badjena
Department of Metallurgical and Materials Engineering, Veer Surendra Sai University of Technology, Burla, Odisha 768018, India
e-mail: nksahu_mme@vssut.ac.in

S. K. Badjena
e-mail: skbadjena_mme@vssut.ac.in

S. Dwibedi
Department of Mechanical Engineering, Indian Institute of Technology Guwahati, Guwahati, Assam 781039, India
e-mail: swagatsudhadwibedi@iitg.ac.in

S. Jhavar (✉)
School of Mechanical Engineering, VIT-AP University, Amaravati, Andhra Pradesh 522237, India
e-mail: suyog.jhavar@vitap.ac.in

P. Bhingole et al. (eds.), *Advances in Materials Engineering*, Lecture Notes in Mechanical Engineering, https://doi.org/10.1007/978-981-97-7114-1_18

1 Introduction

Dissimilar welding, particularly between austenitic stainless steel (ASS) and Inconel, has garnered significant interest due to the unique properties offered by each material and their complementary applications [1, 2]. However, the welding of these dissimilar materials poses challenges owing to their distinct metallurgical characteristics, coefficient of thermal expansion, and susceptibility to intermetallic formation [3–5]. Various welding techniques are employed to join austenitic stainless steel and Inconel, GTAW, GMAW, and LBW processes. GTAW produces a clean weld bead with little contamination. Moreover, the GTAW is less expensive and portable. However, GTAW is slow and less productive. In order to improve productivity, flux-assisted GTAW was developed by Paton Institute, Ukraine [6, 7]. Several flux compositions were developed to improve the depth of penetration for dissimilar joints [8]. The lower heat input in GTAW resulted in a higher cooling rate, thus finer grain size and lesser Nb segregation are obtained [9]. Flux-assisted GTAW resulted in improved DOP, lower distortion, more resistance to cracking, and lesser residual stress in the weldments [10].

Generally, Inconel interlayer is used in 9Cr1Mo steel and ASS [11]. Therefore, for the dissimilar joints of ASS/Inconel, Inconel filler metal is preferred owing to its better mechanical properties [12]. Moreover, Inconel filler metal provides an intermediate coefficient of thermal expansion between the two different materials [13]. The weld zone of dissimilar joints with Inconel filler metal exhibited fully austenite grains with the formation of niobium carbide and titanium carbide in the interdendritic regions [14]. However, long-term exposure to ASS cause embrittlement, which severely deteriorates the mechanical properties whereas, no embrittlement was observed in Inconel alloys [15].

From the above discussion, it is clear that mechanical properties at room temperature and elevated temperature are necessary to achieve better joint quality. Therefore, the current experimentation is primarily involved with evaluating the characterization of dissimilar joints. Further, the joint strength using the Charpy impact test at elevated temperatures is also investigated.

2 Experimentation

Coupons ($70 \times 80 \times 6\ mm^3$) of Alloy 800 and SS316L with thicknesses of 6 mm are selected for dissimilar specimens. The chemical composition of alloy 800 contains 23% Cr, 35% Ni, 0.1% C, 0.15% Al, 0.6% Ti, and 39.5% Fe. In addition, SS316L contains 18% Cr, 14% Ni, and 3% Mo. The filler metal used is Inconel 82 to join the dissimilar metals. Grooves were prepared with a grinding machine at a total angle of 60°. The welding was performed by employing 105 A, and 12–15 V. Pure argon gas was used during the welding, and the gas flow rate was kept at 10 lpm. In the present study, a DCEN was used to weld the dissimilar alloys. The voltage was kept between

12 and 15 V and the current was kept constant at 105 amperes. A 1.6 mm lanthanide tungsten electrode was used in this experiment. After the welding was performed the specimens were cut along the transverse direction and polishing was done. After the polishing with different grit sizes, alumina polishing was performed to get a mirror finish. Then the etchant for the base metal was prepared by using ferric chloride (2.1 gm), cupric chloride (0.6 gm), ethanol (30 ml), HCl (30 ml), and HNO_3 (1.5 ml). The base metal sample was dipped in the etchant for 2 min. The etchant for the weld zone was prepared by using cupric chloride (1 gm), HCl (20 ml), and ethanol (20 ml). The weld zone sample was dipped in the etchant for 4 min. hardness test was carried out at 10 different places on the sample. For the Charpy test, V notches were made in 8 samples with a depth of 2 mm and a V angle of 45°. Specimens were heated to 700 °C for 15 min for a hot impact test.

3 Results and Discussion

3.1 Metallurgical Characterization

Figure 1 shows the macrostructure of dissimilar welded joints between SS316L and Alloy 800. The microstructure of two as-received base metals SS316L and Alloy 800, are shown in Fig. 2a and b respectively. The microstructure of SS316L contains smaller austenite grains, whereas in Alloy 800, austenite grains were more prominent in size. The microstructures of the different zones of the welded joints are shown in Fig. 3. Figure 3a shows the SS316L side of HAZ and weld zone whereas, Fig. 3b shows the Alloy 800 side of HAZ and weld zone. The weld zone is the mixture of filler metal (Inconel 182) and two dissimilar base metals. Due to high peak temperatures and slow cooling near the weld zone boundary, coarser grains were found near the high temperatures in the heat-affected zone. No defects or porosity were observed in the weld zone from the microstructural analysis.

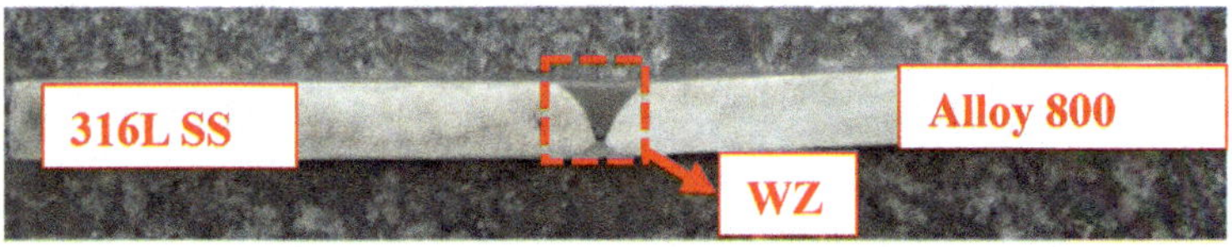

Fig. 1 Macrostructure of dissimilar welded joints between SS316L/Alloy 800

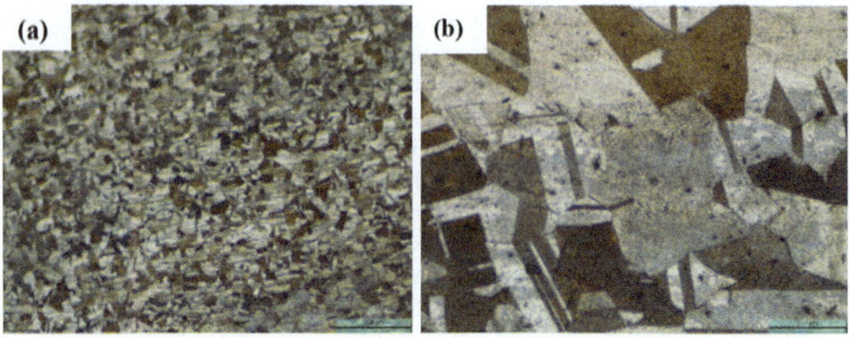

Fig. 2 Microstructure of as-received base metals **a** SS 316L and **b** Alloy 800

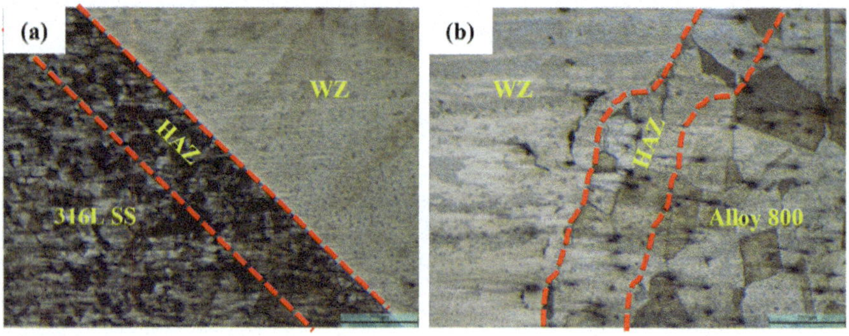

Fig. 3 Microstructure of welded joints **a** SS316L and **b** Alloy 800

3.2 Hardness Test

The Vickers hardness test was conducted at different locations, such as HAZ of SS316L, WZ, and HAZ of Alloy 800. For this test, the load was selected as 1 kg. The hardness data for different regions is shown in Fig. 4. The hardness of WZ is highest, followed by HAZ of alloy 800; the least hardness was observed in HAZ of SS316L. The higher hardness of WZ might be attributed to finer grain size in the WZ. The cooling rate in the WZ is faster, which results in the finer grain size. As the Alloy 800 is more robust and has more alloying elements than SS316L, the hardness of the former is higher than the latter.

3.3 Hot Impact Test

The dissimilar welded joints are widely used in power plants that operate under high temperatures. However, the strength of the alloys decreases at high temperatures. The

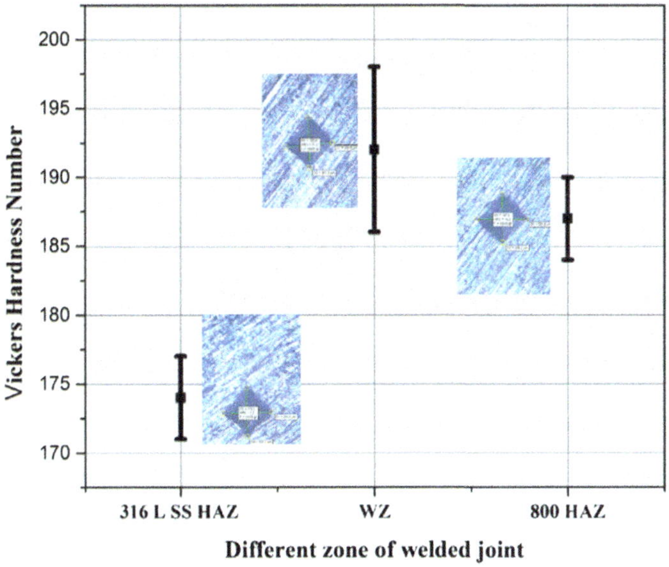

Fig. 4 Vickers hardness of welded specimen at different zone

Charpy impact test of welded joints both at room temperature and 700 °C evaluated. The Charpy impact data at room temperatures and 700 °C temperature is shown in Fig. 5. The room temperature charpy impact value was found to be 350 J/cm^2 in contrast at 700 °C, it decreased to 295 J/cm^2. Figure 5 shows the fractured samples after the charpy impact test. The decrease in impact energy at high temperatures might be due to more oxide inclusions [16] (Fig. 6).

Fig. 5 Fractured weldment specimens after the Charpy impact test

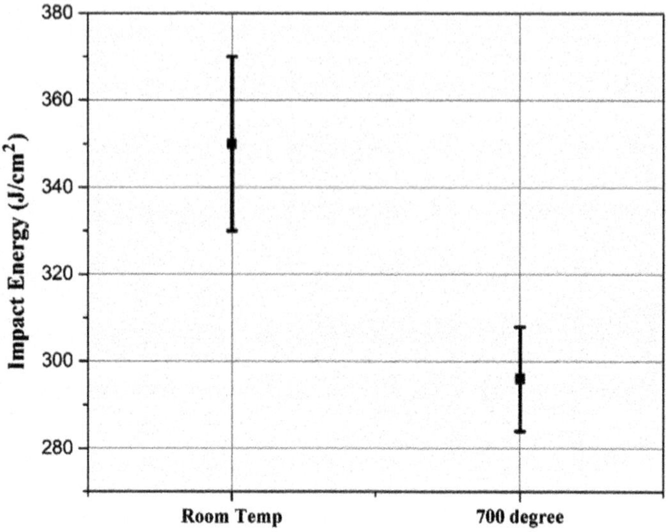

Fig. 6 Charpy impact test energy value of weldment performed at room temperatures and at 700 °C

4 Conclusion

The present work demonstrates the joining of SS316L and Alloy 800. The significant outcomes are mentioned below:

a. The microstructural characterization revealed distinct differences between the base metals SS316L and Alloy 800, with varying grain sizes and distribution. No defects or porosity were observed in the weld zone, indicating a sound welding process.
b. Vickers hardness tests conducted at different zones of the welded specimens demonstrated variation in hardness values, with the weld zone exhibiting the highest hardness. Finer grain size and faster cooling rate in the weld zone. The higher hardness of Alloy 800 compared to SS316L is consistent with its more robust and more alloyed composition.
c. Charpy impact tests were performed at both room temperature and 700 °C to determine the impact energy of the welded joints. The values showed a decrease in impact energy at elevated temperatures, potentially due to oxide insertions.

References

1. Lundin CD (1982) Dissimilar metal welds-transition joints literature review emphasis is on carbon migration, the stress/strain state of welds, and transition joint failure mechanisms

2. Dwibedi S, Bag S (2024) Critical assessment on the influence of process parameters in micro-plasma arc welding process: a review. Weld Int 38:583-617. https://doi.org/10.1080/09507116.2024.2396917
3. Bhaduri AK, Venkadesan S, Rodriguez P, Mukunda PG (1994) Transition metal joints for steam generators-an overview. Int J Press Vessels Pip 58:251–265. https://doi.org/10.1016/0308-0161(94)90061-2
4. Dwibedi S, Bag S (2021) Development of micro-plasma arc welding system for different thickness dissimilar austenitic stainless steels. J Inst Eng India Ser C 102:657–671. https://doi.org/10.1007/s40032-021-00696-6
5. Dwibedi S, Bag S (2022) Influence of process parameters on microstructural evolution, solidification mode, and impact strength in joining of stainless steel thin sheets. Adv Mater Process Technol 8:1089–1104. https://doi.org/10.1080/2374068X.2021.1927645
6. Huang HY, Shyu SW, Tseng KH, Chou CP (2005) Evaluation of TIG flux welding on the characteristics of stainless steel. https://doi.org/10.1179/174329305X48329
7. Sahu N, Panda D, Badjena SK, Sahoo SK, Kar PK (2023) Multicomponent flux for improved penetration and metallurgical properties using A-GTAW. J Mater Eng Perform 32:4237–4248. https://doi.org/10.1007/s11665-022-07383-2
8. Sahu N, Barik BK, Sahoo S, Badjena SK, Sahoo SK (2021) Studies on metallurgical and corrosion characteristics of dissimilar GTAW welding of alloy 800 and SS316L using multi-component activated flux. In: Materials today: proceedings. Elsevier, pp 2533–2536. https://doi.org/10.1016/j.matpr.2020.12.607
9. Kourdani A, Derakhshandeh-Haghighi R (2018) Evaluating the properties of dissimilar metal welding between Inconel 625 and 316L stainless steel by applying different welding methods and consumables. Metall Mater Trans Phys Metall Mater Sci 49:1231–1243. https://doi.org/10.1007/s11661-018-4469-7
10. Sharma P, Dwivedi DK (2021) Study on Flux assisted-Tungsten inert gas welding of bimetallic P92 martensitic steel-304H austenitic stainless steel using SiO2–TiO2 binary flux. Int J Press Vessels Pip 192:104423. https://doi.org/10.1016/j.ijpvp.2021.104423
11. Vidyarthy RS, Kulkarni A, Dwivedi DK (2017) Study of microstructure and mechanical property relationships of A-TIG welded P91–316L dissimilar steel joint. Mater Sci Eng A. https://doi.org/10.1016/j.msea.2017.04.038
12. Sayiram G, Arivazhagan N (2015) Microstructural characterization of dissimilar welds between Incoloy 800H and 321 austenitic stainless steel. Mater Charact 102:180–188. https://doi.org/10.1016/J.MATCHAR.2015.03.006
13. Sayyar N, Shamanian M, Niroumand B (2018) Arc weldability of Incoloy 825 to AISI 321 stainless steel welds. J Mater Process Technol 262:562–570. https://doi.org/10.1016/J.JMATPROTEC.2018.07.020
14. Naffakh H, Shamanian M, Ashrafizadeh F (2009) Dissimilar welding of AISI 310 austenitic stainless steel to nickel-based alloy Inconel 657. J Mater Process Technol 209:3628–3639. https://doi.org/10.1016/j.jmatprotec.2008.08.019
15. Sireesha M, Shankar V, Albert SK, Sundaresan S (2000) Microstructural features of dissimilar welds between 316LN austenitic stainless steel and alloy 800. Mater Sci Eng A 292:74–82. https://doi.org/10.1016/S0921-5093(00)00969-2
16. Ramkumar KD, Kumar BM, Krishnan MG, Dev S, Bhalodi AJ, Arivazhagan N, Narayanan S (2015) Studies on the weldability, microstructure and mechanical properties of activated flux TIG weldments of Inconel 718. Mater Sci Eng A 639:234–244. https://doi.org/10.1016/J.MSEA.2015.05.004

Influence of Post Bond Heat Treatment on IN718 Joints: An Experimental Study

Uttam Kumar Tarai, Dipankar Saha, Swagat Dwibedi, and Suyog Jhavar

Abstract This study investigates the microstructural and mechanical improvement in a TLP-bonded specimen with the application of PBHT. The microstructure analysis of the TLP-bonded shows an athermal solidified zone at the centre due to inadequate solidification time, while the DAZ contained blocky and needle-shaped Nb-Mo-Cr borides. PBHT results in a more uniform microstructure enhances the amount of γ' and γ'' precipitates and reduces the boride fraction in the DAZ and second-phase precipitates ISZ. DAZ exhibited the highest hardness due to boride precipitates, while ISZ displayed lower hardness from insufficient alloying elements. PBHT improved hardness across zones by promoting γ' and γ'' precipitates. Notably, tensile strength in the TLP-bonded specimen significantly improved by 80% with PBHT. These findings underscore the potential for optimizing TLP-bonding processes to enhance mechanical properties.

Keywords TLP-bonding · Isothermal solidification · PWHT · IN718 · Homogenization

U. K. Tarai
Department of Mechanical Engineering, DRIEMS University, Cuttack, Odisha 752023, India
e-mail: uttamtarai@driems.ac.in

D. Saha · S. Dwibedi
Department of Mechanical Engineering, Indian Institute of Technology Guwahati, Guwahati, Assam 781039, India
e-mail: dipankar.saha@iitg.ac.in

S. Dwibedi
e-mail: swagatsudhadwibedi@iitg.ac.in

S. Jhavar (✉)
School of Mechanical Engineering, VIT-AP University, Amaravati, Andhra Pradesh 522237, India
e-mail: suyog.jhavar@vitap.ac.in

P. Bhingole et al. (eds.), *Advances in Materials Engineering*, Lecture Notes in Mechanical Engineering, https://doi.org/10.1007/978-981-97-7114-1_19

1 Introduction

The evolution in the manufacturing field has been centred around developing novel welding techniques, particularly for gas turbine engines. These engines find extensive use in aircraft and land-based power generators of various shapes and sizes, rendering them a significant area of focus. The components of gas turbine engines endure severe operating conditions over extended periods at high temperatures. The components undergo demanding operational conditions and necessitate resistance to high-temperature creep, along with high tensile strength, ductility, and oxidation resistance, in order to endure the applied loads. In order to meet the requirements of hot gas path components, nickel-based superalloys are predominantly utilized as raw materials for fabricating gas turbine components. This preference is attributed to their outstanding mechanical strength, corrosion resistance, and oxidation resistance at elevated temperatures [1–3], primarily owing to precipitation strengthening by γ' precipitates and carbides. Unlike other nickel-based superalloys, Inconel 718 (IN718) demonstrates significant resistance to strain age cracking when subjected to welding or PBHT [4]. IN718 is a nickel-based superalloy hardened by γ'' precipitation [5]. Over the last six decades, the superalloy has been the mainstay in the jet gas turbine industry, constituting a substantial portion of superalloy production and utilization [6]. IN718 has been utilized in various high-temperature applications, including turbine disk blades, tubes, space shuttle rocket engines, cases, seals, stators, shafts, compressors, and fasteners owing to its outstanding mechanical strength and corrosion resistance at high temperatures (up to 650 °C) [1, 4, 7]. In order to fulfil the continuously growing need for enhanced engine efficiency, the utilization of advanced processing techniques and alloy designs plays a critical role. However, to increase the turbine efficiency, the geometrical complexity of the engines has to be increased. Previously, the parts are fabricated by various casting processes as a single component.

Fabrication of a single cast component has generally poorer properties and inhomogeneous microstructure [8]. Hence, there is often a need to manufacture engine components in smaller segments and then assemble them using suitable bonding techniques. The fabrication methods employed for producing these components significantly impact the lifespan of the IN718 components. Moreover, as the aero-engine components undergo high creep and fatigue loading at severe operating conditions, the degradation of engine components is often observed. Economic necessity requires refurbishment and repairing of the parts rather than complete replacement of expensive service-damaged components. Several techniques for joining turbine components have been developed historically. These encompass fusion welding methods such as TIG, PAW, EBW, and LBW [1, 4, 6, 7], traditional brazing [9], and solid-state diffusion bonding [10]. Nevertheless, when considering welding aspects, IN718 exhibits disadvantages, particularly micro-fissuring in the HAZ and liquation cracking, which restrict the application of fusion welding in aero-engine applications. To address these issues, several researchers have introduced the TLP bonding process. Taking advantage of solid-state diffusion bonding and brazing, a new viable alternate technique for joining difficult-to-weld superalloys was developed called transient liquid phase bonding [11]. In this method, an interlayer containing a melting point depressant element, such as B, Si, or P, is positioned between the substrate and heated to a temperature exceeding the melting point of the interlayer, where it is held for an adequate duration. After the interlayer melts, the BM dissolves, causing the liquid zone to widen. Equilibrium is reached between the solid and liquid phases at the interface, leading to the inter-diffusion of the alloying elements between the liquid interlayer and BM. This process causes an elevation in the liquidus point of the liquid phase at the interface, initiating iso-thermal solidification. As isothermal solidification advances, the width of the liquid diminishes, and the solid–liquid interface moves towards the middle of the bond. Given adequate time, the assembly undergoes complete isothermal solidification of the liquid. Subsequent holding at an elevated temperature leads to the homogenization of the joint with the BM. However, insufficient holding time may result in the transformation of residual liquid at the joint into eutectic-type brittle phases, potentially compromising the mechanical properties of the bond [12]. One of the primary advantages of the process is the attainment of a uniform microstructure in the joint, closely resembling that of the BM. The mechanical properties of the bond closely match those of the BM, and the melting temperature of the bond exceeds the bonding temperature.

The bond properties of TLP-bonded joints primarily depend on factors such as bonding temperature, isothermal solidification time, interlayer thickness, and applied pressure. Elevating the bonding temperature and extending the isothermal holding time enhances the strength of the bond [13]. Nevertheless, the interlayer thickness must remain sufficient to adequately wet the surfaces intended for joining. However, increasing the interlayer thickness may increase the isothermal holding time and if not held for sufficient periods, then poor strength is observed. Higher pressure increases the bond properties; however, after some pressure, the property may decrease as it may cause a spill out of the molten interlayer material. Numerous studies have been undertaken to explore the effects of bond properties. However, only a few investigations are available on improving the bond properties. Therefore, this investigation was carried out with the objective of enhancing the bonding properties of the IN718 superalloy. Furthermore, a systematic examination of the microstructure of the TLP-bonded IN718 joint was conducted.

2 Experimental Procedure

The IN718 BM was acquired from Shanti Metal Corporation. It was supplied in cylindrical rod form with a diameter of 10 mm. A wire-cut EDM was used to extract 70 mm height right cylinders from the as-received material. The BM faying surface was ground flat before the bonding. The silicon carbide sandpapers with grit sizes 220–1000 were used to ground the specimens. After the grinding operation, each BM specimen underwent a 10-min ultrasonic cleaning in acetone to eliminate any remaining impurities or debris (such as SiC particles), thus ensuring a clean faying

Table 1 Composition of BM and interlayer (wt%)

Elements	Ni	Fe	Cr	Nb	Mo	B	Si
IN718	55.7	17.7	16	5.1	2.0	–	0.2
BNi-2	82.3	3.0	7.0	–	–	3.2	4.5

Table 2 Experimental process parameters

Temperature (°C)	Time (min)	Interlayer thickness (μm)	Load (Kg)
1050	90	120	15.7

surface. In this study, the interlayer used is a 40 μm thick and 10 μm width commercially available amorphous BNi-2 foil interlayer obtained from Alibaba. Table 1 displays the composition of both the interlayer material and the BM. The experiments encompassed variations in interlayer thickness, bonding pressure, bonding temperature, and bonding time. The detailed process parameters are represented in Table 2. TLP-bonding was conducted using an induction-heating machine with a self-made fixture, as depicted in Fig. 1. An optical microscope and FESEM were used to study the morphology of the TLP-bonded joint. Tensile tests were conducted using a servo-controlled, hydraulically operated UTM, adhering to the standards set by the ASTM E8M-08 for specimen preparation. The microhardness values were measured at three different zones, viz., ISZ, DAZ, and BM zone, by using a Vicker micro-hardness tester. To improve the joint properties of the IN718 bonded specimen, two types of heat treatments were applied across three stages. Initially, the specimens underwent solution annealing at 1050 °C, followed by rapid water quenching. Following that, the specimens underwent an aging process at 760 °C for a duration of 8 h. Finally, precipitation hardening was conducted at 660 °C for 8 h, after which the material was cooled to room temperature in ambient air.

3 Results and Discussion

3.1 Microstructural Analysis

Figure 2a, b depicts the optical microstructure of the IN718 superalloy both before and after TLP-bonding. The figure shows austenitic grains embedded with fine precipitates. The EDS results of the precipitates reveal them to be metallic carbides. The EDS results of the carbide particle are depicted in Fig. 2c; the figure reveals that the carbides are rich in Nb. The line intercept process determined the grain size of the as-received and bonded specimens, and it was found to be 15 and 54 μm, respectively. The thermal exposure during TLP-bonding causes an increase in the grain size of the TLP-bonded specimens. Figure 3a shows the FESEM micrograph of the TLP-bonded specimen. The micrograph indicates 4 different zones in the TLP-bonded specimen,

Fig. 1 Illustrates TLP-bonding arrangement

namely ASZ, ISZ, DAZ and BM zone. The BM zone is the unaffected BM zone, which is not affected by the diffusion process. In this zone, the composition is the same as the BM. The DAZ is formed due to the interdiffusion of the melting point depressant elements into the substrate. The enlarged view of the DAZ is presented in the Fig. 3b. The figure reveals that a large number of precipitates are present in the DAZ. During TLP-bonding, the fast diffusing MPD elements diffuse from the liquid into the substrate. When the concentration of these MPD elements exceeds the solubility limits, the excess elements get rejected, and intermetallic compounds are formed. As B is the main diffusing element in the BNi-2 interlayer, borides are formed in the DAZ. The EDS results of the precipitate, labelled as A in Fig. 3b depicted in Fig. 3c. The results reveal that these precipitates are rich in Nb, Mo, and Cr. As these above-mentioned elements are boride-promoting elements, borides of these elements are formed. The formation of the interfacial solid zone (ISZ) occurred as a result of the isothermal solidification (IS) process of the liquid interlayer. During IS, the interdiffusion of the atom between the substrate and the liquid interlayer occurs. During this process, fast, diffusing elements like B and Si diffuse from the liquid interlayer into the substrate, and the main alloying elements, such as Al, Ti, Nb, Fe, and Mo, diffuse from the substrate into the bond zone. Due to this interdiffusion of elements, the composition of the interface changes, and IS occurs. If the assembly is cooled before the complete IS of the whole liquid, the remaining liquid solidifies by athermal solidification. Which forms intermetallic compounds in the bond zone.

The line scan mapping of the ASZ is depicted in Fig. 4. The line scan results indicate that the secondary phases present in the ASZ are rich in Ti, Al, Cr, Nb, Si and Mo. Which indicates interdiffusion of the elements. Initially, Ti, Al, Nb, and Mo elements were not present in the interlayer. That means these elements have diffused from the substrate into the bonded region. These intermetallic decrease the strength of the bond. These have to be reduced in order to obtain good strength.

In order to mitigate the negative impact of the precipitates in the ASZ, a PBHT was conducted. Figure 5 depicts the microstructure of the PBHT specimen. The figure reveals these are three zones instead of the four zones that were observed in the as-bonded specimen. The ASZ in the as-bonded specimen is now absent in the PBHT specimen. During PBHT, the specimens are solution treated at 1050 °C and double

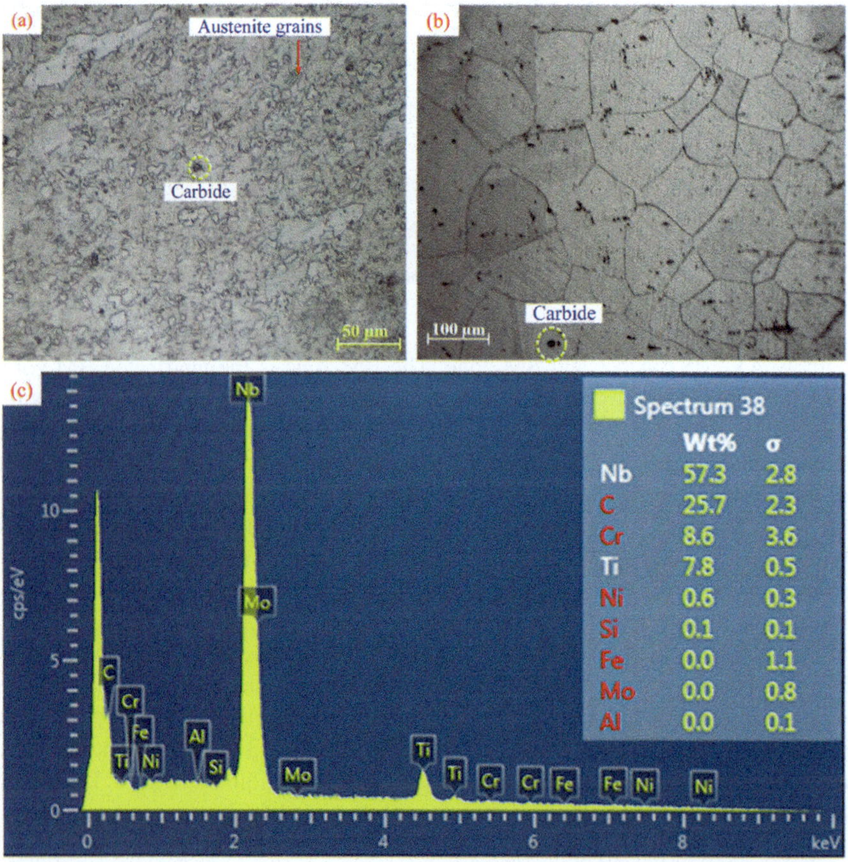

Fig. 2 **a** Microstructure of **a** as-received IN718 and **b** IN718 after TLP-bonding, and **c** EDS of carbide particle

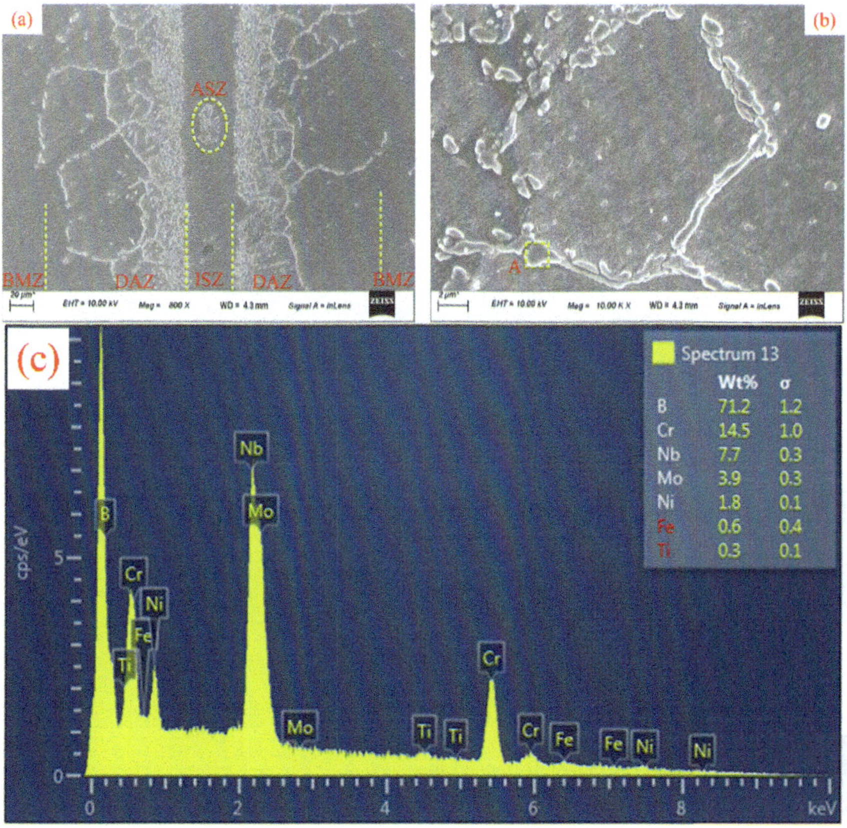

Fig. 3 FESEM microstructure of the **a** as-bonded specimen, **b** magnified image of DAZ and **c** EDS analysis result of the precipitate labelled as A

aged at 750 and 650 °C. During these processes, the interdiffusion of the elements occurs between the precipitate and the Ni-matrix. As the precipitates are enriched with Ti, Al, and Nb, these elements diffuse from the precipitates to the Ni-matrix. Which leads to the formation of the strengthening phases in the matrix. As a result, a uniform microstructure is produced at the ISZ. The magnified view of the DAZ of the PBHT specimen is depicted in the Fig. 5b. In the DAZ of the PBHT specimen, a number of precipitate of blocky, and needle shape was also observed. The microstructure of the BM of the PBHT specimen is depicted in Fig. 5c. The microstructure shows the presence of spherical and disk-shaped precipitates throughout the microstructure. The precipitates are γ' and γ''. These precipitates are formed during double aging of the PBHT specimen. These precipitates improve the strength of the PBHT specimen by precipitation strengthening mechanism. The EDS analysis of the precipitate levelled as B in Fig. 5b is depicted in Fig. 5d. The figure indicates that the precipitates are also rich in Nb, Cr, Mo and B. However, fragmentation of the precipitates is present in the PBHT specimen. Due to the diffusion of elements from the borides,

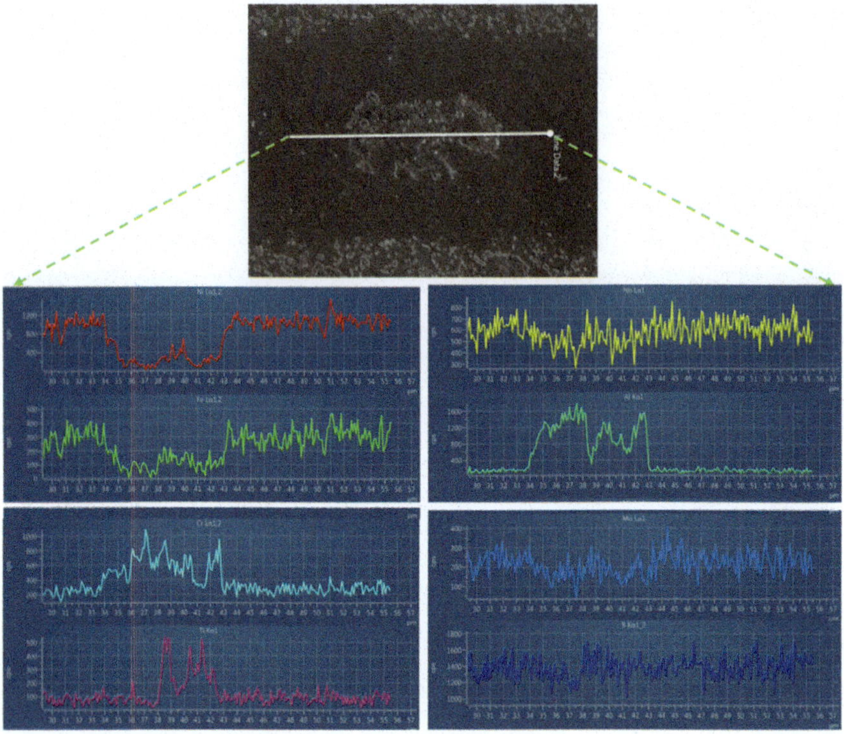

Fig. 4 Line scan results of ISZ of as-bonded specimen

the fragmentation of the borides occurs, which decreases the amount of the boride in the DAZ of the PBHT specimen.

3.2 Mechanical Analysis

The stress–strain curve of the BM, as-bonded specimen and PBHT specimens are shown in Fig. 6a. The graph reveals that the strength of the as-bonded specimen is lower than that of the BM zone. As precipitates are present in the ASZ, this reduces the strength of the as-bonded specimen. These precipitates are brittle in nature; during plastic deformation, cracks are initiated from these precipitates, which reduces the strength of the bond. The graph also reveals that the strength of the PBHT specimen has improved significantly. Through PBHT 80% increase in UTS was observed. This increase in strength can be related to the formation of the additional γ' and γ'' strengthening phase during PBHT. In addition to this, more alloying elements such as Nb, Cr, and Mo have also diffused from the BM zone into the ISZ, which improves the strength of the bond through the solution-strengthening process. In the as-bonded

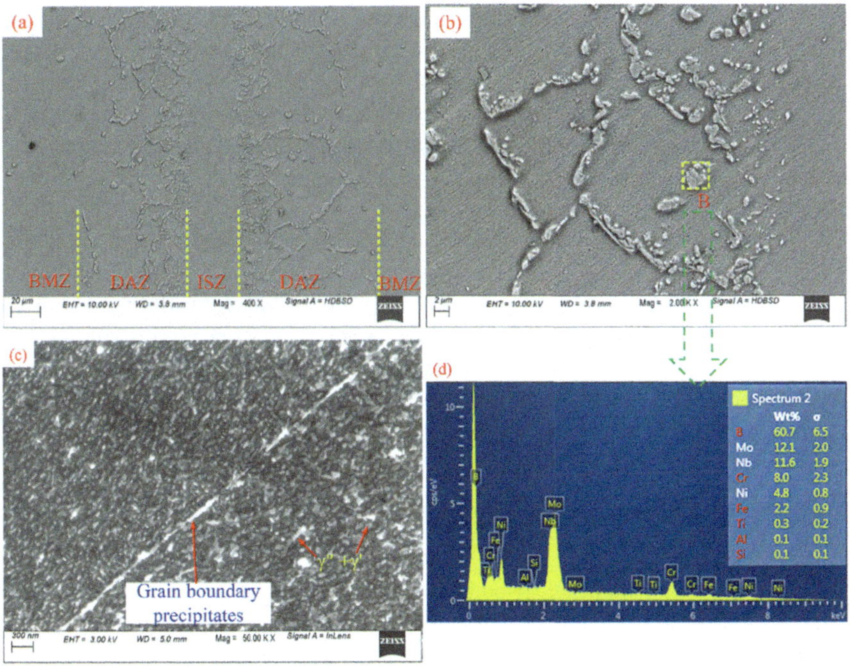

Fig. 5 Illustrates **a** PBHT specimen, **b** magnified view of DAZ, **c** High magnification microstructure of the BM zone of the PBHT specimen, and **d** EDS result of the precipitate indicated as B

specimen and the PBHT specimen, there is a reduction in ductility. This reduction can be attributed to the development of the more γ' and γ'' strengthening phase in three specimens.

Figure 6b illustrates the variation in hardness among the different zones of both the as-bonded and PBHT specimens. In the as-bonded specimen, among all the zones

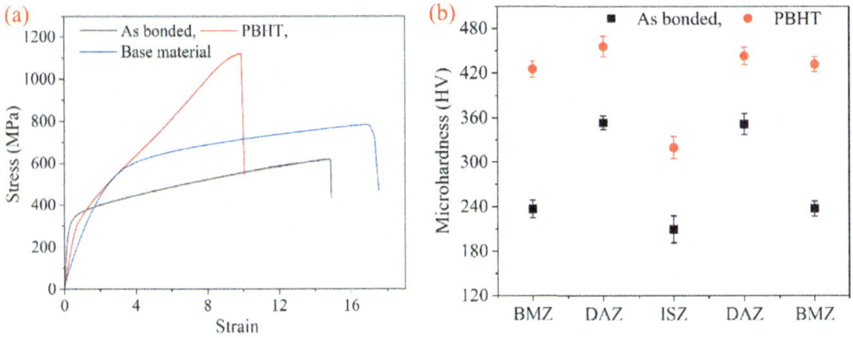

Fig. 6 **a** The stress–strain diagram of the as-received IN718, as-bonded specimen and PBHT specimen, **b** variation of microhardness in various zones

the DAZ hardness is highest and the ISZ hardness is lowest. The presence of borides in the DAZ contributes to its highest level of hardness among all zones. The hardness of the ISZ depends mainly upon the amount of the solution-strengthening elements. As the concentration of these elements (Mo, Ti, Al, and Nb) in ISZ is lower than BM, the hardness of the ISZ is reduced. In case of the PBHT specimen, the hardness of all zones has improved significantly. This phenomenon primarily occurs due to the precipitation of additional γ' and γ'' precipitates during the PBHT process, coupled with the presence of a uniformly distributed solution-strengthening phase across all zones. During PBHT, due to the interdiffusion of elements, uniform microstructure is achieved and more γ' and γ'' phases precipitate. Which increases the bond hardness significantly. In the PBHT specimen, the ISZ hardness is shown lower among other zones. This is due to less availability of the Nb and Al elements in the ISZ for the development of the γ' and γ'' precipitates.

3.3 Micro-fractography Analysis

The fractographs of both the as-bonded specimen and the specimen subjected to PBHT are depicted in Fig. 7a, b. The fractograph of the as-bonded specimen shows a cleavage facet and micro cracks. It also shows intra-granular fracture facets. In addition to this small amount of dimples are observed. This indicates that a mixed mode of failure occurs in the as-bonded specimen. However, in the PBHT specimen, large cleavage facets are observed to indicate brittle fracture failure. In the PBHT specimen, the presence of a large amount of γ' and γ'' precipitates reduces the ductility of the bond.

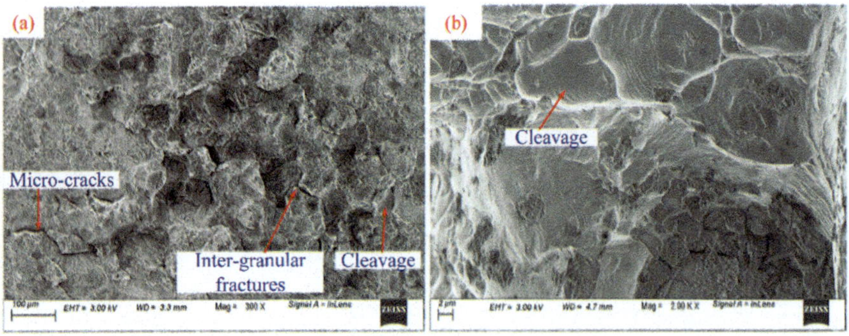

Fig. 7 Fracture surface of the **a** as-bonded specimen and **b** PBHT specimen

4 Conclusion

IN718 was successfully bonded using the TLP-bonding technique with a BNi-2 interlayer. Subsequently, PBHT was conducted on the bonded specimens. The conclusions drawn from this study are as follows:

a. Evidence of ASZ was noted in the TLP-bonded specimen, attributable to inadequate isothermal solidification time. Within the DAZ, blocky and needle-shaped Nb-Mo-Cr borides were observed.
b. PBHT led to a more uniform microstructure. This treatment increased the amount of γ' and γ'' precipitates while simultaneously decreasing the presence of borides in the DAZ and the secondary-phase precipitates in ISZ.
c. The DAZ exhibited the highest hardness value due to a significant presence of boride precipitates, whereas the ISZ displayed lower hardness attributed to insufficient alloying elements such as Nb, Ti, and Mo.
d. Tensile strength notably improved in the TLP-bonded specimen with PBHT, showing an 80% increase in the ultimate tensile strength (UTS). This treatment led to a rise in the amount of γ' and γ'' precipitates, along with a reduction in the presence of borides in the DAZ and the secondary-phase precipitates in ISZ.

References

1. Dwibedi S, Bag S (2024) Critical assessment on the influence of process parameters in micro-plasma arc welding process: a review. Weld Int 38:583–617. https://doi.org/10.1080/09507116.2024.2396917
2. Gruber K, Stopyra W, Kobiela K, Madejski B, Malicki M, Kurzynowski T (2022) Mechanical properties of Inconel 718 additively manufactured by laser powder bed fusion after industrial high-temperature heat treatment. J Manuf Process 73:642–659
3. Koff BL (2004) Gas turbine technology evolution: a designers perspective. J Propuls Power 20:577–595. https://doi.org/10.2514/1.4361
4. Wang Z, Gui Z, Wu J, Zhang Q, Wu X, Lin S, Tian J, Guo C (2023) Microstructure and high-temperature mechanical properties of Inconel 718 superalloy weldment affected by fast-frequency pulsed TIG welding. J Manuf Process 89:338–348
5. Singh SR, Mote RG, Mishra SK (2022) The effect of microstructures and precipitates (γ', γ'', δ) on machinability of Inconel-718 nickel-based superalloy in turning process. J Manuf Process 82:374–389
6. Gasson PC (2008) The superalloys: fundamentals and applications. RC Reed Cambridge University Press, The Edinburgh Building, Shaftesbury Road, Cambridge, CB2 2RU, UK, 2006. 372 pp. Illustrated.\pounds 80. ISBN 0-521-85904-2. Aeronaut J 112:291–291 (2008)
7. Cao X, Rivaux B, Jahazi M, Cuddy J, Birur A (2009) Effect of pre- and post-weld heat treatment on metallurgical and tensile properties of Inconel 718 alloy butt joints welded using 4 kW Nd:YAG laser. J Mater Sci 44:4557–4571. https://doi.org/10.1007/s10853-009-3691-5
8. Oshobe OE (2012) Fiber laser welding of nickel-based superalloy Inconel 718. University of Manitoba (Canada)
9. Huang X, Miglietti W (2012) Wide gap braze repair of gas turbine blades and vanes—a review
10. Kumar DP, Kumaran S (2020) Mechanical and corrosion behaviour of Inconel 718 processed through SPS diffusion bonding and double aging treatment. Mater Today Proc 27:2582–2585

11. Cook GO, Sorensen CD (2011) Overview of transient liquid phase and partial transient liquid phase bonding. J Mater Sci 46:5305–5323. https://doi.org/10.1007/s10853-011-5561-1
12. Tarai UK, Pal S, Robi PS (2023) Reduction of ASZ size and enhancement of mechanical properties of TLP bonded IN718 joint using post bond heat treatment process. Int J Adv Manuf Technol 129:1425–1442. https://doi.org/10.1007/s00170-023-12386-x
13. Tarai UK, Pal S, Robi PS (2023) Kinetics of isothermal solidification during transient liquid phase bonding of Inconel 718 superalloy by differential scanning calorimetry. Proc Inst Mech Eng Part J Mater Des Appl 238(7). https://doi.org/10.1177/14644207231218113

Engineering Applications of Superplasticity of Metals: Review

Amruta Mahajan and Vishvesh Badheka

Abstract Today's demand for lightweight as well as unitized structures focuses attention on various superplastic metals. Superplasticity is a remarkable material property that enables materials like some metals to exhibit exceptional elongation and deformation without fracturing. This unique characteristic is vital for numerous industries such as automotive, aerospace, biomedical, and manufacturing as it allows for intricate shaping of components and structures, reducing the need for complex assembly processes. Though a large variety of materials are used in the manufacturing industry, some of the metals like Aluminium, Magnesium, Titanium, Steel are widely being used due to their desirable properties. Their properties can be modified by using different superplastic deformation methods including modern techniques. Unique design features can be offered by Superplastic Forming and hence it supports creativity in new design and manufacturing methods. This paper summarizes various techniques for achieving superplastic conditions in the metals and using these techniques various applications are also summarized in this paper.

Keywords Superplasticity · Superplastic forming · Grain size

1 Introduction

Lightweight metals with ultrafine grains have been catching the attention of automotive, aerospace industries over the past few decades. Enhanced mechanical properties of metals particularly tensile strength always serve great advantages during the manufacturing of complex forms in a single action phase. The unique property of the materials that provide the strength to get extreme elongation or deformation

A. Mahajan (✉)
Department of Mechanical Engineering, School of Technology, Pandit Deendayal Energy University, Gandhinagar, GJ 382007, India
e-mail: amruta.mphd21@sot.pdpu.ac.in

V. Badheka
School of Technology, Pandit Deendayal Energy University, Gandhinagar, GJ 382007, India
e-mail: vishvesh.badheka@sot.pdpu.ac.in

P. Bhingole et al. (eds.), *Advances in Materials Engineering*, Lecture Notes in Mechanical Engineering, https://doi.org/10.1007/978-981-97-7114-1_20

without failure is known as superplasticity. Superplasticity is the utmost requirement for superplastic forming. But SPF requires some preliminary steps before forming, for example, preparation of fine grain structure in the material used. And this is achieved through some particular deformation processes, known as severe plastic deformation processes (SPD) [1].

2 Superplasticity

2.1 Concept of Superplasticity

Superplasticity is the remarkable material property that enables materials like some metals to exhibit exceptional elongation and deformation without fracturing (refer Fig. 1). So superplasticity is in demand for every manufacturing industry including aerospace, automobile, aircraft, biomedical, and sheet metal forming industry. Superplasticity is defined as the ability of a polycrystalline material to get uniformly elongated by more than 200% by very large elongations under tension at temperatures generally above the 0.5 Tm (where Tm is the melting point temperature of the material on an absolute scale) [2–4].

Simply, the basis of Superplastic Forming (SPF) lies in developing superplasticity behaviors in the materials with increased temperature. The capability of any metallic material to get enhanced tensile elongations of more than 200% through plastic deformation is called superplasticity [5].

Basically, superplastic behavior is of two types: one is internal superplasticity and the second one is fine structure superplasticity.

Fine structure superplasticity is well known for the deformation of metals. Generally slower strain rates are preferred to get superplasticity as compared with conventional forming methods [4]. There is also one commercially accepted high strain rate superplasicity which means superplasticity at strain-rates more than or equal to 10^{-2} S^{-1} [6]. The grains refined by severe deformation processes show high strength. The Hall–Petch relation holds good for ultra-fine-grained metals [7]:

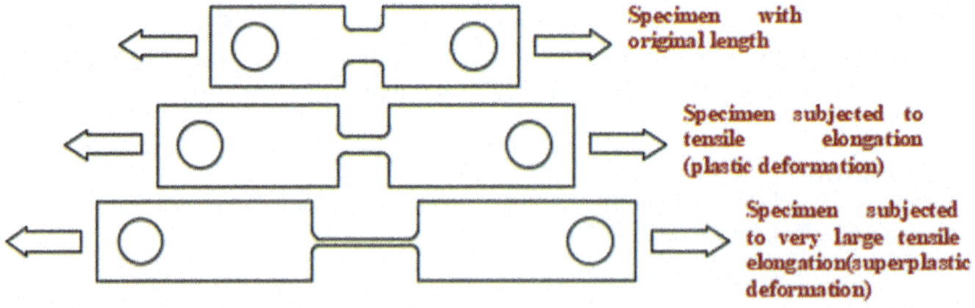

Fig. 1 Schematic illustration of superplastic elongation

$$\sigma_y = \sigma_o + Ad^{-1/2} \tag{1}$$

where σ_y = Yield stress of polycrystalline metals, σ_o = Friction stress, A = constant and d = Grain diameter.

It means that the yield strength of metal increases with the reduction in grain's size. Grain refinement increases the metal strength [7, 8].

2.2 *Superplastic Conditions*

Superplastic conditions are essential to achieve superplasticity in the material, those conditions are as follows:

1. Fine grain size (<10 microns) and grain shape: Basically, Grain boundary sliding in metals takes place because of equiaxed small size grain's presence and such grain's size should be below 10 microns. Also, the strain rate is normally inversely proportional to grain size [6].
2. Appearance of Secondary phase particles: These are evenly distributed so which prevents the fast grain growth.
3. High-angular grain boundaries [2, 6, 9].

All these conditions play a very important role in gaining required superplastic deformation. One of the important parameters is strain rate sensitivity, m. It is the measurement of how the strain rate changes in relation to the change in flow stress. This strain rate sensitivity is responsible for the elongation of grains while resisting necking [5]. Superplastic deformation occurs due to grain boundary sliding (GBS). Hence, GBS is the predominant deformation mechanism in superplastic deformation [2, 4, 9].

Severe Plastic Deformation (SPD). The term "processing by SPD" describes the capacity to apply a very high strain without significantly altering the component's overall dimensions. This basically disallows more conventional processing methods like extrusion, rolling, and drawing, despite the fact that these processes might involve extreme straining [10]. SPD process is the metal forming process that produces large plastic deformations to produce the ultra-fine-grained metal. Hence, it can help in producing lightweight parts using high-strength metals for the reliability and safety of microparts and for environmental safety [7] (Table 1).

Many superplastic deformation behaviors such as high strain rate superplasticity (HSRS), low temperature superplasticity (LTSP), and microcrystalline structure superplasticity individually playing important roles in deformation.

(1) For microcrystalline superplasticity, high deformation temperature should be more than 50% of the melting point temperature of metal, also having a relatively low strain rate between 10^{-4} s^{-1} and 10^{-1} s^{-1}.
(2) HSRS improves the superplastic deformation efficiency and specifies an elongation of materials beyond 200% when the strain rate rises above 1×10^{-2} s^{-1}.

Table 1 Importance of SPD over conventional forming processes [7]

Metal forming processes	Types of forming processes	Imposed plastic strain	Effect of the value of plastic strain	Reason
Traditional metal forming processes	Rolling, extrusion, and forging	<2	Not suitable for structural parts	Due to successive multiple processes deformation causes change in shape (in diameter and thickness)
SPD processes	ECAP, ARB, HPT, RCS, CEC, STS,	>2	Potential for usage as a safe, environmentally friendly metal with extremely high strength	During deformation there is no shape change occur

Refined grains and high-angle grain boundaries are favorable conditions for HSRS.

(3) Low temperature superplasticty of noted at a temperature lower than 50% of the melting point temperature of the material [11].

To work upon this, many researchers explored various severe plastic deformation (SPD) methods to achieve the superplastic characteristic in some metals and alloys. Superplasticity can be obtained through refinement of grains. Materials can achieve even greater refinement in the nanoscale range of less than 100 nm, or sub-micrometer range of 100–1000 nm, by applying SPD methods which, in comparison to those made using traditional bulk-deformation methods, are radically smaller.

There are conventional SPD methods that are known as Standard SPD techniques. But only these methods are not practically possible for mass production in the industry. These methods should be modified and then can continuously manufacture the required production hence such modified methods are known as modified SPD techniques.

Conventional SPD methods. These methods are Equal Channel Angular Pressing (ECAP), High Pressure Torsion (HPT), Accumulative Roll Bonding (ARB) [12], Multi Axial Forging (MAF), Friction Stir Processing (FSP) [4].

Above SPD methods are represented by schematic figures (refer Figs. 2, 3, and 4).

(1) ECAP is such a processing technique that gives metal a severe plastic straining in a simple shear way maintaining consistent cross-section of the sample. So, this method results into ultrafine grains in the processed metal. Multi-pass ECAP processing can produce remarkable total straining into the samples. Also, ultrafine grains find the scope towards attaining superplasticity at different strain rates [13].

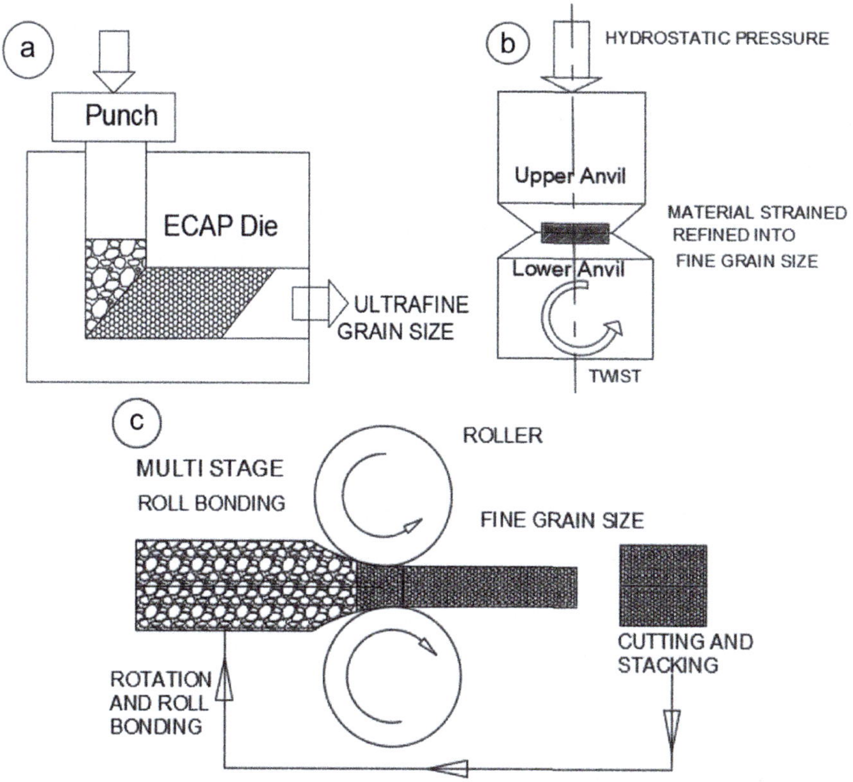

Fig. 2 Diagrammatic depiction of SPD processes **a** ECAP, **b** HPT, **c** ARB (Reproduced from Ref. [12])

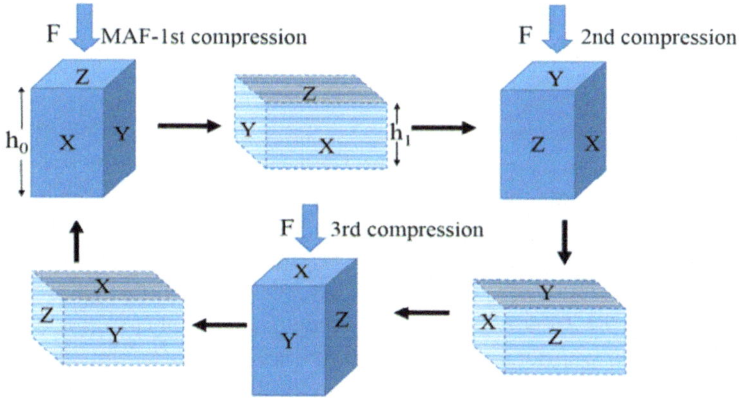

Fig. 3 Schematic of multi axial forging process [17]

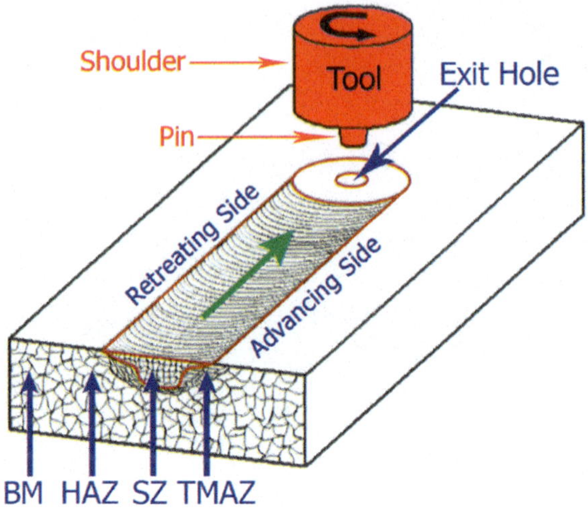

Fig. 4 Schematic of friction stir processing with different microstructural regions [4]

(2) HPT is very effective technique to produce considerable strains and achieve grain refinement by using high hydrostatic pressure [14].
(3) ARB is a multi-stage cyclic roll bonding processing method in which the progressively increasing deformation in between stages is helpful in attaining finer grain size [15].
(4) MAF produces sequential deformation in the material (refer Fig. 3) and causes grain refinement. In addition to its effect on refining grains, MAF may also encourage the weakening of texture in magnesium alloys [16].
(5) FSP technique was invented by Mishra in 1999. This new severe plastic deformation technique, FSP is correlated to the working principle of friction stir welding (FSW) (refer Fig. 4). Basically, it is known as solid state processing. After FSP, the microstructure of materials transformed into refined, homogenized, and densified structures, which improved the materials' strength and plasticity [11, 18, 19].

Modified SPD techniques. Standard SPD methods need to be modified for their actual possible use in industrial production. These modified SPD methods are a blend of traditional metal-forming processes and modern SPD methods. Such new developed SPD methods are generally used in industry e.g., ECAP-Conform, ECAP with rolls, Equal channel angular drawing (ECAD), Incremental ECAP, porthole-equal channel angular pressing (P-ECAP), porthole-equal channel angular pressing (P-ECAP), conshearing, Continuous cyclic bending (CCB), caliber rolling (CAROL), Forward extrusion with torsion deformation process, high-pressure sliding (HPS), continuous high-pressure torsion (CHPT), severe torsion straining (STS), KOBO process, Cryo-rolling [20] (Table 2).

Also, combination of two or more SPD processes is also preferred in industrial applications. For instance, ECAP and HPT, ECAP and ARB. Due to blend of SPD processing techniques expedite microstructures that are different from those obtained

Table 2 Summary of modified SPD techniques [20]

Modified SPD techniques	Description
Integrated extrusion and ECAP process	This is the process that combines ECAP and conventional extrusion in one processing step. There are Conical section and two parallel passages that produce long continuous ultra-fine grain bars
ECAP-conform	Frictional force is used to force workpiece continuously through the die in order to produce UFG samples
Equal channel angular drawing (ECAD)	Two intersecting dies at 135° position are used to pull the workpiece through the die. This is used for multipass processing
ECAP with rolls	Rollers are being part of ECAP used in the deformation process to produce an increased volume of UFG materials
Incremental ECAP	Used to process large size billets and eliminate friction during feeding supply material
Porthole-equal channel angular pressing (P-ECAP)	Blend of extrusion process and quad-channel ECAP. This combined process produces UFG tubes with four-seam welding
Continuous confined strip shearing	Guide roll and feeding rolls are used as feeding apparatus. Multipass operations are possible in continuous confined strip shearing
Conshearing	Material is passed through central roll and number of satellites rolls to produce the large extrusion force
Continuous cyclic bending (CCB)	This is a straining technique for sheet metal industry which produces a very high stretch on the surface and low stretch inside
Caliber rolling (CAROL)	It can make large quantity of UFG rods
Forward extrusion with torsion deformation process	Material first passed through forward extrusion die and then torsional shearing strain has been endured on previously extruded materials so that the microstructure and texture will be further modified. This method helps in improving the performance of processing the Mg and modifying microstructure and texture
High pressure sliding technique (HPS)	This technique was used for processing of metal sheets
Continuous high pressure torsion technique (CHPT)	Processing of metallic sheets is continuously made possible in this CHPT

(continued)

Table 2 (continued)

Modified SPD techniques	Description
Severe torsion straining (STS)	In this method, there is a localized heat zone in which torsional strain developed by providing rotation to the other end. Thus, it produces severe strain to the sample
KOBO process	This process causes very large strain at low temperature. Metallic billet causes reversible plastic straining before entering the cross-section reducing die

by application of an individual SPD technique. This causes the grains to become even more refined and it also causes variations in structure, size, and proportions of second phase precipitates, dislocation structure and density, crystallographic texture and micro texture, and misorientations of grain boundaries. Such refined grains and modified nanostructure after using complex SPD processes commonly result into enhanced mechanical properties [21].

Superplastic Forming (SPF). Superplastic forming is applicable to higher ductility metal alloys. It enables the formation of substantial and intricate feature components without the need for rivets or joints in a single operation. Hence Superplastic forming has the effect of reducing waste, weight, and manufacturing costs. Unique design features can be offered by the SPF and hence it supports creativity in new design and manufacturing methods. For instance, "Avions Marcel Dassault," a French manufacturer, used this SPF technique to create a Ti-6Al-4V alloy wing for the Mirage 2000. This wing weighed 2.8 kg when it was constructed by SPF and was made up of 4 primary and 6 supporting elements. Additionally, the wings, which included 10 primary and 155 attaching elements, weighed 3.2 kg when they were made using the traditional forming process. Hence it confirms that this Superplastic forming technique reduces overall weight as well as number of parts in the assembly. Superplastic forming (SPF) method produced walls with consistent thickness, excellent internal structural symmetry, and exceptional shape control. This increases demand for SPF processing techniques in many industries like automobile, aerospace, aviation, etc. [22].

Light weight metals are very smartly handled during SPF and miraculously opens new possibilities for trending the use in sheet metal industries. The automotive industry always focuses on weight reduction as it helps in reducing the fuel consumption of vehicles and hence strengthening the environmental sustainability approach. Many lightweight alloys used in the industries show superplasticity, for example Ti6Al4A titanium alloy, 5083 Al alloy, and AZ31 Mg alloys.

Superplasticity-oriented SPF technique shows some special advantages over conventional forming processes and that are:

(1) SPF have ability to produce complex shaped components which is impossible by conventional techniques.

(2) Some difficult to form metals (e.g., Titanium and Magnesium) can be easily deformed in required shapes by SPF technique. Ti and Mg alloys have poor ductility due their HCP crystal structure.
(3) Only a single die is required to form the component hence low die cost.
(4) SPF process is completed in single step that produces either finished or semi-finished parts.
(5) As complicated shapes are produced in single piece hence overall quantity of parts gets reduced so the requirement for number of fasteners is also reduced. This offers more safety in the applications e.g. aerospace [1].

3 Industrial Applications

Applications of Aluminum. Only by the application of superplastic shaping, the world's first superplastic bicycle with a bonded frame assembly was manufactured, Also sports car external body panels as well as inner structural panels are superplastically formed from SP5083 [23].

Hefti reported that SPF5083 is considered a commercial Boeing airplane material and hence being used in various parts in commercial airplanes. Airplane Parts can be produced for less cost and weight with SPF process rather than as a casting. Overall, it appears that this material is well-suited for specific applications where its unique combination of properties results in both cost and performance benefits. It's likely a valuable choice in industries where lightweight, structurally sound components are required, such as aerospace manufacturing [24].

Application of Titanium (Ti). There is an increasing need for artificial implantable devices due to the fast growth in the overall human population which results in a larger elderly population. This demographic shift has led to a greater demand for medical implants for various parts like dental implants, hip, and knee. Also, for orthopedic applications various screws, plates, and meshes are required. These medical implants have historically been made of a variety of materials, including cobalt-chrome alloys, titanium alloys, and stainless steel. The choice of material for a specific implant depends upon many factors such as the implant's intended use, mechanical properties, and biocompatibility. Because of their superior fatigue, good formability, high strength, fracture performance, corrosion resistance, and low toxicity, biocompatible metallic alloys like titanium and cobalt-chrome alloys are favored for use in surgical implants. These properties make them suitable for use in physiological conditions within the human body. Titanium has been considered as one of the most bioinert metal, so selected for implants in medical field [21].

Nanostructured titanium alloy is being used for manufacturing of high strength bolts. These bolts are commonly demanded in aircraft and automobile industries, also have excellent requirements in medical applications. High angle grain boundaries and homogeneous structure are the essential conditions for the actual usage of nanostructured structural materials. This can be achieved through one of the SPD methods, i.e., ECA pressing of rods. Nanostructures produced by ECA pressing increase strength and ductility. The threaded bolts made of VT1-0 alloy underwent the ECA pressing process. Using a rolling machine, plastic forming was used to create these threads [25].

The brand name "Timplant" refers to the Nano implant, a new generation of dental implants made by the "Timplant" firm in Ostrava, Czech Republic. With an intraosseal diameter of 2.4 mm, these nanoscale implants are as strong as traditional 3.5 mm diameter implants. Many Nano implants have been implanted on an immediate basis, showing the excellent primary stability of these over other implants [26].

Titanium and its alloys have gained widespread acceptance for use in implantable devices. This trend has been supported by research and industry studies, which have recognized the potential of these materials to meet the growing demand for medical implants, especially as the aging population continues to expand.

Commercially pure titanium (CP Ti) has lower mechanical properties, including lower strength, compared to titanium alloys like Ti-6Al-4V. As a result, CP Ti might be unsuitable for load-bearing implants used for applications in orthopedics where more mechanical strength is needed to sustain the forces and stresses applied to the implant. Due to its lower strength, CP Ti is often limited to use in dental implants and some non-load-bearing applications.

Orthopedic implants, such as hip and knee replacements, typically require materials with higher strength and wear resistance to withstand the mechanical stresses and strains associated with everyday movements. Titanium alloys like Ti-6Al-4V offer improved mechanical properties and are commonly used for such load-bearing orthopedic implants. It's essential for medical professionals and materials scientists to carefully select implant materials based upon particular application requirements to ensure their safety along with their effectiveness. The choice of material depends on factors such as the implant's intended use, location within the body, biocompatibility, and the mechanical properties needed to support the load and function properly [27].

Application of Magnesium (Mg). To focus on the possible utility of Mg and its alloys many methods should be used such as the design and development of new alloys, magnesium-based composites. There is a scope to have surface modification and new processing technologies to build for practicability of Mg. So, they can have optimal Mg implant degradation. Magnesium due to its good biodegradable properties catches the attention of the medical field and finds alternatives for conventional orthopedic and cardiovascular implants. Commonly implemented SPD techniques for biocompatible magnesium alloys are ECAP, HPT, MDF, CEC, and RSW. These techniques are generally used to refine the grains which promote enhancement of the mechanical strength and corrosion resistance. And hence Mg finds clinical use in the medical industry. "Syntel-lix" is the pioneer producer of biodegradable magnesium screws made of the Mg-Y-Re-Zr alloy. Surgery for Hallux valgus involves the use of these screws [28].

It is investigated that after the 4-passes of ECAP process of ZFW MP Magnesium alloy, fine grains can be seen surrounding coarse grains because of DRX. With the increasing passes of ECAP, as surface oxide layers break up, there is a uniform distribution in the matrix takes place. Hence, multiple passes of ECAP process have amplified the strength of the Mg (ZFW MP) alloy by refining the microstructure and decreasing the weak joints among powder particles [29].

Adnan et al. investigated the superplastic forming of Mg alloy from industry viewpoint with an emphasis on the adaptability of the superforming technique. Superplastic forming of AZ31Mg alloy at the 400 °C from the industrial point of view is being successfully tested and results obtained show that there is higher range of ductility achieved at the lower strain rates. The deformation raises amount of thinning to nearly 4% at the lower strain rate 1×10^{-3} S^{-1} and maximum deformation [30].

Application of FSP in Additive Manufacturing. This is the latest manufacturing technique in which material is deposited layer by layer to manufacture the desired shape of the product. The MELD manufacturing company included FSP in additive manufacturing [31].

4 Conclusion

An SPD technique can create a microstructure with a fine grain distribution and orientation, but the strength and distribution of the grains will depend on how many passes or cycles are used. The initial grain size of Material and its texture before processing influence the number of passes required.

Superplastic materials can produce complicated deformations. Simultaneously maintaining the properties of material up to the mark. So ensures safety and environmental sustainability. Conventional processing methods along with modifications or inclusion of forming processes can be implemented in industry for continuous production.

5 Future Scope

In commercial applications, a wide scope is available for superplastic forming of different metals through grain refinement. This is the challenging field for existing applications to be replaced by more efficient metal and its alloys with the proper deformation techniques so far discussed above. So many parameters of various processes are still under research zone for the effective optimum output results for advanced applications. Investigation of optimum parameters for SPD processes like FSP with multipass, with cooling assistance, for nonferrous alloys like Mg are finding scope for advanced research.

References

1. Abu-Farha F, Khraisheh M (2008) An integrated approach to the superplastic forming of lightweight alloys: towards sustainable manufacturing. Int J Sustain Manuf 1(1–2):18–40
2. Harwani D, Badheka V, Patel V, Li W, Andersson J (2021) Developing superplasticity in magnesium alloys with the help of friction stir processing and its variants: a review. J Mater Res Technol 12:2055–2075
3. Khan F MD, Panigrahi S (2018) Achieving excellent superplasticity in an ultrafine-grained QE22 alloy at both high strain rate and low-temperature regimes. J Alloys Compd 747:71–82
4. Mirzadeh H (2021) High strain rate superplasticity via friction stir processing (FSP): a review. Mater Sci Eng A 819(May):141499
5. Parwani A, Abhishek K, Yadav S (2020) Recent advances in mechanical infrastructure
6. Pérez-Prado M, Kassner M (2009) Superplasticity. In: Fundamentals of creep in metals and alloys, 2nd ed, pp 135–152
7. Azushima A et al (2008) Severe plastic deformation (SPD) processes for metals. CIRP Annals Manuf Technol 57:716–735
8. Valiev R, Langdon T (2006) Principles of equal-channel angular pressing as a processing tool for grain refinement. Progr Mater Sci 51:881–981
9. Patel V, Badheka V, Kumar A (2016) Friction stir processing as a novel technique to achieve superplasticity in aluminum alloys: process variables, variants, and applications. Metallogr Microstruct Anal 5(4):278–293
10. Kawasaki M, Langdon T (2019) The contribution of severe plastic deformation to research on superplasticity. Mater Trans 60(7):1123–1130
11. Wang W et al (2019) Friction stir processing of magnesium alloys : a review. Acta Metall Sin (English Lett no. 0123456789)
12. Rakshith M, Seenuvasaperumal P (2021) Review on the effect of different processing techniques on the microstructure and mechanical behaviour of AZ31 magnesium alloy. J Magnes Alloy 9(5):1692–1714
13. Furukawa M, Horita Z, Nemoto M, Langdon T (2001) Processing of metals by equal-channel. 6:2835–2843
14. Edalati K, Horita Z (2010) Continuous high-pressure torsion, pp 4578–4582
15. Wang Q, Xiao X, Hu J, Xu W, Zhao X, Zhao S (2007) An ultrafine-grained AZ31 magnesium alloy sheet with enhanced superplasticity prepared by accumulative roll bonding. J Iron Steel Res Int 14(5):167–172
16. Li T, Zhang K, Du Z, Yuan J, Li X (2014) Production of fine-grained and weak texture structure in an Mg-7Gd-5Y-1Nd-0.5zr alloy by multi-axial forging. Appl Mech Mater 633–634:120–124
17. Salandari-Rabori A, Zarei-Hanzaki A, Fatemi S, Ghambari M, Moghaddam M (2017) Microstructure and superior mechanical properties of a multi-axially forged WE magnesium alloy. J Alloys Compd 693:406–413
18. Del Valle J, Rey P, Gesto D, Verdera D, Ruano O (2012) Friction stir processing of the magnesium alloy AZ61: grain size refinement and mechanical properties. Mater Sci Forum 706–709:1823–1828
19. Patel V, Li W, Vairis A, Badheka V (2019) Recent development in friction stir processing as a solid-state grain refinement technique: microstructural evolution and property enhancement. Crit Rev Solid State Mater Sci 377–425
20. Faraji G, Kim S, Kashi T (2018) Severe plastic deformation for industrial applications. 1:165–186
21. Valiev R, Straumal B, Langdon T (2022) Using severe plastic deformation to produce nanostructured materials with superior properties
22. Padmanabhan K, Prabu S (2018) Superplasticity: common basis for a near-ubiquitous phenomenon
23. Barnes A (2001) Industrial applications of superplastic forming: trends and prospects. Mater Sci Forum 357–359:3–16

24. Hefti L (2007) Commercial airplane applications of superplastically formed AA5083 aluminum sheet. J Mater Eng Perform 16(2):136–141
25. Valiev R, Islamgaliev R, Alexandrov I (2000) Bulk nanostructured materials from severe plastic deformation. Progr Mater Sci 45
26. Bagherpour E, Pardis N, Reihanian M (2018) An overview on severe plastic deformation: research status, techniques classification, microstructure evolution, and applications
27. Mishnaevsky L et al (2014) Nanostructured titanium-based materials for medical implants: modeling and development. Mater Sci Eng R Rep 81(1):1–19
28. Sedighi M, Hashemi R (2022) Severe plastic deformation (SPD) of biodegradable magnesium alloys and composites: a review of developments and prospects. J Magnes Alloys 10:938–955
29. Gholami-Kermanshahi M, Neubert V, Tavakoli M, Pastorek F, Smola B, Neubert V (2018) Effect of ECAP processing on corrosion behavior and mechanical properties of the ZFW MP magnesium alloy as a biodegradable implant material. Adv Eng Mater 1800121:1–12
30. Zaid AIO, Al-Matari MAA, Nazzal MAH (2016) Forming of a super plastic sheet metal made of MgAZ31 alloy. In: 14th international symposium on advanced materials 2016. IOP Conf Ser Mater Sci Eng 146
31. Mishra RS, Haridas RS, Agrawal P (2022) Friction stir-based additive manufacturing

Mechanical Properties of Concrete by Partial Replacement of Cement with Dolomite

Toshan Singh Rathour, Fiza Khan, Khyati Kodopi, Sanjana Sahu, and Somprabha Sonkale

Abstract Concrete, a ubiquitous construction material globally, relies on natural resources such as lime, aggregates, and water. The surge in cement production has led to a significant rise in CO_2 emissions, contributing to environmental pollution. In underdeveloped countries like India, enormous amounts of dolomite waste are generated annually. Owing to the large amount of area needed for disposal and the low amount of waste that is recovered or reused, dolomite waste disposal is a very serious issue. To mitigate these effects, researchers have explored alternative materials like Dolomite Powder to replace or supplement cement. This study delves into the potential of Dolomite Powder as a substitute for cement in concrete, presenting a cost-effective and strength-enhancing solution. Various replacement percentages (0, 5, 10, and 15% by weight of cement) were examined. The research evaluated compressive strength, flexural strength, split tensile strength, and workability revealing that incorporating dolomite powder improves these mechanical properties in concrete, suggesting a promising avenue for further exploration and application in the construction industry.

Keywords Dolomite powder · Compressive strength · Flexural strength · Split tensile strength · Workability

1 Introduction

Recent substantial developments in concrete technology represent a significant breakthrough in the realm of concrete, a crucial material in civil engineering. Concrete is the second most often used substance in the world after water, and it serves as the foundation for many different types of construction, including residential buildings and roads [1]. Because of its tremendous mechanical strength, ordinary Portland cement is thought to be the ideal construction material. It is not strong enough to

T. S. Rathour · F. Khan · K. Kodopi (✉) · S. Sahu · S. Sonkale
Department of Civil Engineering, Shri Shankaracharya Institute of Professional Management & Technology, Raipur 492015, India
e-mail: khyati.kodopi@ssipmt.com

P. Bhingole et al. (eds.), *Advances in Materials Engineering*, Lecture Notes in Mechanical Engineering, https://doi.org/10.1007/978-981-97-7114-1_21

withstand chemical attacks or be durable. Cement production is also acknowledged as one of the energy-intensive and greenhouse gas-emitting industries, accounting for roughly 7% of the world's carbon dioxide emissions [1–4].

The increased demand for materials with low embodied energy and a low carbon footprint has driven research into the development of innovative [5], sustainable alternatives or blended cement with low embodied energy [3]. This is the reaction to the expected doubling of cement and concrete demand by the year 2050 [2], which is being driven by the rapid construction of buildings and infrastructure in developing countries [6].

The mineral dolomite, which is made up of calcium magnesium carbonate ($CaMg(CO_3)_2$) [7], has gained attention as a possible binding agent. Dolomite powder is a type of limestone powder that has different ratios of $CaCO_3$ and $MgCO_3$. It is highly valued for its density, surface hardness, and exceptional resistance to weathering [8]. Dolomite is preferred for use as a filler material in asphalt and concrete applications because of its superior strength and hardness [9]. According to the study, dolomite is substantially less expensive than cement and is easily accessible locally. It also has a far lower cost of procurement [2].

During the research, we get to know numerous favorable properties of dolomite powder that not only contribute to mitigating global warming but also offer an economical alternative for individuals who face constraints in constructing their homes due to the high cost of building materials. By replacing a portion of the cement content with dolomite powder, concrete becomes more economical without compromising its strength. This substitution also enhances the workability of concrete, making it more durable [10], especially in high-temperature conditions, without the need for expensive plasticizers and admixtures [2].

Enhanced durability of concrete is achieved, and operational parameters in elevated temperature environments are attainable without the need for expensive plasticizers and admixtures. M25-grade concrete samples were formulated, incorporating substitutions of 0, 5, 10, and 15% of cement with dolomite powder [2]. The compressive, split tensile, and flexural strength of the specimens were assessed on the 28th day, facilitating the identification of the optimal dolomite replacement percentage [2].

2 Materials

2.1 Cement

Cement serves as a crucial binding agent in construction work, uniting all the components within concrete. The cement utilized was in a dry state, devoid of any lumps, and was appropriately stored [2, 11, 12]. The study comprises of cement grade 43 employed, adhering to the standards specified in IS 2720 Part 3 [13] for the evaluation of specific gravity and fineness modulus in IS 4031 part 1 [14], as shown in Table 1.

Table 1 Physical properties of cement

Properties	Observed value	Recommended value
Specific gravity (kg/m^3)	2.9	2.9–3.15
Fineness modulus (%)	2.66	<10

Fig. 1 Dolomite powder

Table 2 Physical properties of dolomite

Properties	Observed value	Recommended value
Specific gravity (kg/m^3)	2.8	2.7–2.9
Fineness modulus (%)	6	<10

2.2 Dolomite Powder

Dolomite powder came from the manufacturing factory in Bilaspur in powdered form and was sieved using an 90-micron sieve, before being added to concrete [2, 15]. The carbonate material dolomite is made up of calcium magnesium carbonate ($CaMg(CO_3)_2$) [1, 16], which often results from diagenesis and has a stoichiometric ratio showing 50% or higher magnesium content instead of calcium [12, 17]. In this experiment, Dolomite is replaced by 5, 10, 15% by weight of cement in concrete (Fig. 1 and Table 2).

2.3 Coarse Aggregates

Coarse aggregate refers to chemically inert materials that [18], when combined with cement paste form concrete [3]. Aggregates have a significant impact on the strength of concrete [8]. In experiments, 20 mm-sized aggregates are utilized [18]. The quality of concrete is directly linked to the properties of its components; hence, the aggregates integrated into a concrete mix should possess characteristics such as hardness, strength, density, durability, and freedom from any clay, loam, vegetable matter, or other foreign particles [3]. The presence of such impurities hinders the adhesion of

Fig. 2 Coarse aggregate

Table 3 Physical properties of coarse aggregate

Properties	Observed value	Recommended value
Specific gravity (kg/m^3)	2.92	2.5–3.0
Fineness modulus (%)	7.38	6.75–8.0
Water absorption (%)	0.8	<2%

cement to the aggregate surface, consequently diminishing the overall strength of the concrete [2].

Selecting the proper rocks and sand is crucial for creating durable concrete. The strength of the concrete is significantly influenced by these ingredients. The rocks that are used should be clean and free of filth, and they should be strong and devoid of any weak points or cracks [2] (Fig. 2 and Table 3).

2.4 Fine Aggregate

Fine aggregates are defined as the material that has passed through an I.S. Sieve of size 4.75 mm [19]. They serve the purpose of densifying concrete by occupying the voids within coarse aggregates. Experimental work utilizes natural sand for this purpose. Fine aggregate is expected to possess durability, cleanliness, and freedom from organic contaminants. The primary role of fine aggregate resides in facilitating the generation of workability and consistency within the concrete mix [2]. In this study, river sand served as a fine aggregate [2, 9] (Fig. 3 and Table 4).

2.5 Water

In addition to actively participating in chemical reactions with cement, water is an essential component of concrete. Pure, uncontaminated water free of acids, alkalis,

Fig. 3 Fine aggregate

Table 4 Physical properties of fine aggregate

Properties	Observed value	Recommended value
Specific gravity (kg/m^3)	2.64	2.5–3.0
Fineness modulus (%)	3.9	2–4
Water absorption (%)	0.6	<3

and oil is required for use in the concrete mix. The optimal pH range for the water used to produce concrete is between 6 and 8. In this investigation, the concrete mixing procedure used locally accessible potable water [6, 20].

3 Objectives

- To find out if adding Dolomite Powder in concrete can make it stronger and easier to handle, by testing different proportions of dolomite powder in concrete mixes.
- To determine if using Dolomite Powder instead of cement in concrete can help the environment by reducing CO_2 emission from cement production.

4 Methodology

The study involved the formation of M-25 concrete mix using Pozzolana Portland cement by the concrete mix proportioning requirements specified in IS:456-2000 [21] and IS10262-2019 [22]. The mixing process was conducted at room temperature, and the prepared mixtures were cured in water for 28 days at room temperature as well. To find the compressive strength and flexural strength, standard-sized concrete cubes of dimensions 150 × 150 × 150 mm were prepared. These tests were conducted precisely on the 28th day after the concrete preparation [2, 3, 5] (Fig. 4).

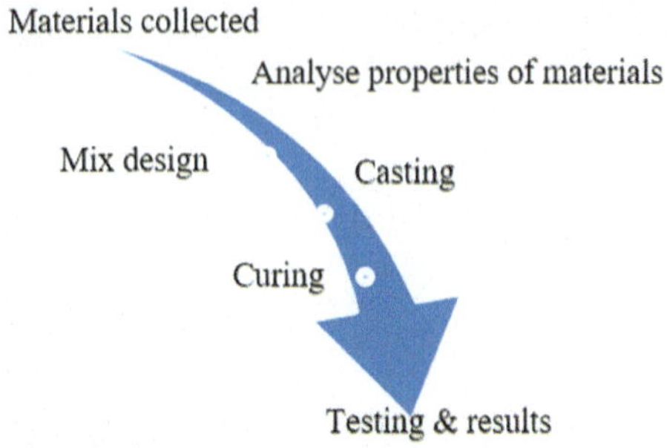

Fig. 4 Flow chart of methodology

By testing the mechanical properties of materials of concrete containing dolomite in place of cement. The mix design of dolomite concrete is prepared and the ratio of proportion is given in the Table 5.

Mix Design. The M-25 concrete mix design is prepared, and the quantity of materials required to cast cubes is calculated [5, 8]. The concrete mix is designed as per the code provisions of IS 10262-2019 [22] (Fig. 5).

Table 5 Case study of dolomite powder

Case ID	Description	Cement (kg/m^3)	Dolomite powder (kg/m^3)
D0	Dolomite powder replaced by 0% by weight of cement	319.30	0
D5	Dolomite powder replaced by 5% by weight of cement	303.34	15.97
D10	Dolomite powder replaced by 10% by weight of cement	287.40	31.90
D15	Dolomite powder replaced by 15% by weight of cement	271.41	47.89

Fig. 5 Dolomite concrete cube

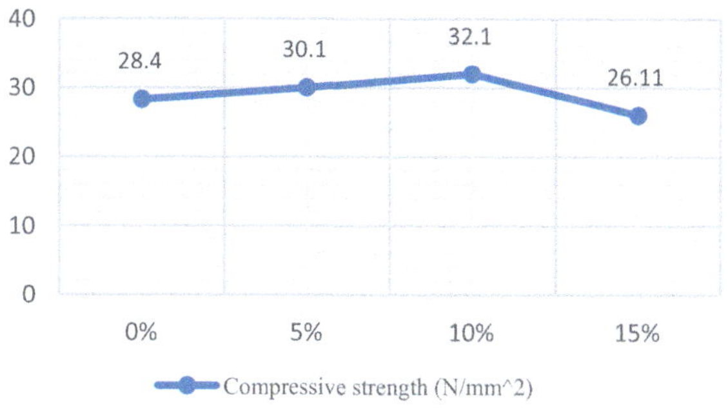

Fig. 6 Compressive strength of concrete

5 Testing and Discussion

5.1 Compressive Strength

This parameter serves as an indicative measure of the comprehensive quality of the concrete [11, 13]. The average of the three results was considered as the compressive strength. The 28 days compressive strength of standard concrete is 28.4 N/mm^2, while the maximum compressive strength of 32.1 N/mm^2 is obtained by replacing 10% of cement with Dolomite waste (Fig. 6).

5.2 Flexural Strength

To foresee future cracking, the flexural tensile strength of concrete components must be determined. Flexural strength was measured using prisms measuring 100 mm × 100 mm × 500 mm. During the testing, a Universal Testing Machine (UTM) with a capacity of 10,000 KN was used with a two-point loading approach [9]. Standard concrete has a flexural strength of 5.32 N/mm^2 after 28 days, while the maximum Flexural strength of 7.38 N/mm^2 is obtained by replacing 10% of cement with Dolomite waste (Fig. 7).

5.3 Workability

The recommended ranges for concrete workability, as measured are given in IS-1199-1959 [23], workability test by slump cone on fresh concrete was tested at Concrete and test results are given in Fig. 8.

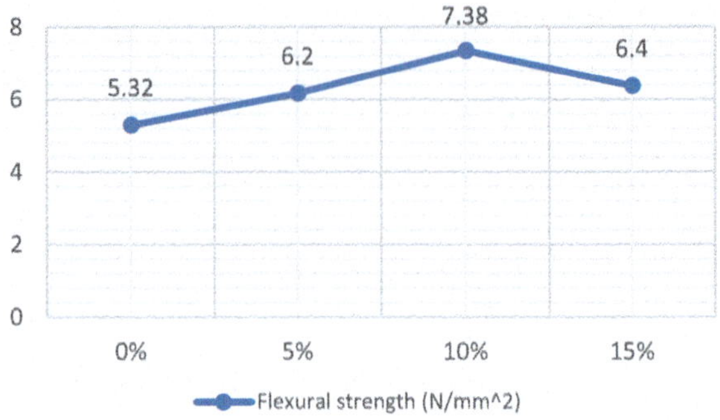

Fig. 7 Flexural strength of concrete

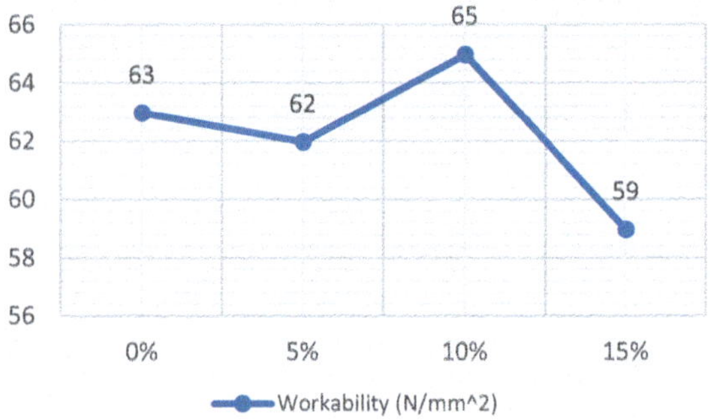

Fig. 8 Workability of concrete

5.4 Split Tensile Strength

To determine the tensile strength, 150 mm diameter and 300 mm height cylindrical specimen conforming to IS: 10086:1982 [24] were cast. The cylinders were placed horizontally on the specimen's perimeter, and the load was delivered along the longitudinal axis [25]. After 28 days, the concrete specimen's split tensile strength with cement (the control specimen) was determined to be 3.95 N/mm^2. According to the testing findings, the M25 Concrete mix's split tensile strength increases when 10% dolomite powder is replaced for cement when compared to conventional concrete. Further addition of dolomite powder decreases the strength of concrete [7] (Fig. 9).

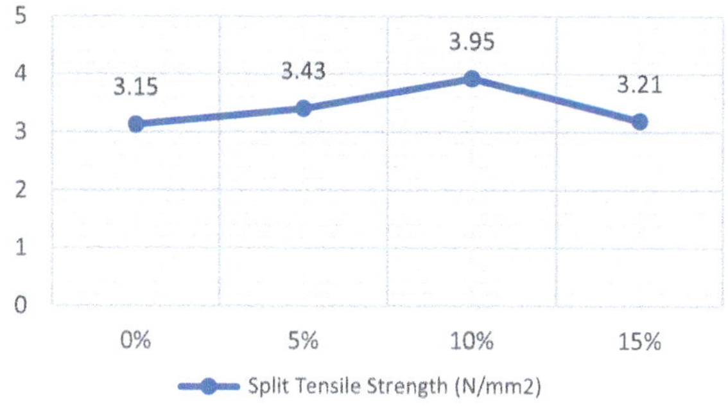

Fig. 9 Split tensile strength of concrete

6 Conclusion

The use of dolomite powder up to a 10% replacement by weight of cement increased the compressive strength, flexural strength, split tensile strength, and workability of concrete. Any addition of dolomite powder, however, has been seen to result in a decrease in strength. During the research, it is determined that 10% is the best amount for substituting cement with dolomite powder in cubes.

After performing a compression test, flexural test, split tensile test, and workability test (slump test). The following data has been collected:

- Substituting cement with dolomite up to a certain percentage leads to a notable increase in strength, demonstrating the efficiency and cost-effectiveness of this approach.
- The compression strength was found 13% higher than the nominal concrete by 10% replacement of cement with dolomite.
- Split tensile strength was found 25% higher than the nominal concrete by 10% replacement of cement with dolomite.
- Similarly, the maximum flexural strength was observed 38% higher than the nominal concrete by 10% replacement of cement with dolomite.
- The workability was found 3.1% higher than the nominal concrete by 10% replacement of cement with dolomite.

By utilizing readily available and cost-effective dolomite powder, it is possible to effectively manage construction costs.

References

1. Bhusare PVP (2019) A review paper on experimental study of effect on concrete properties with partial replacement of cement with dolomite powder. Int J Res Appl Sci Eng Technol 7(7):966–969
2. Patwari VG, Gund KC (2020) Effect on concrete strength due to partial replacement of cement in concrete with dolomite powder. Int Res J Eng Technol 7(12):1654–1657

3. Sugathan A (2017) Experimental investigation on partial replacement of cement with marble powder. Int J Innov Res Sci Eng Technol 6(7):12748–12756
4. Mikhailova O, Yakovlev G, Maeva I, Senkov S (2013) Effect of dolomite limestone powder on the compressive strength of concrete. Procedia Eng 57:775–780
5. Deepthi CG, Baby S (2016) Study on compressive strength of concrete with dolomite powder and crushed tiles. Int J Innov Res Sci Eng Technol 5(9):16584–16590
6. Kumar LR, Kiran J, Rangarajan P (2017) Properties of concrete incorporating dolomite powder. IOSR J Mech Civ Eng 14(2):78–80
7. Prajapati K (2019) Experimental investigation on strength of concrete by partial replacement of cement by dolomite powder. Int J Res Appl Sci Eng Technol 7(5):2611–2616
8. Harish UL, Harikrishna G (2021) Partial replacement of cement with dolomite powder. Int J Mod Trends Sci Technol 7:7
9. Mahesh Yelbhar DR, Megha S, Suraj S, Rohit F (2020) Experimental study of performance of concrete with partial replacement of cement by dolomite. Int Res J Mod Eng Technol Sci J Adv Res Innov Ideas Educ 6(3):38–42
10. Anand MR (2022) An experimental study on partial replacement of cement by GGBS. Int J Sci Res Eng Manag 6(5)
11. Rahul Rollakanti C, Venkata Siva Rama Prasad C, Poloju KK, Juma Al Muharbi NM, Venkat Arun Y (2021) An experimental investigation on mechanical properties of concrete by partial replacement of cement with wood ash and fine sea shell powder. Mater Today Proc 43(10):1325–1330
12. Basha SA, Services IC (2018) Impact of quarry dust and dolomite powder on compressive strength of concrete. J Emerg Technol Innov Res 5(6):38–42
13. IS 2720 Part-3 (1980) Methods of test for soil, Part-3: Deytermination of specific gravity, 1st ed. BIS, New Delhi
14. IS 4031 Part-1 (1996) Method of physical tests for hydraulic cement, 2nd edn. Bureau of Indian Standards, New Delhi
15. Devu K, Sreerath S (2019) Experimental investigation on partial replacement of cement with fly ash and glass powder. Lecture notes in civil engineering, vol 46, pp 73–82
16. Preethi G, Arulraj P (2015) Effect of replacement of cement with dolomite powder on the mechanical properties of concrete. IJISET Int J Innov Sci Eng Technol 2(4):1083–1088
17. Kumar NS, Sirish RD, Amit R, Gillasrinivas (2018) Experimental study on partial replacement of sand with m-sand and cement with dolomite powder in cement concrete. Anveshana's Int J Res Eng Appl Sci 3(5):326–330
18. Kaushal Prajapati PDJS, Varghese V (2019) Experimental investigation on strength of concrete by partial replacement of cement by dolomite powder. Int J Res Appl Sci Eng Technol 7(5):2611–2616
19. Kaushal Prajapati PDJS, Varghese V (2019) Experimental investigation on strength of concrete by partial replacement of cement by dolomite powder. IJRASET 7(4):146
20. Gusain I, Sharma S, Debarma S, Kumar Sharma A, Mishra N, Prakashrao Dahale P (2023) Study of concrete mix by adding dolomite in conventional concrete as partial replacement with cement. Mater Today Proc 73:163–166
21. IS 456-2000 (2007) Plain concrete and reinforced, 4th edn. Bureau of Indian Standards, New Delhi
22. IS 10262-2019 (2019) Concrete mix proportioning, 2nd edn. Bureau of Indian Standards, New Delhi
23. IS 1199-1959 (2012) Methods of sampling and analysis of concrete, 11th edn. Bureau of Indian Standards, New Delhi
24. Kisan M, Sangathan S, Nehru J, Pitroda SG (2008) IS 10086-1982 Indian Standard specification for moulds for use in tests of cement and concrete, 3rd edn. BIS, New Delhi
25. Jyothirmai A, Anuboyina HR (2022) An experimental analysis on behaviour of concrete with copper slag and dolomite. Mater Today Proc 62(4):2214–7853

Cellular Mg-Based Biodegradable Bone Implant: Recent Development to Future Prospects

Dhaval Makwana, Bhingole Pramod, and Manan Shah

Abstract Magnesium is the most attractive biomaterial due to its unique bioresorbable nature with good mechanical properties and low density. Recently developed advance manufacturing techniques used to fabricate biodegradable porous Mg bone implant materials, its mechanical and biological performance have been reviewed. Advantages as well as limitations related to mechanical properties, in-vivo performance, corrosion, and tribological requirements are discussed. Future perspectives of closed-cell Mg-based biomedical implants, their benefits, and challenges are highlighted.

Keywords Cellular magnesium · Open cell foam · Closed cell foam · Bone implant

1 Introduction

Magnesium-based bioresorbable materials are much more focused on the researcher due to its low density, high strength and biodegradable properties in bone implant applications [1]. Commercially available Ti-Co and stainless steel-based bone implants are bio-inert materials that have the main disadvantage of long term corrosion, and required second surgery for implant removal after bone healing [2]. Furthermore, high density and elastic modulus compared to natural bone may create a mechanical imbalance in skeletal structure and stress-shielding respectively, which is painful to the patients and sometimes causes failure in bone regeneration and suggests lifetime permanent implant [3].

Mg is the most attractive element in the field of biodegradable bone implants compared to bio-inert materials due to advantageous properties of high strength-to-weight ratio, modulus of elasticity nearly equal to natural bone, excellent tissue

D. Makwana · B. Pramod (✉) · M. Shah
Mechanical and Aero-Space Engineering, Institute of Infrastructure Technology Research and Management, Ahmedabad, Gujarat 380026, India
e-mail: pramod_bhingole@yahoo.com

P. Bhingole et al. (eds.), *Advances in Materials Engineering*, Lecture Notes in Mechanical Engineering, https://doi.org/10.1007/978-981-97-7114-1_22

response in the body and good biocompatibility [4]. However, commercial use of Mg-based bone implants is still limited due to various challenges like rapid degradation rate and hydrogen evolution [5]. In-vivo studies reported in literature showed when Mg is implanted in the body, corrosion products formed mainly consist of magnesium carbonate, magnesium hydroxide, and magnesium phosphate [4]. Magnesium chloride also formed due to the presence of chloride ions in blood plasma [6]. All biodegradation products formed due to Mg corrosion were found biocompatible and no harmful effects were reported [5]. Hydrogen released due to magnesium corrosion has an important role in the implantation study of biodegradation. Research reported that amount of H_2 gas released was absorbed by the body and it depends on the hydrogen evolution rate [5]. Also, local gas cavities may be fixed by incision or puncturing. However, after 45 days of implant, H_2 gas cavities were not observed and no toxic effects found [5]. Further studies are required in H_2 gas evolution in large implant degradation, long term corrosion rate, and its effects due to local pH in the body environment to control from it bone healing area. Also, there was no infectious effects observed based on in-vivo implantation studies as reported in literature [7, 8].

Developing cellular Mg structures are much attraction in biodegradable bone implants as cellular structure permit bone ingrowth. Recently, various production methods implemented to fabricate open pore Mg-based structures to fulfill mechanical support and bone regeneration [9]. However, mechanical support requires long term sustainability after its degradation until complete bone healing, which make it challenging. Additive manufacturing (AM) and powder metallurgy (PM) are most explored advanced techniques to elaborate porous Mg-based bone implant [9].

Wen et al. introduced a new powder metallurgical (PM) process to develop open pore Mg and Ti foam using ammonium hydrogen carbonate and carbamide as space particles with obtained porosity of 50%, compressive strength of 2.33 MPa and young's modulus of 0.35 GPa [10]. Hao et al. also used carbamide as space particles to prepare open cell Mg foam through PM route and described mechanical properties with 40–80% porosity and 3–9 MPa compressive strength were achieved [11]. Vahid et al. fabricated open cell Mg-Nb and Mg-Ta foam using ammonium hydrogen carbonate space holder to obtain 30–70% porosity [12], Yang et al. prepared Cu-Mg alloy closed cell foam by PM method using gas releasing $CaCO_3$ powder including 50% porosity and compressive strength of ~24 MPa achieved [13]. Apart from various powder metallurgical methods developed by researchers, the required porosity for bone implants can be achieved and proposed as a biodegradable implant. Poor mechanical properties, cost of powder processing and its handling make it more challenging to overcome its commercial applications.

Additive manufacturing (AM) of magnesium-based porous implants is one of the most favorable techniques reported as recently advanced technology. Selective laser melting (SLM) is generally used in processing to fabricate porous Mg structures. This process required fine Mg powder with a special Ar gas inert SLM environment chamber [14]. AM methods satisfied all requirements for long term mechanical integrity with short term implant support and biocompatibility [15, 16].

2 Future Prospects

Apart from many promising techniques developed recently to fabricate porous Mg implants as reported, it is difficult to achieve all consideration to use in commercial applications. Overall cost and processing methods have an important role in implant development and suitable methods need to be identified for further studies. However, some technical challenges were identified based on the recent development. Mg has a primary challenge of rapid degradation compared to bone healing rate. Moreover, open cell construction has more surface contact area to degrade the entire implant volume rapidly. It is suggested to use closed cell geometry, which may be useful to improve tissue growth on cellular surface and prevent internal degradation that can provide long term mechanical integrity. Degradation rate can be controlled by further degradation of outer cell walls and tissue growth may increase on cell structure to maintain mechanical integrity according to degradation rate of implant with bone healing. Figure 1 represents the major aspects of mechanical, biological, corrosion, and tribological with concentrating overall cost of developing porous Mg-based implant.

On focusing overall cost of porous Mg implant, four major aspects of mechanical, biological, corrosion and tribological properties need to be considered. On the

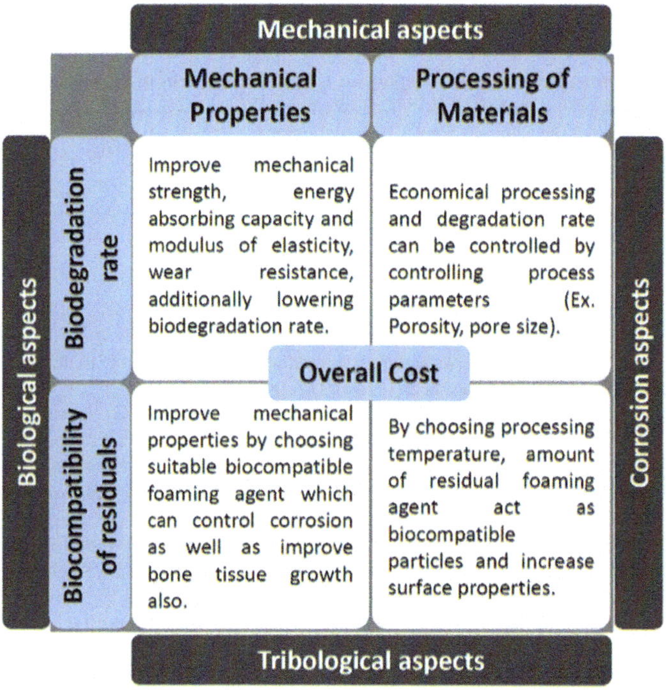

Fig. 1 Major aspects for considering to develop Mg-based biodegradable bone implant

perspective to closed cell Mg foam, mechanical properties may obtain a great combination with strength to weight ratio, energy absorbing capacity, and modulus of elasticity. Same point of view, biocompatibility, controllable corrosion rate, and tribological properties may achieved by controlling process parameters [17] and selecting suitable alloying and foaming agent. Overall cost of liquid metal processing is also economical. Controlling process parameters and porosities, detailed in-vitro and in-vivo studies are required and give great potential to achieve future challenges in cellular Mg-based biodegradable bone implant.

3 Conclusion

Apart from many development methods for porous Mg-based implant, two most explored method of powder metallurgy (PM) and additive manufacturing (AM) having great potential in biodegradable porous Mg-based implants are discussed. Its advancement in mechanical integrity and biocompatibility are highlighted. To overcome technical challenges faced in powder-based techniques, closed cell Mg structures have been suggested for bone implant. Achieving the required porosity, pore shapes, and size with combined advantages of mechanical, corrosion, biocompatibility, and surface properties by further in-vitro and in-vivo detailed study makes great potential for commercial use.

Acknowledgements This work was supported by the Department of Science and Technology; Science and Engineering Research Board, Government of India (Grant No. ECR/2016/001518). Authors acknowledge the Siemens Centre of Excellence; Centre of Advanced Defence Technology (CADT), and IITRAM (Project No: IITRAM/COE/CADT/22-23/IRP05) for support.

References

1. Gu X, Zheng Y (2010) A review on magnesium alloys as biodegradable materials. Front Mater Sci Chin 4:111–115. https://doi.org/10.1007/s11706-010-0024-1
2. Dargusch MS, Balasubramani N, Yang N, Johnston S, Ali Y, Wang G, Venezuela J, Carluccio J, Lau C, Allavena R, Liang D, Mardon K, Ye Q (2021) In vivo performance of a rare earth free Mg–Zn–Ca alloy manufactured using twin roll casting for potential applications in the cranial and maxillofacial fixation devices. Bioact Mater 12:85–96. https://doi.org/10.1016/j.bioactmat.2021.10.026
3. Yu Y, Liu W, Chen J, Quan L, Zheng X, Liu L (2019) No need to routinely remove titanium implants for maxillofacial fractures. J Oral Maxillofac Surg 77:783–788. https://doi.org/10.1016/j.joms.2018.10.022
4. Kavyani M, Ebrahimi GR, Ezatpour HR, Jahazi M (2020) Microstructure refinement, mechanical and biocorrosion properties of Mg-Zn-Ca-Mn alloy improved by a new severe plastic deformation process. J Magnes Alloys 6(10):1640–1662. https://doi.org/10.1016/j.jma.2020.11.013
5. Witte F (2010) The history of biodegradable magnesium implants: a review. Acta Biomater 6:1680–1692. https://doi.org/10.1016/j.actbio.2010.02.028

6. Atrens A, Johnston S, Shi Z, Dargusch MS (2010) Viewpoint - understanding Mg corrosion in the body for biodegradable medical implants. Scripta Mater 154:92–100. https://doi.org/10.1016/j.scriptamat.2018.05.021
7. Li Y, Liu G, Zhai Z, Liu L, Li H, Yang K, Tan L, Wan P, Liu X, Ouyang Z, Yu Z, Tang T, Zhu Z, Qu X, Dai K (2014) Antibacterial properties of magnesium in vitro and in an in vivo model of implant-associated methicillin-resistant staphylococcus aureus infection. Antimicrob Agents Chemother 58:7586–7591. https://doi.org/10.1128/AAC.03936-14
8. Rahim MI, Rohde M, Rais B, Seitz JM, Mueller PP (2016) Susceptibility of metallic magnesium implants to bacterial biofilm infections. J Biomed Mater Res Part A 104:1489–1499. https://doi.org/10.1002/jbm.a.35680
9. Vahidgolpayegani A, Wen C, Hodgson P, Li Y (2017) Metallic foam bone 25–82. https://doi.org/10.1016/B978-0-08-101289-5.00002-0
10. Wen CE, Mabuchi M, Yamada Y, Shimojima K, Chino Y, Asahina T (2001) Processing of biocompatible porous Ti and Mg (45):1147–1153
11. Ling G, Fu HÆ, Han S, Dong ÆW (2009) Processing and mechanical properties of magnesium foams 251–256. https://doi.org/10.1007/s10934-008-9194-y
12. Vahid A, Hodgson P, Li Y (2017) New porous Mg composites for bone implants. J Alloy Compd 724:176–186. https://doi.org/10.1016/j.jallcom.2017.07.004
13. Yang D, Guo S, Chen J, Lu J, Wang L, Jiang J, Ma A (2018) Fabrication of Cu-Mg alloy foam with close pore structure by gas release reaction powder metallurgical approach. J Alloy Compd 766:851–858. https://doi.org/10.1016/j.jallcom.2018.06.365
14. Kopp A, Derra T, Müther M, Jauer L, Schleifenbaum JH, Voshage M, Jung O, Smeets R, Kröger N (2019) Influence of design and post processing parameters on the degradation behavior and mechanical properties of additively manufactured magnesium scaffolds. Acta Biomater 98:23–35. https://doi.org/10.1016/j.actbio.2019.04.012
15. Li, Y., Zhou, J., Pavanram, P., Leeflang, M.A., Pouran, B., Weinans, H., Jahr, H., Zadpoor, A.A.: Additively manufactured biodegradable porous magnesium (67), 378–392 (2017) https://doi.org/10.1016/j.actbio.2017.12.008.
16. Putra NE, Mirzaali MJ, Apachitei I, Zhou J, Zadpoor AA (2020) Multi-material additive manufacturing technologies for Ti-, Mg-, and Fe-based biomaterials for bone substitution. Acta Biomater 109:1–20. https://doi.org/10.1016/j.actbio.2020.03.037
17. Makwana D, Bhingole P, Chaudhari P (2021) Statistical analysis for cost effective process parameters and localized strain hardening behavior of Al-7050 foam. Mater Lett 298:129998. https://doi.org/10.1016/j.matlet.2021.129998

Behaviour Analysis of Functionally Graded Sandwich Plate for Free Vibration by Finite Element Method Solution

Charmi Vidja and Supen Kumar Sah

Abstract This study shows the free vibration response of functionally graded metal-ceramic sandwich plates for the validation of the frequency parameter according to classical plate theory (CPT) using ANSYS software by considering the finite element solution approach. The faces of the functionally graded sandwich plate have layers, which are considered to be isotropic. Volume fraction, modulus of elasticity, density, and Poisson's ratio of the different faces of the functionally graded material (FGM) plate are presumed to differ with power law distribution. For this study, the Functionally graded plate is symmetric from the middle plane. The core layer or the middle layer is ceramic and has an isotropic and homogenous nature. The model chosen for the analysis is 1:1:1 having same thickness ratio for top, core, and bottom, and frequency variation with different volume index is obtained for the clamped boundary condition considering classical plate theory. Impacts of change in aspect ratio, volume fraction index over frequency have been studied. Variation of frequency for different mode shapes is studied. Also, the impact of nodes and elements on frequency is investigated.

Keywords Functionally graded materials · Free vibration · Power law · Classical plate theory · FGFM sandwich plate

1 Introduction

The FGM is considered by a gradual variation in composition and structure, resulting in properties that alter smoothly from one end to the other, either from top to bottom or bottom to top. FGM has a smooth transition from one extreme to another in the quality and composition of material. A typical composite material is made up

C. Vidja · S. K. Sah (✉)
School of Mechanical Engineering, Vellore Institute of Technology Bhopal, Bhopal, Madhya Pradesh, India
e-mail: supen146@iitkgp.ac.in

Department of Mechanical Engineering, SRM University, Amaravati, Andhra Pradesh, India

P. Bhingole et al. (eds.), *Advances in Materials Engineering*, Lecture Notes in Mechanical Engineering, https://doi.org/10.1007/978-981-97-7114-1_23

of two or more than two components that have different material properties at the layer interface while FGM is made in such a way that can have smooth and gradual transitions of material properties, which avoids the chances of interlaminar stress concentration. It shows a gradual shift from metal at one end to ceramic at the other with a change in thermal and elastic properties. Applications of FGM include the aerospace industry, automobiles, biomedicine, heat-resistant systems, cutting tools, etc.

Abrate, Serge [1] demonstrated the proportionality of natural frequencies of FGM plate to homogeneous isotropic plates by utilizing classical plate theory. Zenkour [2] for the response of FG sandwich plate which is simply supported used sinusoidal shear deformation theory. He examined that there are deflection and stresses in several modes of FGM sandwich plates for classical plate theory, sinusoidal plate theory, first-order and third-order theory. Additionally, Zenkour [3] investigated the impact of core thickness on critical buckling stress in proportion to the complete thickness of the plate, Additionally, the possessions of the aspect ratio, side-to-thickness ratio, and volume fraction index are considered. Sah and Ghosh [4] conducted a study on the free vibration and buckling response of FGM plates by means of the inverse trigonometric shear deformation theory (ITSDT). Their intention was to derive equilibrium and stability equations for several boundary conditions. By employing finite element-based numerical techniques based on ITSDT, the study addresses the frequency and critical buckling load under uniaxial and biaxial compressive stresses. Sah and Ghosh [5] explored the influence of porosity distribution on the free vibration and buckling behaviour of FGM sandwich plates. These plates were supported and functionally graded in multiple directions. Nguyen et al. [6] employed a novel inverse trigonometric shear deformation theory to evaluate the free vibration, static, and buckling characteristics of FG sandwich plates. The study considered various plate models subjected to different boundary conditions and investigated their behaviour in relation to aspect ratio variations. Using first-order shear deformation theory, Lanhe, Wu [7] obtained the equilibrium and stability comparison for thermal buckling of thick rectangular FG sandwich plates underneath simply supported conditions. Sah and Ghosh [8, 9] considered power law based on perfect and porous FG plate to investigate buckling analysis under thermal loads. Reddy [10] researched an isotropic rectangular FGM plate consisting of two materials distributed through the thickness, and varying elastic modulus following power law and used third-order shear deformation concept along with Navier's solution of rectangular plates for a response of FGM plates. Gasik et al. [11] designed and simulated symmetric FGM plates and disks using ABAQUS. Utilizing ANSYS, Helal and Shi [12] investigated how the material gradient of rectangular FGM plates varied. The modified sigmoid function is the ideal material gradient for FG rectangular plates, according to modelling and simulation conducted using the finite element method. Tabatabaei et al. [13] designed a single-layer FGM plate consisting of ceramic (Al_2O_3) and metal (stainless steel) in ABAQUS Software Coupled with FORTRAN to study modal study by means of a finite element approach for simply supported (SSSS) and clamped (CCCC) boundary condition for various volume fraction. Ramu and Mohanty [14] developed a MATLAB code to analyze a 3D model of an FGM plate. The code calculates the frequencies & mode shapes of the plate by the FEM under diverse boundary conditions. Additionally, research is done on how power law and volume fraction indices affect natural frequency and mode shape variations. Tabatabaei et al. [15] studied the decrement in the frequency of FG plate with an increment in figures of n (volume fraction index), which results in a decrease in the ceramic portion and an increase in the metallic portion which shows a decrease of stiffness of the plate. Al_2O_3 and stainless steel 304 are combined to create the functionally graded material (FGM) plate, which was studied using ABAQUS. The results obtained for natural frequency and mode shapes are under fully clamped conditions. Yaghoobi and Yaghoobi [16] examined the effects of heat, mechanical action, and thermo-mechanical action on sandwich plates that were symmetrical. These plates were elastically supported and had face sheets made of material that was functionally graded. Shear deformation theory, first-order, was applied to the analysis. Free vibration investigation of thermally loaded FG plates was explored by Burlayenko et al. [17] utilizing ABAQUS in conjunction with UMAT and UMATHT.

The primary goal of the current work is to conduct a free vibration response of rectangular FGM plates with finite element analysis (FEA). FGM model is formed via FUSION 360, and the free vibration analysis has been carried out using FEM with ANSYS. In the study material characteristics of FG sandwich plate are classified along thickness direction following power law distribution method. The variable mode shapes and resonance frequencies for distinct volume fraction indices are observed using modal analysis. This study also describes how the thickness ratio affects the deviation of frequency characteristics.

Previous studies have utilized various theories to explore the free vibration response of FGM plates. In this study, the examination of sandwich plates is conducted using different software. Specifically, the free vibration investigation of FGM plate is carried out by means of FE solution with ANSYS utilizing classical plate theory. To analyze the FGM plate, the initial step involves determining values of volume fraction of ceramic and metal (V_c and V_m) for different volume fraction indices (n), which are calculated numerically once vertical coordinates (z) are known for each layer. Subsequently, the geometric modelling of the FGM plate is performed in Fusion 360, with the values of a/b and a/h already decided. Finally, the analysis is executed in ANSYS after assigning material properties and applying mesh and boundary conditions to the sandwich plate.

2 Modelling of FGM

This study describes the structure of a plate which is rectangular in shape consisting of FGM. The plate consists of metal in the upper and bottom faces and ceramic in the core layer, all with unvarying thickness. The material characteristics change gradually through the thickness of the plate conferring to a power law distribution approach.

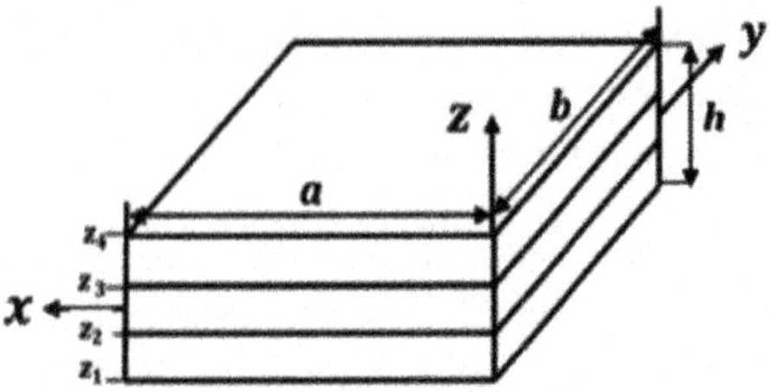

Fig. 1 Functionally graded sandwich plate

In the above plate shown in Fig. 1, a, b, and c represent length, breadth, and height or thickness respectively. The plate utilized in this learning comprises three layers. The uppermost and bottommost faces are composed of metal, while the ceramic makes up the core which is the middle layer. The FG plate used here is represented using the x, y, and z coordinate system. Also, the midplane is placed at $z = 0$ and the top intersections are z_3 and z_4 whereas bottom intersections are z_2 and z_1. 1:1:1 means that the FGM sandwich plate has three equally divided layers. It can be of different ratios like 1:2:1 or 2:2:1 but the present study deals with three equally divided layers only.

2.1 Power Law Formulation

In this study, FGM sandwich plate is composed of metal and ceramic. The summation of volume fraction of metal and ceramic is unity [4]:

$$V_c + V_m = 1 \tag{1}$$

In the above equation V_m and V_c symbolize the volume fraction of metal and ceramic. The FGM sandwich plate comprises top, bottom, and core layers. The volume fraction of ceramic for individual layers of FG sandwich plate [4]:

$$V^{(1)}(z) = \left(\frac{z - z_1}{z_2 - z_1}\right)^n, \quad z \in [z_1, z_2] \tag{2}$$

$$V^{(2)}(z) = 1, \quad z \in [z_2, z_3] \tag{3}$$

$$V(z) = \left(\frac{z - z_4}{z_3 - z_4}\right)^n, \quad z \in [z_3, z_4] \tag{4}$$

$V^{(1)}(z)$, $V(z)$, and $V^{(3)}(z)$ represent the volume fraction of bottom, core, and top layers.

In this study, power law has been utilized for continuous distribution of material characteristics of the plate across the depth of plate using volume fraction of metal

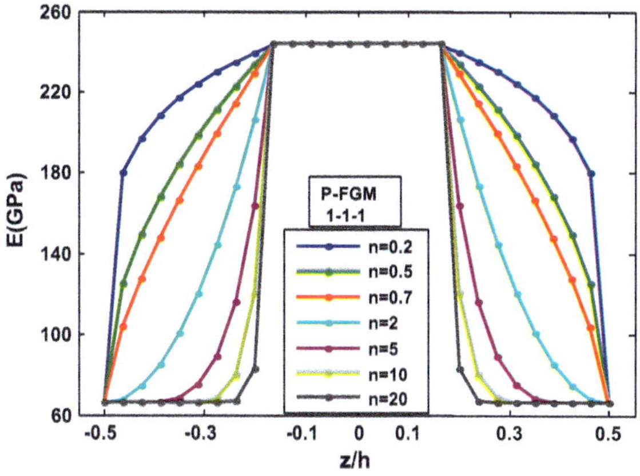

Fig. 2 Variation of modulus of elasticity with *z*/*h* ratio for 1-1-1 FGM sandwich plate using power law distribution

and ceramic. The effective material properties of each layer using volume fraction of FG plate [5]:

$$P(z)^k = P_m + (P_C - P_m)\left(\frac{1}{2} + \frac{z}{h}\right)^n \tag{5}$$

Here k represents layer number, as the bottom layer represents first layer, and the upper layer is represented by the third layer then k will be equal to 1 and 3 respectively for both layers. $P(z)$ represents effective material properties of FGM sandwich plate. P_m and P_C represents metal and ceramic components respectively.

Figure 2 presents variation of modulus of elasticity with z/h ratio for various range of volume fraction index for 1:1:1 FG sandwich plate. It is quite evident from Fig. 2 that there is a smooth variation of modulus of elasticity at the interface of the FGM sandwich plate. It is clearly seen that modulus of elasticity of FGM plate decreases with increasing volume fraction.

2.2 *Kirchoff Plate Theory*

The equation of thin plate can be derived from Kirchoff plate theory by considering the plate as planes perpendicular to the x-axis. According to this theory displacement in x and y directions due to rotation can be written as shown in Eqs. (6) and (7) respectively [2].

$$u = -z * \left(\frac{\partial w}{\partial x}\right) \tag{6}$$

$$v = -z * \left(\frac{\partial w}{\partial y}\right) \tag{7}$$

The in-plane strain and shear strain can be written as [5]:

$$\varepsilon_x = -z\frac{\partial^2 w}{\partial x^2}, \varepsilon_y = -z\frac{\partial^2 w}{\partial y^2} \text{ and } \gamma_{x_y} = -2z * \frac{\partial^2 w}{\partial x \partial y} \tag{8}$$

where ε_x and ε_y represent normal strain along x and y directions. γ denotes shear strain.

The stress equation of plane for isotropic material according to Kirchhoff's theory are [2]:

$$\sigma_x = \frac{E}{1-\theta^2}(\varepsilon_x + \vartheta\varepsilon_y), \quad \sigma_y = \frac{E}{1-\theta^2}(\varepsilon_x + \vartheta\varepsilon_y) \tag{9}$$

where σ_x and σ_y are normal stress along x and y directions. E represents elastic modulus of the material, ϑ denotes the Poison's ratio. Shear stress can be stated as [5]:

$$\tau_{xy} = G\gamma_{xy} \tag{10}$$

where, $G = \frac{E}{2(1+\vartheta)}$.

G represents the shear modulus of the material.

The normal stresses and shear stresses of in-plane act at the edge of plate. Also, transverse shear stress is present but transverse shear deformation is neglected. And the equations for bending moments at the edge of plate are written as [4]:

$$\begin{Bmatrix} M_X = \int_{-h/2}^{h/2} Z * \sigma_x * dz \\ M_y = \int_{-h/2}^{h/2} z * \sigma_y * dz \\ M_z = \int_{-h/2}^{h/2} z * \gamma_{xy} * dz \end{Bmatrix} \tag{11}$$

The differential form of governing equation to carry out free vibration analysis of FGM plate are expressed as [4]:

$$\begin{Bmatrix} \frac{\partial Q_x}{\partial_x} + \frac{\partial Q_y}{\partial_y} + q = 0 \\ \frac{\partial M_x}{\partial_x} + \frac{\partial M_{xy}}{\partial_y} - Q_x = 0 \\ \frac{\partial M_y}{\partial_x} + \frac{\partial M_{xy}}{\partial_y} - Q_y = 0 \end{Bmatrix} \tag{12}$$

Here q is transverse distributed loading whereas transverse shear load is Q_x and Q_y.

Hence partial form of governing equation after substituting and simplification can be written as [4]:

$$D\left[\frac{\partial^4 w}{\partial x^4} + 2 * \frac{\partial^4 w}{\partial x^2 \partial y^2} + \frac{\partial^4 w}{\partial y^4}\right] = q \tag{13}$$

2.3 Finite Element Analysis Formulation

In the present study, it is considered that FGM plate has four nodes and three degrees of freedom. The displacement component perpendicular to the plate surfaces is denoted by W. Rotations in X and Y directions ($x \,\&\, \theta_y$) are articulated in the form of $x = \varepsilon a$ and $y = b\eta$ [2]. Here ε and η are local coordinates.

So, three degrees of freedom can be written as [2]:

$$W, \theta_x = \frac{1}{b}\frac{\partial w}{\partial \eta}, \; \theta_y = \frac{1}{b}\frac{\partial w}{\partial \varepsilon} \tag{14}$$

Therefore, the element possesses twelve D.O.F, and a twelve-degree polynomial is utilized to express the displacement function, as demonstrated below [2]:

$$\begin{aligned} w = {} & \alpha_1 + \alpha_2\varepsilon + \alpha_3\eta + \alpha_4\varepsilon + \alpha_5\varepsilon\eta + \alpha_6\eta^2 + \alpha_7\varepsilon^3 \\ & + \alpha_8\varepsilon^2\eta + \alpha_4\varepsilon\eta^2 + \alpha_{10}\eta^3 + \alpha_{11}\varepsilon^3\eta + \alpha_{12}\varepsilon\eta^3 \end{aligned} \tag{15}$$

Shape function for four nodded elements with three degrees of freedom can be written as [5]:

$$N_j^T(\varepsilon, \eta) = \begin{bmatrix} \frac{1}{8}(1 + \varepsilon_j\varepsilon)(1 + \eta_j\eta)(2 + \varepsilon_j\varepsilon + \eta_j\eta - \varepsilon^2 - \eta^2) \\ \frac{b}{8}(1 + \varepsilon_j\varepsilon)(\eta + \eta_j)(\eta^2 - 1) \\ \frac{a}{8}(\varepsilon_j + \varepsilon)(\eta + \eta\eta_j)(\varepsilon^2 - 1) \end{bmatrix} \tag{16}$$

Element stiffness matrix is written in the form of [2]:

$$[k]_e = \int \frac{h^3}{12}[B]^T[D][B]dA \tag{17}$$

where $[k]_e$ is the element stiffness matrix [5].

$$[B] = \begin{bmatrix} \frac{\partial^2}{\partial x^2} \\ \frac{\partial^2}{\partial y^2} \\ \frac{\partial^2}{\partial x \partial y} \end{bmatrix}[N] = \begin{bmatrix} \frac{1}{a^2}\frac{\partial^2}{\partial \varepsilon^2} \\ \frac{1}{a^2}\frac{\partial^2}{\partial \eta^2} \\ \frac{2}{ab}\frac{\partial^2}{\partial \varepsilon \partial \eta} \end{bmatrix}[N(\varepsilon, \eta)] \tag{18}$$

$[D]$ can be obtained from the stress strain relationship.

Element mass matrix can be stated as [5]:

$$[m]_e = \rho h \int [N]^T[N]dA \tag{19}$$

where $[m]_e$ is the element mass matrix.

Element stiffness matrix and element mass matrix are assembled to get global stiffness and global mass matrix.

The equation of motion for free vibration response of FGM plate [5]:

$$[K] - \omega_{mn}^2[M] = 0 \tag{20}$$

Here $[K]$ and $[M]$ are global stiffness matrix and global mass matrix.

3 Methodology

The plate used for analysis has a combination of two materials which are ceramic (Al_2O_3) and metal (aluminum). Then ANSYS was used for modal analysis after designing the plate in FUSION 360. The modal analysis is performed for rectangular plates under clamped conditions for classical plate theory. Further variation of frequency parameter with aspect ratio is investigated. The modelling and analysis work includes building geometry, varying the properties for core and metal layers according to the thickness of the layer, followed by meshing and adding boundary conditions on the model and lastly solving and getting natural frequency along with mode shapes. Modulus of elasticity and density varies with the thickness path of the plate used here for the research. However, Poisson's ratio is considered to remain steady.

4 Analysis for Free Vibration of FG Sandwich Plate

This section describes the modelling process and the creation of a geometric model, followed by modal analysis to obtain natural frequencies of the FGM plate for clamped boundary condition.

4.1 Geometric Model of FG Sandwich Plate

The sandwich plate used for the examination is a plate in which the core has isotropic ceramic (Al_2O_3) and the faces vary gradually from metal to ceramic The plate is 0.1 m long, 0.1 m wide, and 0.01 m thick. Apart from that, the model chosen here has thickness divided in the ratio of 1:1:1. The top and bottom faces are graded from metal to ceramic and symmetric with respect to the core. So Top, Bottom, and core all are 0.0033 m thick. As the transition from metal to ceramic or ceramic to metal is gradient and smooth, top, and bottom faces are further divided into two equal halves making 0.001667 m each to have different Young's modulus and density. By using this dimension, the plate is sketched, and the geometric model is created in FUSION 360 (Table 1).

Table 1 Length, breadth, and thickness of each layer of FGM sandwich plate

Layers	Length (m)	Breadth (m)	Thickness (m)
Top	0.1	0.1	0.00333
Middle	0.1	0.1	0.00333
Bottom	0.1	0.1	0.00333

Fig. 3 Functionally graded sandwich plate geometric model created in FUSION 360

And for variation in thickness ratio, the dimension of the FGM sandwich modal is changed according to the required *a*/*b* value (Fig. 3).

4.2 *Modal Analysis of FGM Sandwich Plate*

First, add each face's mechanical properties to the engineering data module for the modal analysis. The calculated Young's Modulus and density for each face using power law are added and the Poisson ratio is kept steady, i.e., 0.3 throughout the plate. Hence there will be five material files created with different mechanical properties. Then the geometric model is imported into the geometry module (Table 2 and Fig. 4).

The element size is kept at 5 mm and the element order linear. Also, the number of nodes is 4410 and the number of elements 2000. After importing geometry modal analysis can be carried out in the Model module. Here the added properties of each face can be assigned to the face to which it refers. After assigning properties to each layer meshing is done followed by adding the boundary condition to the sandwich

Table 2 Properties of material used in the FGM sandwich plate

Materials	Properties		
	E, Modulus of elasticity (GPa)	Poisson's ratio (μ)	ρ, Density (Kg/m^3)
Aluminum (metal)	70	0.3	2702
Ceramic (Al_2O_3)	380	0.3	3800

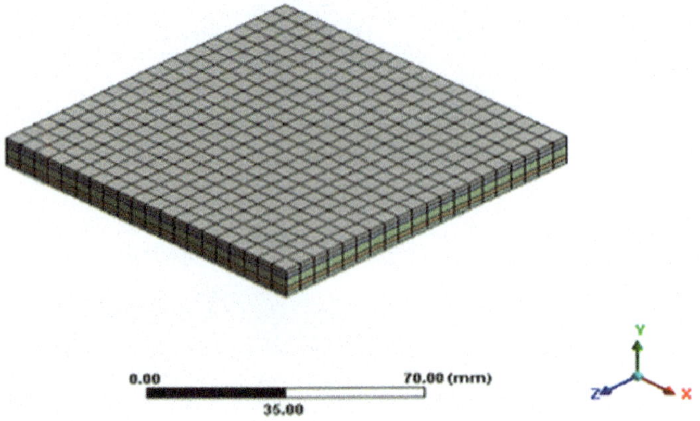

Fig. 4 Meshed model of FG sandwich plate

plate. The type of analysis chosen has clamped boundary conditions to validate classical plate theory results. So, for that purpose displacement is assigned to a particular direction making it free and rest both directions as 0 m. Similarly for other directions too. Hence there will be two displacements as given to both length and breadth. The problem is then solved to determine the resonant frequencies and mode shapes that characterize the vibration of the FG sandwich plate (Fig. 5).

The following formula determines the frequency parameter:

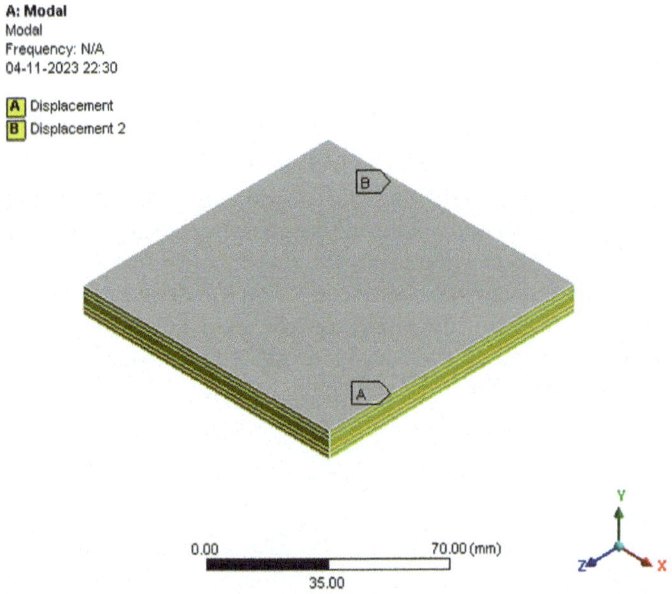

Fig. 5 Applying boundary condition to FGM sandwich plate

$$= \frac{\omega a^2}{h}\sqrt{\frac{P_0}{E_0}} \tag{21}$$

where $E_0 = 1$ GPa, $P_0 = 1$ kg/m^3 and ω is the value of frequency obtained in modal analysis.

5 Results and Discussion

The outcomes of the modal analysis of FG sandwich plate, which consists of Al metal and ceramic, are presented, and discussed using the FEM. The results are verified with Zenkour [3]. To obtain the frequency, modal analysis was performed with the parameters and boundary conditions matching the reference result that they used. A plate of 0.1 m long, 0.1 m wide, and with a height of 0.01 m was used for the study and the results obtained for the frequency are for $a/b = 1$ and $a/h = 10$ for various n values like 0, 0.5, 1, 5 and 10 are shown below. The results are verified with non-dimensional frequency values for classical plate theory mentioned in Zenkour [3].

It can be seen from Table 3 that frequency parameter decreases with an increment in values of volume fraction index (n). The value of n ranges from zero to 10. The frequency results obtained in the study are compared with the existing results. Additionally, the study examines how frequency varies as values of a/b increase for a range of n values.

From the Table 4, we can see that with an increase in a/b value frequency parameter increases. However, frequency parameter reduces with increasing volume fraction index. In this study, value of a/b ranges from 1 to 5 for various volume fraction index values such as 0, 0.5, 1.5, and 10.

Also, the mode shapes for $n = 5$, $b = 1$, and $a/h = 10$ for $a = 0.1$ m, $b = 0.1$ m, and h 0.01 m for FGM Sandwich Plate are shown in Fig. 6.

Table 5 mentions the frequency parameter for distinct volume fraction indexes with varying mode shapes. The table shows that when the n values increase, the frequency value drops. The results include frequency for six mode shapes for various values of n from zero to ten. For particular mode, shape frequency decreases with an increase in the value of n.

Table 3 Result for frequency parameter (rad/s) with varying volume fraction index (n), $\frac{a}{h} = 10$, and $\frac{a}{b} = 1$ for 1:1:1 rectangular FG sandwich plate

n	Present study	Zenkour [3]
0	1.8314	1.87359
0.5	1.5749	1.54903
1	1.3999	1.37521
5	1.0491	1.05565
10	0.9796	1.00524

Table 4 Comparison of frequency parameter with varying for *a*/*b* varying values of volume fraction index, 1:1:1 rectangular FGM sandwich plate

a/b	$n = 0$	$n = 0.5$	$n = 1$	$n = 5$	$n = 10$
1	1.8314	1.5749	1.3999	1.0491	0.9796
1.5	2.9255	2.5249	2.24919	1.69339	1.5666
2	4.4077	3.822	3.41392	2.580012	2.4012
2.5	6.0575	5.4346	4.87013	3.67725	3.4382
3	6.1675	5.7904	5.5355	4.88285	4.6913
3.5	6.1675	5.7904	5.5355	4.88285	4.6913
4	6.1675	5.7904	5.5355	4.88285	4.6913
4.5	6.1675	5.7904	5.5355	4.88285	4.6913
5	6.1675	5.7904	5.5355	4.88285	4.6913

When the element size of the FG plate rises, the number of nodes and elements reduces, Table 6 shows values for n equal to 5. Additionally, frequency increases with mode shapes for specific nodes and element sizes. Frequency increases with an increase in element size along with nodes and element size decrease.

When the n is equal to 0.5, Table 7 displays the frequency variation with nodes and element size. It is quite evident that the resonating frequency of FG sandwich plates rises along with a drop in the number of nodes and elements due to an increase in element size.

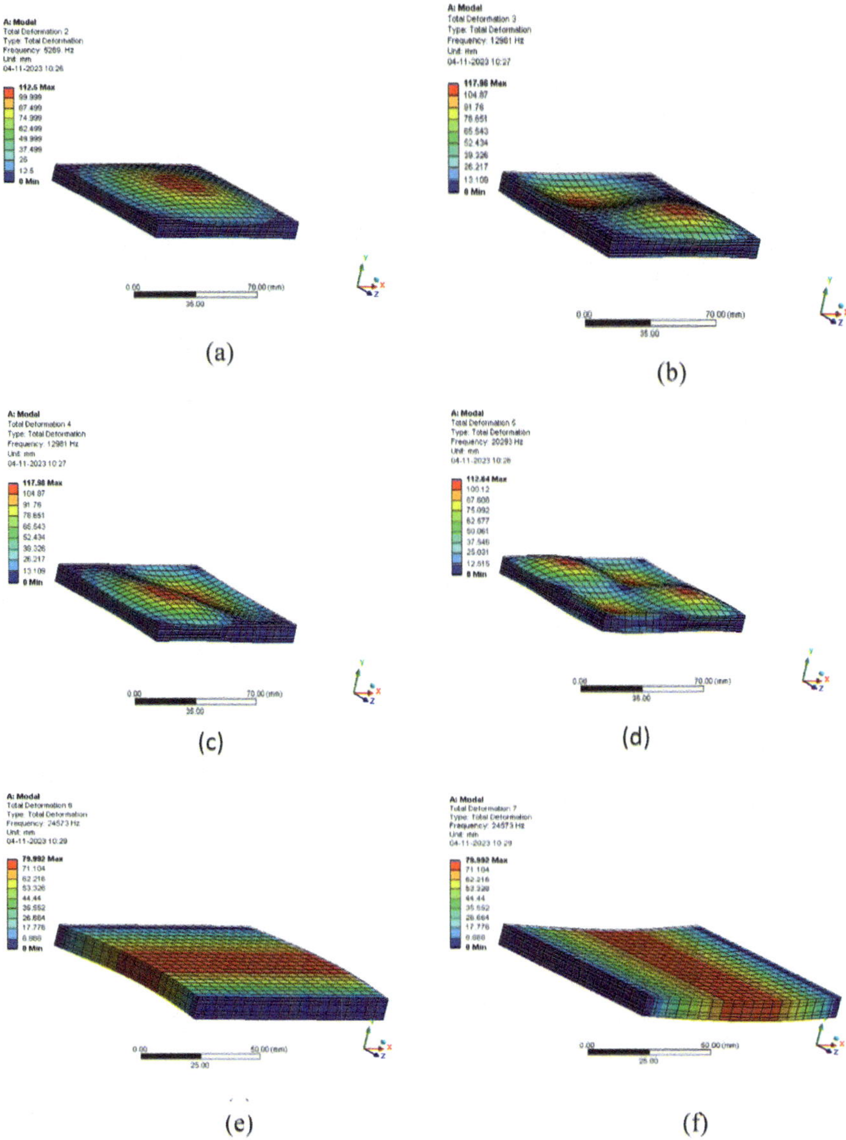

Fig. 6 **a**, **b**, **c**, **d**, **e**, and **f** show mode shapes result for $n = 5$, $a/b = 1$, and $a/h = 10$ of rectangular functionally graded material sandwich plate

Table 5 Comparison of frequency parameter for different mode shapes varying with (n)

Mode shapes	Volume fraction index (n)				
	$n = 0$	$n = 0.5$	$n = 1$	$n = 5$	$n = 10$
1	1.8314	1.5749	1.3999	1.0508	0.9796
2	4.4065	3.8208	3.4129	2.5792	2.4011
3	4.4065	3.8208	3.4129	2.5792	2.4011
4	6.1675	5.7902	5.3038	4.032	3.7501
5	6.1675	5.7902	5.5353	4.8824	4.6758
6	6.7738	5.9134	5.5353	4.8824	4.6758

Table 6 Variation of frequency parameter with no. of elements and number of nodes ($n = 5$)

No. of elements	No. of nodes	Frequency parameter					
		Mode 1	Mode 2	Mode 3	Mode 4	Mode 5	Mode 6
15,000	28,611	1.0486	2.5589	2.5589	3.9992	4.878	4.878
6936	13,475	1.0491	2.5633	2.5633	4.0064	4.879	4.879
3125	6760	1.0498	2.5700	2.5700	4.0169	4.8806	4.8806
2000	4410	1.0508	2.5792	2.5792	4.0320	4.8824	4.8824
1445	3240	1.052	2.5889	2.5889	4.0485	4.8842	4.8842
1125	2560	1.0532	2.5992	2.5992	4.066	4.8862	4.8862
500	1210	1.0605	2.6608	2.6608	4.1749	4.8973	4.8973
245	640	1.0771	2.7938	2.7938	4.4318	4.9184	4.9184
125	360	1.1184	3.1133	3.1133	4.9577	4.9577	5.1306
80	250	1.1184	3.1133	3.1133	4.9577	4.9577	5.1306
80	250	1.1833	3.613	3.613	5.0032	5.0032	6.3255
45	160	1.3899	2.4552	2.4552	5.2989	5.2989	6.3255
45	160	1.3899	2.4552	2.4552	5.2989	5.2989	7.3645
45	160	1.3899	2.4552	2.4552	5.2989	5.2989	7.3645
20	90	2.4552	5.3773	5.3773	7.9959	12.1645	19.2282
20	90	2.4552	5.3773	5.3773	7.9959	12.1645	19.2282

Table 7 Impact of nodes and frequency for varying numbers of nodes and elements for $n = 0.5$

No. of elements	No. of nodes	Frequency parameter					
		Mode 1	Mode 2	Mode 3	Mode 4	Mode 5	Mode 6
15,000	28,611	1.5717	3.7922	3.7922	5.7853	5.7853	5.8691
6936	13,475	1.5724	3.7987	3.7987	5.7865	5.7865	5.8792
3125	6760	1.5734	3.8079	3.8079	5.7882	5.7882	5.8932
2000	4410	1.5749	3.8208	3.8208	5.7902	5.7902	5.9134
1445	3240	1.5765	3.8347	3.8347	5.7926	5.7926	5.9359
1125	2560	1.5782	3.8494	3.8494	5.795	5.795	5.9597
500	1210	1.5885	3.9362	3.9362	5.8081	5.8081	6.104
245	640	1.6101	4.1166	4.1166	5.8329	5.8329	6.4215
125	360	1.6605	4.5299	4.5299	5.8798	5.8798	7.2204
80	250	1.735	5.1449	5.1449	5.9339	5.9339	8.4885
80	250	1.735	5.1449	5.1449	5.9339	5.9339	8.4885
45	160	1.9622	6.0509	6.0509	7.1078	7.1078	8.7384
45	160	1.9622	6.0509	6.0509	7.1078	7.1078	8.7384
45	160	1.9622	6.0509	6.0509	7.1078	7.1078	8.7384
20	90	3.1319	6.3782	6.3782	9.4907	14.4456	20.3210
20	90	3.1319	6.3782	6.3782	9.4907	14.4456	20.3210

6 Conclusion

The finite element approach has been utilized to have a free vibration response of FG sandwich plates of 1:1:1 constituting aluminum (Al_2O_3). The fluctuation in material characteristics is governed by a power-law relationship, and the analysis is conducted to ascertain the natural frequency under clamped boundary conditions. The FGM sandwich plate used has a thickness ratio (a/b) equal to 1 and an aspect ratio of 10. The values obtained for natural frequency and mode shapes are compared to the existing research of Zenkour [3] for classical plate theory. Also, the study investigates how frequency parameters deviate with changes in the volume fraction and thickness variation.

The conclusion of the present study can be outlined as:

- As the value of n or volume fraction index rises, the frequency parameter of FGM sandwich plates varies inversely.
- Frequency varies positively with an increase in thickness ratio.
- After a certain increment in frequency, it remains constant with an increase in thickness ratio.
- The values obtained in results by using the FEM approach show similarity with the available result.
- The frequency of the FG sandwich plate increases for higher mode shapes as the number of elements and nodes decreases.

- The comparison

Present study may be considered as a benchmark on further investigation on free vibration of FG sandwich plate.

References

1. Abrate S (2006) Free vibration, buckling, and static deflections of functionally graded plates. Compos Sci Technol 66(14):2383–2394
2. Zenkour AM (2005) A comprehensive analysis of functionally graded sandwich plates: Part 1—Deflection and stresses. Int J Solids Struct 42(18–19):5224–5242
3. Zenkour AM (2005) A comprehensive analysis of functionally graded sandwich plates: Part 2—Buckling and free vibration. Int J Solids Struct 42(18–19):5243–5258
4. Sah SK, Ghosh A (2021) Free vibration and buckling analysis of FGM plates using inverse trigonometric shear deformation theory. Aircr Eng Aerosp Technol 93(2):298–310
5. Sah SK, Ghosh A (2022) Influence of porosity distribution on free vibration and buckling analysis of multi-directional functionally graded sandwich plates. Compos Struct 279:114795
6. Nguyen V-H et al (2014) A new inverse trigonometric shear deformation theory for isotropic and functionally graded sandwich plates. Compos Part B 66:233–246
7. Lanhe W (2004) Thermal buckling of a simply supported moderately thick rectangular FGM plate. Compos Struct 64(2):211–218
8. Sah SK, Ghosh A (2021) Thermal buckling behavior of power and sigmoid functionally graded material sandwich plates using nonpolynomial shear deformation theories. Eng Comput 39(5):1723–1751
9. Sah SK, Ghosh A (2022) Effect of porosity on the thermal buckling analysis of power and sigmoid law functionally graded material sandwich plates based on sinusoidal shear deformation theory. Int J Struct Stabil Dyn 22(05):2250063
10. Reddy, JN0970 (2000) Analysis of functionally graded plates. Int J Numer Methods Eng 47(1–3):663–684
11. Gasik MM et al (2003) Design and fabrication of symmetric FGM plates. Mater Sci 423:1984
12. Helal WMK, Shi D (2014) Optimum material gradient for functionally graded rectangular plate with the finite element method. Indian J Mater Sci 2014
13. Tabatabaei SJS, Fattahi AM (2022) A finite element method for modal analysis of FGM plates. Mech Based Des Struct Mach 50(4):1111–1122
14. Ramu I, Mohanty SC (2014) Modal analysis of functionally graded material plates using finite element method. Procedia Mater Sci 6:460–467
15. Tabatabaei SJS, Fattahi AM (2016) Finite element method for modal analysis of FGM. Int J Mech Mechatron Eng 10(2):352–356
16. Yaghoobi H, Yaghoobi P (2013) Buckling analysis of sandwich plates with FGM face sheets resting on elastic foundation with various boundary conditions: an analytical approach. Meccanica 48:2019–2035
17. Burlayenko VN, Sadowski T, Dimitrova S (2019) Three-dimensional free vibration analysis of thermally loaded FGM sandwich plates. Materials 12(15):2377

Material Selection and Design

Composite Material for Design and Manufacturing of the Wind Turbine Blades—A Review

Pranoti Shinde, Prashant Sharma, Arun Kumar Sharma, Ashok Soni, and Arpit Shukla

Abstract This article overviews the most current composite materials for designing and producing wind turbine rotor blades. The design of the blade, which displays the cross-section area of the blade and its design requirements, is discussed. The material selection criteria are also essential to consider when selecting a material. The mechanical properties that are important while choosing an appropriate material are discussed. The advantages and disadvantages of plastic and natural fibers demonstrate the material used under circumvent conditions. A brief idea of the technologies and methods used to manufacture the blades using composite material is also provided. The effectiveness of the blades is the only factor influencing the wind turbine's capture efficiency. Therefore, using the best design and durable materials is essential when making wind turbine blades.

Keywords Composite material · Wind turbine blade · Wind energy · Manufacturing technology · Natural fiber · Hybrid composite material

P. Shinde (✉)
COEP Technological University, Pune, Maharastra 411005, India
e-mail: sp18.mech@coeptech.ac.in

P. Sharma
Shri Ram Institute of Science & Technology, Jabalpur, Madhya Pradesh 482001, India
e-mail: pra.sharma30683@gmail.com

A. K. Sharma
Gyan Ganga College of Technology, Jabalpur, Madhya Pradesh 482003, India
e-mail: arunksharma232@gmail.com

A. Soni
Gyan Ganga Institute of Technology and Sciences, Jabalpur, Madhya Pradesh 482003, India
e-mail: ashoksoni@ggits.org

A. Shukla
Takshshila Institute of Engineering and Technology, Jabalpur, Madhya Pradesh 482001, India
e-mail: arpitshukla@gmail.com

P. Bhingole et al. (eds.), *Advances in Materials Engineering*, Lecture Notes in Mechanical Engineering, https://doi.org/10.1007/978-981-97-7114-1_24

1 Introduction

Carbon emission is becoming problematic in every aspect of life, including pollution and ozone depletion. Ozone depletion and a polluted environment led to dangerous and incurable skin diseases. Reducing carbon emissions is vital to the environment and human life. This necessity increases interest in power generation from renewable sources like wind energy. Wind energy is one of the lowest kinds of sustainable energy that may be used indefinitely for no cost at all. As wind energy usage is increasing daily, its size is also growing. The increasing size leads to the use of a massive amount of material. The diameter of wind turbine blades must be expanded due to the larger number of wind turbines.

In the case of wind turbines, the blades are an essential component. The practical design of the turbine's blades determines its efficiency. Aerodynamic performance and the structural characterization of the wind turbine blade are two essential design factors that influence the optimum design process [1]. The wind turbine blade's optimal structural design may be attained by selecting a material that has the highest strength, stiffness, and least amount of weight. As wind turbine blades get bigger, kinetic wind energy causes aero-elastic effects and deflection over the underloaded part of the blade's surface. With the increasing size of wind turbines, advanced materials are available for designing the blade. Figure 1 shows the modern wind turbine designed using advanced composite material.

Advanced composite materials are considerably preferred to minimize the deflection and failure of the blade, which occurs due to cyclic load experienced on the blade

Fig. 1 Modern wind turbine

surface because of the wind speed. The proportion of strength to weight gauges the robust material's aerodynamic and aero-elastic properties.

Emerging composite materials that are lightweight, sustainable, and cost-effective are natural fibers. They exhibit several advantages and are used as promising alternatives to the traditional composite material. Environmental factors play an important role while using nature fiber for the manufacturing of wind turbine blade. The strength-to-weight material is a measure of Aerodynamic and aero-elastic characteristically sturdy material. According to Lolo et al. [2], composite materials account for 95% of the materials utilized in the wind turbine blade industry. The issue of different blade stresses is resolved and structural integrity is provided using fiber-based composite materials.

2 Historical Aspect

A renewable energy source replaces conventional energy resources due to the depletion of fossil fuels and the adverse effects on climate like pollution, greenhouse gas emissions, etc. Windmills have been re-engineered throughout the past to develop more powerful sources of energy production. Researchers also invent new advanced materials, higher technology, and light and aerodynamics to challenge conventional energy sources.

The Panemone windmill, as shown in Fig. 2a, is the first windmill in the fields. The first windmill is manufactured using 6–12 sails covered in reed or cloth matting. The basic design of the windmill is used only to grind grains and draw up water. Windmills are converted into advanced wind turbines to increase the power output, as shown in Fig. 2b. The material already in use for small windmills couldn't be used in the wind turbine because of the easy failure of the wind turbine blade. Because

Fig. 2 **a** European windmill and **b** aerodynamically advanced wind turbine

of the cyclic tension-tension load on the upper surface and cyclic compression-compression load on the blade, the previous material cannot withstand the forces due to high wind speed. To attain the requirement of material stiffness, strength, toughness, and weight, a criterion is described in Eq. (1), which is known as the beam merit index [3, 4], defined as

$$M_b = \sqrt{E}/\rho \tag{1}$$

where ρ is the material's density, and E is the young modulus. The optimum material for creating a contemporary wind turbine is that which meets the beam merit index requirements [5]. Several studies have suggested using composite materials to manufacture turbine components, despite the lack of references to the production of rotating molded blades for small wind turbines. Large wind turbine blades in the context of wind power generation are mostly built of composite materials. Hardwood, steel, aluminum, and composite materials are among the materials utilized in the production of wind turbine blades [6].

2.1 Wind Turbine Blade Design

Wind turbines are manufactured in three different sections. One is the generator part, the second is the towers, and the third is the aerodynamic rotor section. The generator and tower units are generally manufactured using metallic material. The aerodynamic rotor consists of the wind turbine blade. The capture efficiency of the wind turbine is dependent on its rotor blade. In addition, the rotor blade is separated into two portions: the bottom and upper blade portions. Every component is made independently of the other.

The wet-hand lay-up procedure is used in the production of significant blades. Large amounts of blades are produced during this process. A mould of the small and medium blade's upper and lower sections is used for mass manufacture. For the large radius blade. Similar technology is used with the web's adhesive bonding between the upper and lower units of the blade. The other manufacturing techniques applied to design wind turbine blades are Filament winding technology, Vacuum infusion and prepreg technology, and Resin infusion technology. The research on improvement in quality, finishing, and reducing cost of manufacturing composite blades, as well as enhanced technology like automation of lay-up and fiber placement, is proliferating.

The NREL S818 airfoil design used by Otero and Ponta [7] which is generated by Somer [8], is shown in Fig. 3, which displays an example of a large-size blade design with webs allocated between the upper and lower surfaces made of balsa-core skin. Spar caps on both sides of the blade support the forward and aft shear web. The large blade size requires a tremendous amount of material, increasing the cost of manufacturing the rotor blade. The area where webs are adhesively bonded between surfaces experiences maximum load due to kinetic wind energy. The cost-efficient manufacturing of the blade, these web supports are manufactured using expensive

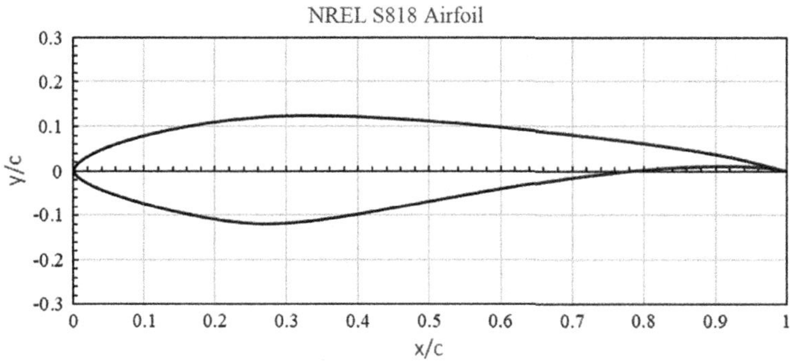

Fig. 3 Commercial blade design

material with higher strength, stiffness, and minimum mass, like carbon fiber. Furthermore, the blade's surface is manufactured using glass fiber reinforcement, which is inexpensive compared to carbon fiber reinforcement.

2.2 Modal Analysis Technique for Composite Material Used in Wind Turbine Blade

According to studies conducted by Cui et al. [9], modal stacking may be employed to achieve any movement of free vibration in a generic multi-degree-of-freedom structural system.

Developing a modal model for numerical calculations is known as finite element modal analysis. The following matrix calculation may represent the system's non-damping-free vibration as shown in the Eq. (2) [9]:

$$[M]\{\ddot{u}\} + [K]\{u\} = \{0\} \quad (2)$$

[M] and [K] are real symmetric matrices in the linear structural system Eq. (2). The solution of simple harmonic motion to the equation is as follows:

$$\{u(x, y, z, t)\} = \{\phi(x, y, z)\}e^{iwn} \quad (3)$$

where w_n denotes the angular frequency of the simple harmonic motion and $\phi(x, y, z)$ is the displacement vector of the amplitude, which defines the spatial distribution of the displacement vector (u). Equations (3) and (2) [9] can be used to obtain the following formula related to u.

$$[k - \omega_n^2 M]\{\phi\} = \{0\} \quad (4)$$

Since Eq. (4) [9] is a typical real eigenvalue issue, if the value of its coefficient determinant is zero, then $\{\phi\}$ has a non-zero solution.

Three primary kinds of fiber-reinforced material are analyzed by examining the fundamental structure of composite materials.

1. Single-layer composite material.
2. The laminated composite material.
3. Short-fiber composite material.

The research and creation of composite materials are becoming more dependent on vibration parameters studies due to the widespread application of multi-layer composite materials in high-tech industries such as aerospace engineering and wind energy.

2.3 Multi-objective Optimization Trade-Off Parameters of Composite Material

Each material used for the analysis has been considered a trade-off of the three primary material and process parameters, i.e., material strength, stiffness, and cost. The author Griffin and Ashwill [10], lists the possible combination of trade-off parameters. There are 72 material/process parameters listed in Table 1, of which 21 combinations could be modelled for study.

Furthermore, blades must also be built with a suitable clearance from the structure to have the necessary stability and fatigue strength. If all the additional design variables are the same, greater stain levels inside the blade's alloy will lead to increased

Table 1 Material or process combination matrix considered in trade-off analysis

Parameters	Combination/values considered
Fiber material	• Fiberglass 0° and ±45° fibers • Carbon 0° fibers with fiberglass ±45° • Carbon 0° and +45° fibers
Fabrics used in spar cap construction	• Woven unidirectional fabric combined with stitched biaxial fabric (dry) • Stitched unidirectional and biaxial fabrics (dry) • Prepreg unidirectional and biaxial fabrics
Laminate fiber volume fraction	• 0.4 • 0.5
Percentage of unidirectional laminate in spar cap (by volume)	• 70% • 80%
Cure temperature	• Unheated molds with room temp. cure • Heated molds with elevated Temperature post cure

displacement at certain twisting strains. This may reduce a portion of the heaviness and money saved, particularly when using materials with more tip displacement.

2.4 Methodology Assessment

The total efficacy and cost-effectiveness of wind energy generation are largely dependent on the lifespan and durability of wind turbine blades. However, issues with wind turbine blade longevity and dependability still exist, even with advances in materials science and production processes. Scholars have recognized the urgent need to improve wind turbine blade performance and durability, especially considering the materials that go into making them. The goal of extending the life of wind turbine blades has brought composite polymer materials to light because of their high strength-to-weight ratio and resistance to erosion. The wind energy sector is seeing a rise in the use of these materials. However, there are many obstacles in the way of recovering and recycling waste materials from wind turbine blades, especially those composed of composite polymers. Furthermore, using synthetic fibers like glass and carbon raises questions regarding their long-term sustainability and potential effects on the environment.

Research collaborations between academia and industry are still essential to addressing these issues. These initiatives aim to comprehend and improve wind turbine blade reliability, with a particular focus on flow characterization, defects in manufacturing, and the investigation of natural, biodegradable, and ecologically friendly materials. The objective also includes developing more sustainable solutions for the materials used in wind turbine blades and optimizing the expenses related to the reliability of the wind turbine system.

In the end, the renewable energy industry is constantly striving to enhance the functionality, economy, and environmental impact of the materials used in wind turbine blades. In order to improve the sustainability and dependability of wind energy generation, it is imperative that we keep researching and creating new materials and recycling techniques. The current state of research indicates that a significant research gap exists in the areas of optimizing composite polymer recycling processes and developing natural, biodegradable, and environmentally friendly materials for wind turbine blades [11]. Strong approaches are desperately needed to assess these new materials' performance across a range of operational scenarios, in addition to material development and recycling optimization.

3 Manipulation of Advanced Composite Material Throughout the Study

Advanced composite materials, known as ACMs, combine fibers and matrices, giving unusually high stiffness and strength. This material also occupies a large volume with a very low density [12]. The material that inherits these extraordinary mechanical properties is an advanced composite material compared to other materials like reinforced concrete. Some materials in polymer matric composite, ceramic matrix composite, and metal matrix composite are progressive composites.

Nowadays, composite materials are broadly utilized in airplane manufacturing and aerospace industries, as well as in sports equipment, building infrastructure, etc., because they can withstand high temperatures and exhibit stability towards chemicals [12]. Advanced composite materials widely used in manufacturing wind turbine blades are glass and carbon fiber reinforced with the desired matrix. An example of two kinds of glass fibers is shown in Fig. 4. The glass fiber is a borosilicate glass commonly known as E-glass or electric-glass fiber because it has high electrical resistance.

To achieve the best performance of the glass fiber in wind turbines, 75% of glass is designed in glass fiber reinforced epoxy composite. The chemical composition of some glass fibers is shown in Table 2. S-glass is known as silica glass, whereas C-glass is known as corrosive glass. Because it exhibits 15% more stiffness and 30% more strength than E-glass fiber, Windstrand™ glass fiber is widely used in producing wind turbine blades [13].

Carbon fiber is rapidly replacing glass fiber in usage. The benefits of carbon fiber over glass fiber are its higher strength and stiffness, yet it is considered a lighter material. Carbon fiber has some disadvantages: it is insatiable to the misalignment

(a) (b)

Fig. 4 **a** Glass fiber that is consistent in length and **b** glass fiber that is cut into pieces

Table 2 Some standard glass fiber's chemical composition

Compound	E-glass	S-glass	C-glass
SiO_2	55.2	65.0	65.0
Al_2O_3	8.0	25.0	4.0
CaO	18.7		14.0
MgO	4.6	10.0	3.0
Na_2O	0.3	0.3	8.5
K_2O	0.2		
Li_2O	7.3		5.0

of fiber, which can reduce strength in terms of fatigue and compression. The solution to this predicament is combining two fibers according to the strength, stiffness, and weight requirement. All the forms of carbon fiber available in the market selected to fulfill different purposes can be seen in Fig. 5a. The combination of fibers in a composite material is known as a hybrid material. The hybrid material will help reduce weight even more and show good mechanical properties.

Natural fibers like jute, flax, hemp are also accessible in the market and propose upright mechanical characteristics, are affordable, and are environmentally friendly [14]. Timber and bamboo composite fibers have high strength, reliability, and durability in manufacturing wind turbine blades. The mechanical properties of natural fibers make them appropriate for manufacturing small wind turbine blades. Natural fibers have a higher basic water absorption rate and worse heat resilience than polymer compound fibers. Figure 5 shows the key factors affecting the mechanical properties of natural fiber composite [15]. Table 3 summarizes the various works done for wind

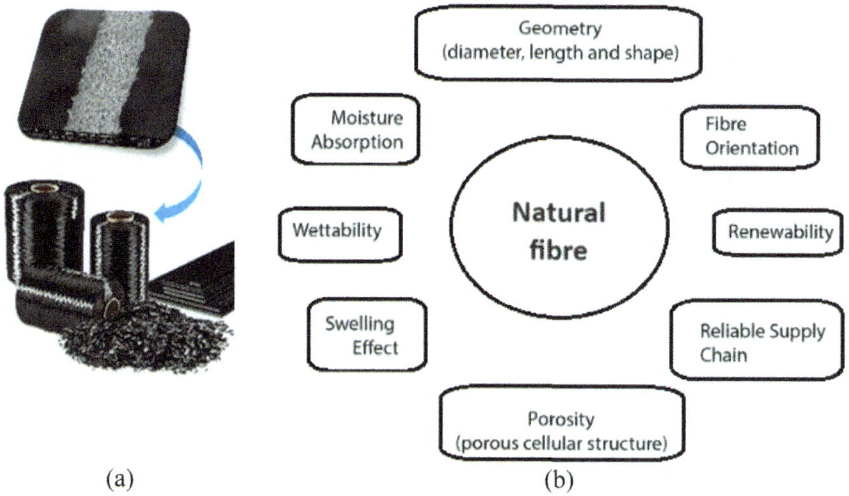

Fig. 5 **a** Various forms of carbon fiber and **b** natural fiber composite

turbine blade analysis for different composite materials by various researchers using various methods.

Thermosets and thermoplastics are incorporated with composite fibers as a matrix material in composite material. Epoxy is classified as a thermoset and is usually mixed with fibers to create composite substances. Polyester is another material utilized as a matrix in compounds besides epoxy. Thermosets are easier to infuse and cure more quickly than thermoplastics. Polypropylene, polyvinyl chloride, polystyrene, and polyethylene (LDPE and HDPE) are popular thermoplastics [22].

4 Critical Review and Research Scope in Selection of Material in Blade Manufacturing

Composites made of glass fiber, carbon fiber, and thermoplastic are the main elements utilized to create huge blades in the production of wind energy. According to García-Arrieta et al. [23], there is a tendency towards the utilization of composite materials because of their advantageous attributes, including low weight, resistance to corrosion, strength, endurance, and stability in dimensions. Researchers have been concentrating on improving the design and materials used in wind turbine blades as energy efficiency and sustainability become more and more important. Wind turbine technology has advanced considerably over the last ten years, with an increasing tendency toward the use of composites for the production of blades.

To advance the sustainability and environmental impact of blades, additional research into alternative materials and manufacturing processes is needed, as the critical evaluation indicates apparent. Researchers and industry professionals are asked to look at the integration of eco-friendly and biodegradable materials into the creation of blades to reduce the dependency on petroleum-based resources, given the growing emphasis on renewable energy and sustainable practices. Moreover, the literature highlights the need for efficient repair techniques to resolve problems and extend the service life of wind turbine blades and suggests a more thorough examination of the production procedures and issues related to composite materials. Research should keep focusing on creating dependable repair methods and investigating naturally occurring, biodegradable materials that may provide improved resistance to damage and capability for repair.

With an emphasis on composite materials specifically, the literature review as a whole delivers an excellent overview of the materials utilized to manufacture wind turbine blades. Nonetheless, to guarantee the precision and dependability of the data offered, it is crucial to assess the sources and their procedures attentively.

The review of the literature highlights the benefits of using composite materials, including their high strength, low cost, and particular stiffness, in the production of blades. Understanding the long-term ecological effect and recyclability of these composite materials is still lacking in studies. Future studies could fill the Research gap by analyzing the environmental effects of the life cycle assessments of composite

Table 3 Summarizes the various works done for wind turbine blades

S. no	Material	Blade	Technique	Parameters	Turbine	Objective	Author
1	Glass fiber/carbon fiber	Blade length 75 m	Ex-situ 3D XCT technology with in situ transilluminated white light imaging (TWLI)	Width, thickness, and flexural rigidity	10 MW DTU turbine	Fatigue damage and stiffness degradation	Mikkelsen [16]
2	Steel, gel coat, CSM 300, unidirectional laminate	5 m composite wind turbine blade	White light imaging and ex-situ X-ray computed tomography were employed in the investigation	Density, thickness, material mechanical properties	RUZGEM composite wind turbine	To determine static and fatigue strength	C. Muyan and D. Coker (Muyan & Coker, 2020)
3	Glass fiber reinforced plastic	Large scale blade	A FEA obtains the inherent frequency range of the blade, including its modal characteristics	Tensile modulus of elasticity, shear modulus of elasticity, Poisson's ratio, density	Large-scale wind turbine	Modal characteristics and the blade's inherent spectrum of frequencies	Cui et al. [9]
4	E-glass/epoxy composite blade	30-m blades length with NACA 64–618 at the tip with DU & EU airfoil at the root	FEA and composite laminate concept	Layup thickness, layup numbers, and layup angle	0.0012 GW turbine with a blade	Maximise structural stiffness and strength	Lanting [17]
5	E-glass/epoxy	23.3 m in overall span-wise length/ NACA 63(6)-0xx airfoil	Finite element method for structural analysis	Aerodynamic loads, ice-related loads, hygro-thermal loads, and mechanical loads	Horizontal axis wind power setup with 750 kW category	Blade's endurance under fatigue	Kong et al. [18]

(continued)

Table 3 (continued)

S. no	Material	Blade	Technique	Parameters	Turbine	Objective	Author
6	Carbon fiber cloth/ epoxy (C-E) composite, glass fiber/ vinyl ester (G-V), and foam core (FC)	Blades consist of six airfoils and have a length of 61.5 m	Maximum Tsai-Wu failure criterion using the finite element method	The ratio of ±Xn, the stacking angle, with the sequence of stacks	5 MW setup of the NREL	The highest tension applied to the blade	Zhang et al. [19]
7	Glass–epoxy	Blade length 4.86 mm/	Finite element analysis using ANSYS	Stacking sequence and ply thickness	1.2 MW vertical axis wind turbine	Maximize strength-to-weight ratio	Hameed et al. [20]
8	Glass fiber-reinforced plastic (GFRP) and carbon fiber-reinforced plastic (CFRP)	Blade length 40 m/ root-circles and rear part- NREL S-809	Finite element methods together alongside an evolving approach	The thickness, form of material, along with orientation angle of the laminate layer	2 MW rated power horizontal axis wind turbine	An analysis of overall weight as well as blade value	Hu et al. [21]

materials used in wind turbine blades, from production to disposal. Additionally, studies might focus on creating environmentally friendly wind turbine blade end-of-life possibilities, looking into possible recycling or repurposing strategies that complement circular economy ideas and sustainable manufacturing techniques [24]. Furthermore, the wind turbine needs lightweight turbine blades, and fiber-reinforced composites are a logical alternative for this. 'Disposing' of huge composite structures will become a worry as end-of-life approaches, even though composites are long-lasting. Waste from the production process is an immediate concern. The Composites UK paper outlines the many recycling alternatives for composite materials and illustrates how each recycling method affects the environment.

5 Discussion and Conclusions

In the present scenario of depletion of fossil fuels, dependency on renewable energy sources is increasing expeditiously. Wind energy is an unflattering source of energy that is castoff anywhere nowadays. Advancement of wind turbines is accelerating the use of the technology in every wind condition. Wind turbine usage in any weather condition highly depends on its rotor blade's design. Design and manufacturing of the rotor blade play an energetic part in this framework. Selecting suitable materials according to space requirements and environmental conditions is essential. The key points for the research direction and development of materials used for the manufacturing of the wind turbine blade:

1. Environmental Impact and Recyclability: The growing usage of composite materials in the manufacturing of blades has highlighted the necessity for in-depth studies on the environmental effects and recyclability of these materials over the long run. The research now in publication has called attention to the urgent need to emphasize the benefits of particular rigidity, low cost, and high strength associated with composite materials. Investigating possible wind turbine blade recycling or repurposing techniques can make a big difference in supporting a circular economy and sustainable manufacturing practices. The development of novel methods for the effective recovery and recycling of waste materials from abandoned wind turbine blades could be the main goal of research activities.
2. Methods used for development and repairing: The creation of dependable and affordable repair techniques for composite materials used in wind turbine blades is another essential field for further study. The production, manufacture, transportation, and usage of composite materials will inevitably result in faults or damages. However, cutting-edge repair techniques can extend the life of wind turbine blades while cutting waste.
3. Innovations of Testing Techniques: It is recommended that future research endeavors incorporate sophisticated testing methodologies, such as non-destructive testing techniques, to precisely evaluate the structural integrity and performance of composite materials. The overall efficiency of wind turbines can

be improved by using these strategies to gather useful data for improving the safety and dependability of blades.

The complex design of the blade is achieved by using the desired composite material. The material should withstand the cyclic load and environmental impact exhibited on the blade. The mechanical properties of carbon fiber satisfy the significant conditions that can withstand the impact and loads on the rotor blade. The combination of two plastic fibers is employed in most industries. The carbon fiber could be reused using carbon fiber conversion technology. The combination of plastic fiber and natural fiber is recommended for manufacturing a rotor blade as a composite material. The disadvantages of natural fiber discussed in Sect. 3 could be stabilized using reusable carbon fiber. The strength and durability of the hybrid composite can be increased on the nano level due to damage in the material used for the wind turbine rotor blade at the micro level. The lifecycle of the wind turbine could also be enhanced by using nanoengineered hybrid material with reusable capabilities.

References

1. Torregrosa AJ, Gil A, Quintero P, Cremades A (2022) On the effects of orthotropic materials in flutter protection of wind turbine flexible blades. J Wind Eng Ind Aerodyn 227:105055. https://doi.org/10.1016/J.JWEIA.2022.105055
2. Lolo K, Ouedraogo S, Attipou K, Tiem S (2019) Mechanical characterization of ronier wood for its use in the design of windmill blades. Int J Adv Res (Indore) 7:600–609. https://doi.org/10.21474/IJAR01/9872
3. Ashby MF, Shercliff H, Cebon D (2007) Materials: engineering, science, processing and design. 514
4. Ashby M, Shercliff H, Cebon D (2007) Strategic thinking: matching material to design. Mater Eng Sci Process Des 29–46
5. Ziane K, Ilinca A, Karganroudi SS, Dimitrova M (2021) Neural network optimization algorithms to predict wind turbine blade fatigue life under variable hygrothermal conditions. Eng 2:278–295. https://doi.org/10.3390/ENG2030018
6. Rajendran I, Vijayarangan S (2001) Optimal design of a composite leaf spring using genetic algorithms
7. Otero AD, Ponta FL (2010) Structural analysis of wind-turbine blades by a generalized timoshenko beam model. J Solar Energy Eng Trans ASME 132:0110151–0110158. https://doi.org/10.1115/1.4000596/475878
8. The S816, S817, and S818 Airfoils October 1991 – NREL. https://www.yumpu.com/en/document/read/36061951/the-s816-s817-and-s818-airfoils-october-1991-nrel. Last accessed 15 December 2023
9. Cui Y, Shi L, Zhao F (2010) Modal analysis of wind turbine blade made of composite laminated plates. In: Asia-pacific power and energy engineering conference, APPEEC. https://doi.org/10.1109/APPEEC.2010.5448923
10. Griffin DA, Ashwill TD (2002) Blade system design studies volume I: composite technologies for large wind turbine blades. https://doi.org/10.2172/800994
11. Mekonnen BY, Mamo YJ (2020) Tensile and flexural analysis of a hybrid bamboo/jute fiber-reinforced composite with polyester matrix as a sustainable green material for wind turbine blades. Int J Eng 33:314–319. https://doi.org/10.5829/IJE.2020.33.02B.16
12. Pilato LA, Michno MJ (1994). Adv Compos Mater. https://doi.org/10.1007/978-3-662-35356-1

13. Ashwill TD (2009) Materials and innovations for large blade structures: research opportunities in wind energy technology
14. Kalagi GR, Patil RJ, Nayak M (2016) Natural fiber reinforced polymer composite materials for wind turbine blade applications
15. Lau K, Hung P, Zhu MH, Hui D (2018) Properties of natural fibre composites for structural engineering applications. Compos B Eng 136:222–233. https://doi.org/10.1016/J.COMPOSITESB.2017.10.038
16. Mikkelsen LP (2020) The fatigue damage evolution in the load-carrying composite laminates of wind turbine blades. In: Fatigue life prediction of composites and composite structures, pp 569–603. https://doi.org/10.1016/B978-0-08-102575-8.00016-4
17. Lanting Z (2012) Research on structural lay-up optimum design of composite wind turbine blade. Energy Procedia 14:637–642. https://doi.org/10.1016/J.EGYPRO.2011.12.988
18. Kong C, Bang J, Sugiyama Y (2005) Structural investigation of composite wind turbine blade considering various load cases and fatigue life. Energy 30:2101–2114. https://doi.org/10.1016/j.energy.2004.08.016
19. Zhang Z, Qiao Y, Wang S (2022) Damage analysis of wind turbine blade with different stacking parameters. J Phys Conf Ser 2276. https://doi.org/10.1088/1742-6596/2276/1/012001
20. Hameed MS, Afaq SK, Shahid F (2015) Finite element analysis of a composite VAWT blade. Ocean Eng 109:669–676. https://doi.org/10.1016/j.oceaneng.2015.09.032
21. Hu W, Han I, Park S-C, Choi D-H (2012) Multi-objective structural optimization of a HAWT composite blade based on ultimate limit state analysis. J Mech Sci Technol 26:129–135. https://doi.org/10.1007/s12206-011-1018-3
22. Asim M, Jawaid M, Saba N, Ramengmawii, Nasir M, Sultan MTH (2017) Processing of hybrid polymer composites—a review. In: Hybrid polymer composite materials: processing, pp 1–22. https://doi.org/10.1016/B978-0-08-100789-1.00001-0
23. García-Arrieta S, Benito IL, García M, Bonaiti G, Escudero OO, Elizetxea C (2022) Use case 2: thermal recycling of long fibers, pp 323–344. https://doi.org/10.1007/978-3-031-22352-5_16
24. Jiang SRX, Inkermann D, Vietor T, Shu LH (2018) Generating reuse possibilities for retired wind-turbine blades. https://doi.org/10.25071/10315/35406

Design Evaluation of Splinted and Non-splinted Crown of Dental Implant to Improve Stress Distribution at Bone-Implant Interface Using FEA

Aditya Gujare, Bhargav Hindurao, Pankaj Dhatrak, and Sandipan Roy

Abstract There has been an upsurge in applications of dental implant over last few decades and FEM has played an essential role in rapid development of dental implant. In spite of that, dentists are still doubtful regarding the use of splinted or non-splinted dental implants. Few studies state that splinted implants are better than non-splinted dental implants from a biomechanical viewpoint while few studies state that non-splinted dental implants are better for oral hygiene and retreatment point of view. This study presents a biomechanical aspect by analyzing distribution of stress in splinted as well as non-splinted dental implant based on occlusal loadings. The results state that splinted dental implants cause uniform stress distribution in the mandibular region and thus are favorable from biomechanical standpoint.

Keywords Implant system · Splinted · Non-splinted · Stress distribution · Equivalent stress

1 Introduction

There has been an increase in the use of dental implant in partially or fully edentulous jaws over the past few years because of their high rates of success and feasible applicability [1–4]. Potential of these implants for improved function over more favorable stress distribution makes them a viable treatment option [4]. Even though these implants have many advantages, there are a few imperfections in these systems. Among them, the most serious one is implant failure after prosthetic surgery which occurs due to occlusal loading [5]. Furthermore, because of the emergence of less

A. Gujare (✉) · B. Hindurao · P. Dhatrak
School of Mechanical Engineering, Dr. Vishwanath Karad MIT World Peace University, Pune 411038, India
e-mail: adityagujare1@gmail.com

S. Roy
Department of Mechanical Engineering, SRM Institute of Science and Technology, Chennai 603203, India

P. Bhingole et al. (eds.), *Advances in Materials Engineering*, Lecture Notes in Mechanical Engineering, https://doi.org/10.1007/978-981-97-7114-1_25

intensive therapy planning and anatomic limits of the jaw, short implants have become an alternative therapeutic choice throughout time. As a result, the likelihood of implant failure has increased because short implants provide less Osseo integration area in the posterior part of the jaw bone, where occlusal forces are higher and quality of bone is less agreeable. Therefore, splinting of adjoining dental implants is carried out for better load distribution [1].

In dentistry, unilateral or bilateral joining of teeth for better positional and functional stability is known as splinting. The reasoning for splinting teeth is increased stability, transmission of horizontal forces acting on teeth gets reduced, and compensation for increased crown to root ratio [6–8]. However, there is still an ambiguity among clinicians regarding the use of non-splinted or splinted implants. According to the literature, oral hygiene, occlusion, implant length, abutment connection designs, and the fitting framework must be considered while taking the decision of whether to splint the implant or not [3, 9]. From a biomechanical point of view, few studies mentioned that splinting has a positive effect because it reduces the intensity of stress in adjacent bone tissue due to better load distribution [1, 4, 10]. Based on oral hygiene, and clinical results, non-splinted crowns are preferred [2, 9, 11]. Few studies also assert that there is no significant change in load distribution for splinted and non-splinted restorations [12]. The aim of the present study is to analyze the stress distribution in cancellous and cortical bone for both splinted and non-splinted short dental implant using FEA. The implant, abutment, and crown are meshed in Hypermesh and analyzed in Abaqus.

2 Materials and Methods

2.1 3D CAD Geometry Modelling

The dimensions for implant, abutment, and screw were measured using an optical measuring machine (SWIFT DUO UK). A cuboidal mandible in the first molar and premolar region was used as the basis for the solid model. The cuboidal mandible is split into two layers cortical and cancellous bone. The cortical bone thickness is assumed to be 2 mm. The three-dimensional geometry for dental implant is created using Solidworks CAD software. The images of splinted, non-splinted crown, abutments, and implant referred for creating 3D geometry are shown in Fig. 1.

2.2 Three-Dimensional FE Modelling

The abutment, screw, implant, cancellous and cortical bones were done using C3D10 tetrahedron element with element size of 0.1 mm, whereas S3R/S4R linear tri/ quadrilateral element was used to mesh the crown with a size of 0.3 mm as shown

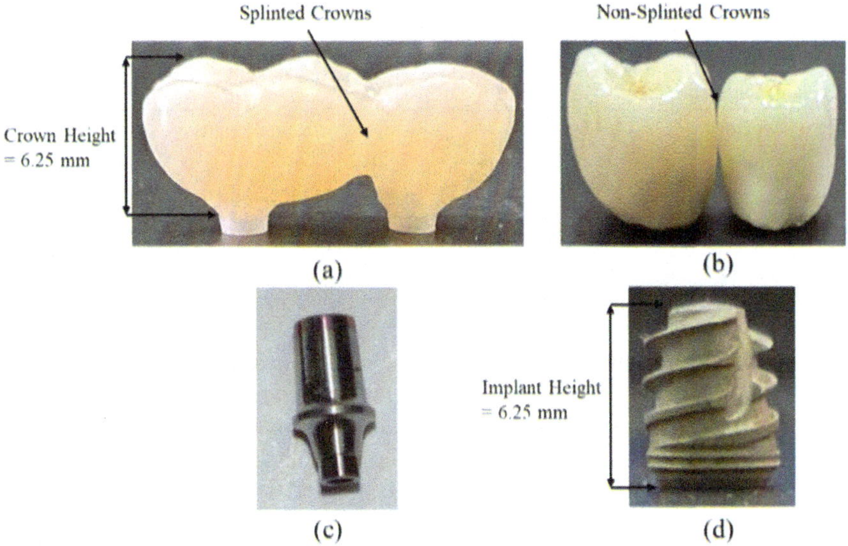

Fig. 1 Images of **a** splinted crowns, **b** non-splinted crowns, **c** abutment, **d** implant

in Fig. 2. To satisfy the convergence criteria, mesh was refined at the bone-implant interface. 3% variation was overserved while meshing the implant prosthetics to achieve mesh convergence [13, 14].

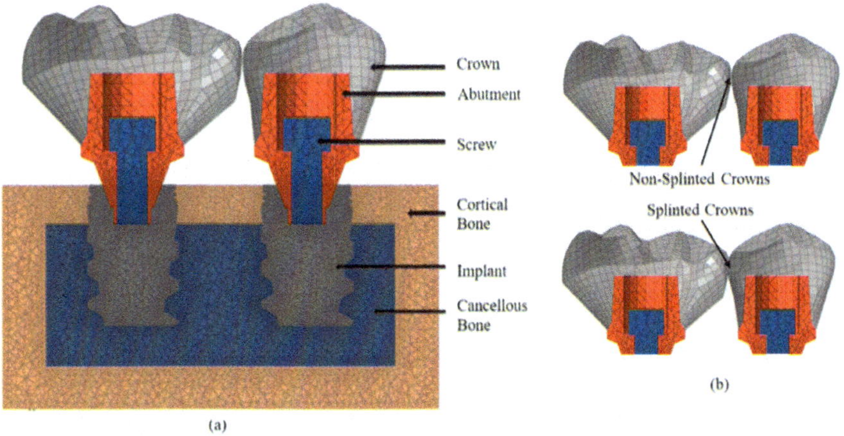

Fig. 2 Cut section of finite element meshing for **a** non-splinted dental implant assembly, **b** non-splinted and splinted crowns

Table 1 Mechanical properties of implant system

Implant system	Materials	Young's modulus	Poisson's ratio	Refs.
Implant, abutment, screw	Titanium	110,000	0.33	[18]
Hard jawbone	Cortical	13,700	0.26	[19]
Soft jawbone	Cancellous	1370	0.31	[19]
Crown	Porcelain	70,000	0.19	[18]

2.3 *Mechanical Properties*

The material used for the implant, abutment, and screw is Grade 4 Titanium alloy (Ti-6Al-4V). The implant crown is made of porcelain as shown in Fig. 1. Depending upon the age and gender, the mechanical properties of these bones vary. The success of a dental implant is based upon a number of factors, one of which is bone density. Siddhi et al. stated that bone density is an important consideration in obtaining predictable osseointegration and implant success [14, 15]. DII type of bone is considered in the present study. The orthotropic material model gives more acceptable results in predicting stress at bone implant interface than the isotropic material model [16]. However many literatures assumed the isotropic material properties. Therefore, the elastic properties of bone are considered to be isotropic as shown in Table 1. Adapted values of Young's modulus for the bone are considered because it is difficult to determine these values experimentally [17].

2.4 *Occlusal Loads and Boundary Condition*

To replicate the masticatory forces, two loading conditions were considered to find the generated equivalent stress at the cortical and cancellous bones. As shown in Fig. 3, both splinted and non-splinted implants were subjected to the masticatory load of 125 N along vertical axis (vertical load) and obliquely at 45°. Both of these forces were applied at the crown surface. In the Finite Element Model, constraints in all directions (Degree of freedom (DOF) is zero) are applied to the outer surface of the cortical bone.

3 Results

Dental implant such as splinted and non-splinted crowns are used by dentists. To determine which type of dental crown to be preferred, this study has been done. The dental implant is subjected to a 125 N load applied vertically and 45° obliquely. The least stress was produced by vertical loading, followed by force applied obliquely at

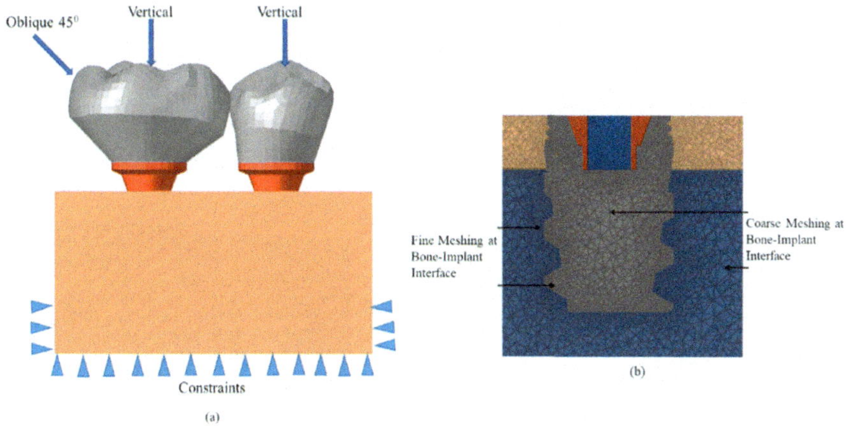

Fig. 3 **a** Loading and boundary conditions, **b** element meshing at bone-implant interface

45°. Vertical loading, on the other hand, produced the least displacement, followed by oblique loading at 45°.

3.1 Cancellous Bone

The vertical load of 125 N was applied, equivalent stress of 3.93 MPa was observed at bottom of cancellous bone for non-splinted crown whereas a stress of 4.07 MPa was observed near the last thread of cancellous bone for splinted crown as shown in Fig. 4. The stress in cancellous bone of splinted crown was increased by 3.44% as compared to non-splinted crown for vertical loading condition. Similarly, for oblique loading the stress is 8.48 MPa and 8.35 MPa in non-splinted and splinted crown respectively. For oblique loading, the stress in splinted crown was reduced by 1.53% as compared to non-splinted crown.

3.2 Cortical Bone

When vertical load of 125 N was applied, equivalent stress (von Mises Stress) of 12.53 MPa was observed in cortical bone of non-splinted crown whereas a stress of 13.15 MPa in splinted crown near the cancellous and cortical interface for both as shown in Fig. 5. For vertical loading, the stress in cortical bone of splinted crown was increased by 4.71% as compared to non-splinted crown. Similarly, for oblique loading the stress was generated at the uppermost part of cortical bone of magnitude 29.16 MPa and 29.02 MPa in non-splinted and splinted crown respectively. The

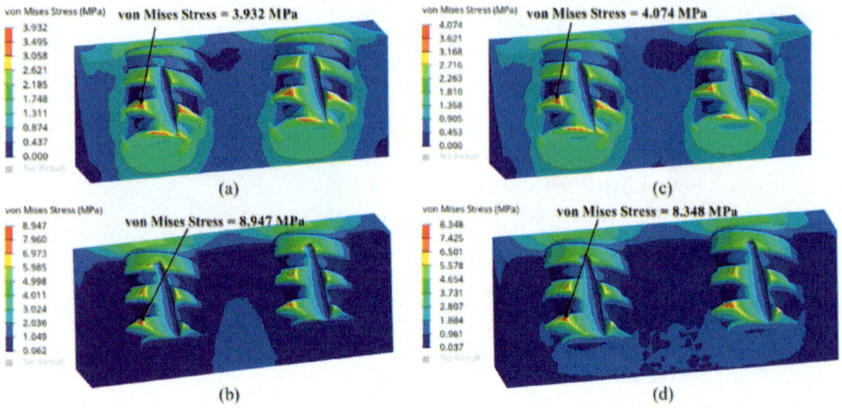

Fig. 4 Equivalent stress plot for cancellous bone **a** non-splinted oblique 45° loading, **b** splinted oblique 45° loading, **c** non-splinted vertical loading, and **d** splinted vertical loading

stress in splinted crown was reduced by 0.48% as compared to non-splinted crown for oblique loading condition.

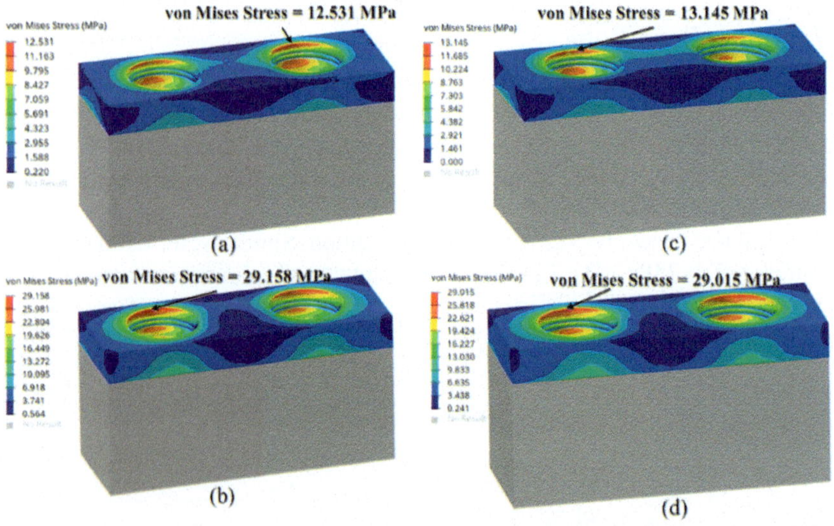

Fig. 5 Equivalent stress plot for cortical bone **a** non-splinted oblique 45° loading, **b** splinted oblique 45° loading, **c** non-splinted vertical loading and **d** splinted vertical loading

4 Discussion

The goal of this study is to examine the distribution of stress in the mandible for both splinted as well as non-splinted short dental implants by use of the finite element method. It should be noted that minor variations can cause bone damage, as confirmed by Cendoya et al. [20]. Furthermore, as Baldisserotto et al. point out, there are significant differences in bite forces based on gender and age. Moreover, this study does not contain bruxism-related bite force variations which analyzes bite forces under only physiologic or normal load forces [21]. Therefore, the occlusal bite force of 125 N is applied vertically and 45° obliquely to the implant to examine the stresses generated in cancellous and cortical bones.

According to the authors (Guichet et al.) [12], although the occlusal adjustment was well-balanced for the dental implant, stress distribution found in individual dental implant was less uniform than splinted dental implant. Vasco et al. [22] used image correlation as well as finite element analysis to show how stresses are reduced in the bones since splinting has better effect. Jofre et al. [20] observed that biomechanical behavior of a splinted dental implant system is better as a result of infrastructure union. This increases bone-implant anchoring area and decreases bone resorption when operational loads are acting on bone [23]. Göran Bergkvist et al. [10] stated that when comparing bone tissue nearby splinted dental implants to non-splinted dental implants, a significant decrease in stress was discovered. Splinting is believed to improve healing post-surgery from a strictly mechanical standpoint. Allahyar Geramy et al. [5] used three-dimensional Finite Element Analysis to investigate the biomechanical state of splinting in implant-supported maxillary overdentures and discovered that implant components which are employing splinted dental implants have lowered stress levels. Higher stresses are created in non-splinted designs due to the way stress is distributed in the crestal bone in a pathway between dental implants. The capacity to withstand the lateral stresses has been increased due to splinting of prostheses as it increases the supporting functional surface area. Splinting not only cement the retention of dental implant but also reduces the possibility of loosening of abutment screw, component fracture, and marginal bone resorption surrounding implants. Göran Bergkvist et al. [10] reported, for vertical loading conditions, stress is reduced by a factor of 0.54 in splinted implants compared to non-splinted implants. In contrast, for oblique loading conditions, splinted implants reduced stress by a factor of 8 when compared to non-splinted implants. This study concluded that splinting an implant can improve implant stability in bone tissue and is recommended as it generates less stress as compared to non-splinted implants. This study follows a similar pattern to that of the aforementioned researchers. As shown in Fig. 6a, b the stress in a splinted implant is greater than in a non-splinted implant under vertical loading conditions. For oblique loading, a splinted implant has less stress than a non-splinted implant. Furthermore, the cortical bone has greater stress than the cancellous bone as Young's Modulus of cortical bone is greater than the cancellous bone [24, 25] (resistance offered by cancellous bone to deformation is less as compared to the cortical bone).

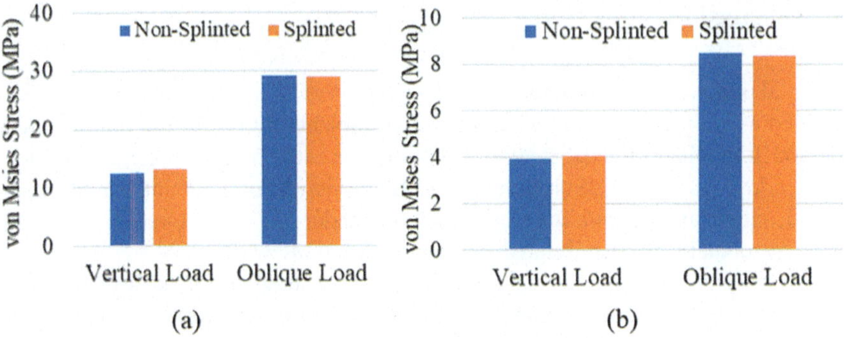

Fig. 6 Stress distribution for vertical and oblique 45° loading in **a** cortical and **b** cancellous bone

5 Conclusion

For oblique loading conditions, splinted implants have better stress distribution than non-splinted implants. Equivalent stress in cancellous as well as cortical in splinted is less as compared to non-splinted implants. Considering the results obtained from analysis, it can be concluded that uniform stress distribution has been seen in the cancellous and cortical bones in splinted implants as compared to non-splinted. The stress generated in splinted implant is low as compared to the non-splinted implant for the oblique loading. So, it is recommended to use splinted implants.

References

1. Meimandi M, Ardakani MRT, Amid R, Motlagh AM, Beheshti S (2018) Comparison of stress and strain distribution around splinted and nonsplinted 6 mm short implants in posterior mandible: a finite element analysis study. Implant Dent 27(1):74–80
2. Ozeki K, Nishio H (2020) Comparison of stress distribution in partially and completely edentulous mandibles around splinted and non-splinted implant prostheses: a finite element study. Biomed Mater Eng 31(1):19–33
3. de Souza Batista VE et al (2019) Should the restoration of adjacent implants be splinted or nonsplinted? A systematic review and meta-analysis. J Prosthet Dent 121(1):41–51
4. Rieger MR, Adams WK, Kinzel GL (1990) A finite element survey of eleven endosseous implants. J Prosthet Dent 63(4):457–465
5. Geramy A, Habibzadeh S (2018) Stress distribution in splinted and unsplinted implant-supported maxillary overdentures: a 3D finite element analysis. Implant Dent 27(1):56–62
6. Grossmann Y, Finger IM, Block MS (2005) Indications for splinting implant restorations. J Oral Maxillofac Surg 63(11):1642–1652
7. Kittur N, Oak R, Dekate D, Jadhav S, Dhatrak P (2020) Dental implant stability and its measurements to improve osseointegration at the bone-implant interface: a review. Mater Today Proc 43:1064–1070
8. Joshi S, Bora U, Karnik N, Bhadri K, Dhatrak P (2022) Current and emerging technologies for resonance frequency analysis-based devices for measuring dental implant stability: a review. Springer Singapore

9. Clelland N, Chaudhry J, Rashid R, McGlumphy E (2016) Split-mouth comparison of splinted and nonsplinted prostheses on short implants: 3 year results. Int J Oral Maxillofac Implant 1135–1141
10. Bergkvist G, Simonsson K, Rydberg K, Johansson F, Dérand T (2008) A finite element analysis of stress distribution in bone tissue surrounding uncoupled or splinted dental implants. Clin Implant Dent Relat Res 10(1):40–46
11. Yilmaz B, Mess J, Seidt J, Clelland NL (2013) Strain comparisons for splinted and nonsplinted cement-retained implant crowns. Int J Prosthodont 26(3):235–238
12. Guichet DL, Yoshinobu D, Caputo AA (2002) Effect of splinting and interproximal contact tightness on load transfer by implant restorations. J Prosthet Dent 87(5):528–535
13. Dhatrak P, Shirsat U, Sumanth S, Deshmukh V (2020) Numerical investigation on stress intensity around bone-implant interface by 3-dimensional FEA and experimental verification by optical technique. Mater Today Proc 39:35–41
14. Sunil S, Dhatrak P (2021) Biomechanical consideration of bone density and its influence on stress distribution characteristics of dental implants. Mater Today Proc 46:478–483
15. Dhatrak P, Girme V, Shirsat U, Sumanth S, Deshmukh V (2019) Significance of orthotropic material models to predict stress around bone-implant interface using numerical simulation. Bionanoscience 9(3):652–659
16. Piotrowski B, Baptista AA, Patoor E, Bravetti P, Eberhardt A, Laheurte P (2014) Interaction of bone-dental implant with new ultra low modulus alloy using a numerical approach. Mater Sci Eng C 38(1):151–160
17. Chen J, Zhang Z, Chen X, Zhang C, Zhang G, Xu Z (2014) Design and manufacture of customized dental implants by using reverse engineering and selective laser melting technology. J Prosthet Dent 112(5):1088–1095.E1
18. Dhatrak P, Shirsat U, Sumanth S, Deshmukh V (2020) Influence of cutting flutes on stress distribution for selected dental implants: numerical studies. Mater Today Proc 38:2680–2686
19. Jofre J, Cendoya P, Munoz P (2012) Effect of splinting mini-implants on marginal bone loss: a biomechanical model and clinical randomized study with mandibular overdentures. Int J Oral Maxillofac Implant 25(6):1137–1144
20. Dúcia Caldas Cosme RSS, Baldisserotto SM, de Andrade Canabarro S (2005) Bruxism and voluntary maximal bite force in young dentate adults. Int J Prosthodont 18(4):328–332
21. Tiossi R et al (2013) Validation of finite element models for strain analysis of implant-supported prostheses using digital image correlation. Dent Mater 29(7):788–796
22. Nimbalkar S, Dhatrak P, Gherde C, Joshi S (2020) A review article on factors affecting bone loss in dental implants. Mater Today Proc 43:970–976
23. Tiossi R, Vasco MA, Lin L et al (2013) Validation of finite element models for strain analysis of implant-supported prostheses using digital image correlation. Dent Mater: Off Publ Acad Dent Mater 29(7):788–796
24. Ramos Xavier Coutinho Nascimento L, Monteiro Torelly G, Nelson Elias C (2023) Analysis of bone stress and primary stability of a dental implant using strain and torque measurements. Saudi Dent J 35(3):263–269
25. Sasaki T, Nakata H, Suzuki A, Hada T, Kasugai S, Kuroda S (2020) Comparison of splinted and non-splinted superstructures of three implants placed in a mandibular distal extension model with missing teeth using modal analysis. J Mech Behav Biomed Mater 112:104050

Performance Evaluation of Powder-Lubricated Journal Bearing Using FDM and CFD Technique

Jijo Jose and Niranjana Behera

Abstract Dry powdered particles have been proposed as suitable lubricant alternatives in hydrodynamic bearings that operate in extreme environments where normal lubricants fail, such as at high temperatures and under extreme working conditions. As oil lubricant was prone to thermal degradation at high temperatures, powder lubricants will not undergo degradation. In the present work, a suitable model has been proposed for predicting the behavior of a powder-lubricated circular journal bearing by using the CFD technique and the results have been validated with the numerical solution of the equations solved using FDM technique. Static and dynamic characteristics have been predicted for powdered particles with a size of 1 and 2 μm and for bearings with *L*/*D* ratios of 1 and 0.58. Using both these techniques the load carrying capacity is found to be increasing with an increase in *L*/*D* ratio, particle diameter, and eccentricity ratio.

Keywords CFD · FDM · Friction · Particle diameter · Powder lubrication

1 Introduction

Dry powdered particles have been proposed as suitable lubricant alternatives in hydrodynamic bearings that operate in extreme environments where normal lubricants fail, such as at high temperatures and under extreme working conditions [1, 2]. Powder lubricants, unlike liquid lubricants, do not deteriorate at high temperatures. Unlike oil lubricants, powder lubricants do not suffer from thermal degradation at high temperatures. Solid lubricant boundary layers, unlike fluid films created in hydrodynamic lubrication, can have a uniform thickness regardless of load, temperature, or speed [3]. The solid lubricant-generated fluid layer has good shearing and

J. Jose
Amal Jyothi College of Engineering, Koovappally, Kerala, India
e-mail: jijojose@amaljyothi.ac.in

N. Behera (✉)
VIT University, Vellore, Tamil Nadu, India
e-mail: niranjanabehera@vit.ac.in

P. Bhingole et al. (eds.), *Advances in Materials Engineering*, Lecture Notes in Mechanical Engineering, https://doi.org/10.1007/978-981-97-7114-1_26

load-carrying capabilities [4]. Granular powders used in bearings work well as lubricants in hot and severe environments [5, 6]. An investigation by Heshmat and Brewe on journal bearings lubricated with two different powders, MoS_2 and WS_2 revealed that the friction coefficient of bearing remained constant throughout the analysis, and stable thermal operation was achieved across the entire range of load and speed combinations [7, 8]. Rahmani et al. found that powder-lubricated journal bearings are more stable than oil-lubricated bearings [9]. As a result, these bearings can be used in high-speed machinery including centrifugal compressors, flywheel energy storage devices, and gas turbine engines [10].

Most of the mathematical models used to conduct numerical investigations into powder lubrication were built using Haff's grain theory [11]. In this concept, powder particles were represented as molecules in granular fluids. McKeague and Khonsari developed a new analytical model that predicts the behavior of slider bearings lubricated with powder based on a set of geometrical and material properties without taking side leakage into account [12]. Several researchers have been working on developing more detailed statistical models for particle lubrication during the last few years. Zhou and Khonsari [13] investigated the performance of powder lubricants sheared in between two plates which are parallel using the fundamental equations provided by Lun et al. [14]. Pappur and Khonsari investigated a powder-lubricated thrust bearing as part of their research [15]. The generalized Reynolds equation for three-dimensional flow was established by Jang and Khonsari, and it specifies the pressure created in a powder-lubricated bearing with side flow [16]. They identified the relationships among pressure, solid volume fraction, and pseudo-temperature. Jose and Behera and Rahmani et al. investigated the dynamic and static characteristics of powder-lubricated non-circular bearings using finite difference technique [17, 18]. They studied the dynamic and static characteristics of the bearing for different L/D ratios, ellipticity ratios, and particle diameters.

Many researchers have adopted commercial computational fluid dynamics (CFD) software in their research due to developments in computer technology. A CFD algorithm has the essential benefit over finite difference methods in that it uses the entire Navier–Stokes equations to solve the flow problems instead of just the Reynolds equation. The outcomes of the two strategies are therefore likely to differ. The primary advantage of the CFD code is that it uses the finite volume method and the complete Navier–Stokes equations to handle flow problems. Therefore, even very complex geometries can be solved using CFD techniques. Shankar et al. analyzed the static characteristics of four-lobe bearing and compared with plain circular bearings using CFD technique [19]. Tauviqirrahman et al. applied the CFD technique for analyzing the tribological properties in heterogeneous rough journal bearings [20]. Manojkumar et al. applied CFD to predict the pressure profile and stress in the bearing [21]. Wang et al. studied the effect of thermal and cavitation in journal bearing using CFD [22]. Kadhim et al. applied CFD for predicting the load bearing capacity in nano-lubricated journal bearings by considering variable viscosity effects [23]. Nie et al. applied CFD to study the impact of surface texture on the load carrying capacity of journal bearing [24]. Tauviqirrahman et al. studied the effect of dimple depth on the load bearing capacity in the textured journal bearing [25].

In this paper, a hydrodynamic powder-lubricated circular bearing with particles of size 1 and 2 microns is assessed using finite difference method (FDM) technique and Computational fluid dynamics (CFD) technique. The dynamic and static characteristics bearing have been investigated.

2 Mathematical Model

2.1 Model Using FDM

This section briefly describes the mathematical model that was developed to carry out the analysis. The following is a list of the assumptions that the model considers.

i. The cohesive force between grain particles is not considered.
ii. The average spacing between the particles is greater than the grain diameter.
iii. The elasticity of grain particles is taken into consideration, and spinning motion is restricted.
iv. Grain particles show fluctuations in many characteristics due to the faster shear flow.
v. When compared to the viscosity of particles due to grain contact, the importance of interstitial fluid (air) is determined to be insignificant.
vi. A constant bulk density was chosen for the granular material.

Static Characteristics of Bearings

The components of fluid film force along the X and Y directions on the journal are [17]

$$F_X = \int_0^L \int_0^{2\pi} pR\cos\theta d\theta dz \quad (1)$$

$$F_Y = \int_0^L \int_0^{2\pi} pR\sin\theta d\theta dz \quad (2)$$

The load (W) that acts on the journal is determined using the following equation [17]

$$W = \sqrt{F_x^2 + F_y^2} \quad (3)$$

The side leakage (Q_1) present in the bearing is determined using the following expression [17]

$$Q_1 = 2\int_0^{2\pi}\int_0^{h} u_z dyRd\theta = 2\int_0^{2\pi}\left(-\frac{1}{6}\frac{t}{a}B\psi h^2\frac{\partial}{\partial z}(\ln p)\right)Rd\theta \quad (4)$$

Bearing frictional force has been determined using the following expression [17],

$$F_s = \int_0^L \int_0^{2\pi} \left(\frac{h}{2R} \frac{\partial p}{\partial \theta} + \eta \frac{U}{h} \right) R d\theta dz \tag{5}$$

Coefficient of friction is determined using the following equation [17],

$$f = F_s / W \tag{6}$$

Dynamic Characteristics of Bearings

The dynamic properties of a journal bearing are used to determine its stability. The stability study determines if the bearing can regain equilibrium after being perturbed. The bearing in question has linear and time invariant equations of motion, and its stability has been determined using Routh-Hurwitz criteria. Figure 1 shows the direct and cross damping and stiffness properties of the powdered fluid film, which are characterized using the Finite perturbation method. The damping and stiffness coefficients are calculated using the calculations below [26].

$$K_{XX} = \frac{\left(\frac{C}{W}\right)\left[F_X(X_0 + \Delta X, Y_0, 0, 0) - F_X(X_0 - \Delta X, Y_0, 0, 0)\right]}{2\Delta X} \tag{7}$$

$$K_{XY} = \frac{\left(\frac{C}{W}\right)\left[F_X(X_0, Y_0 + \Delta Y, 0, 0) - F_X(X_0, Y_0 - \Delta Y, 0, 0)\right]}{2\Delta Y} \tag{8}$$

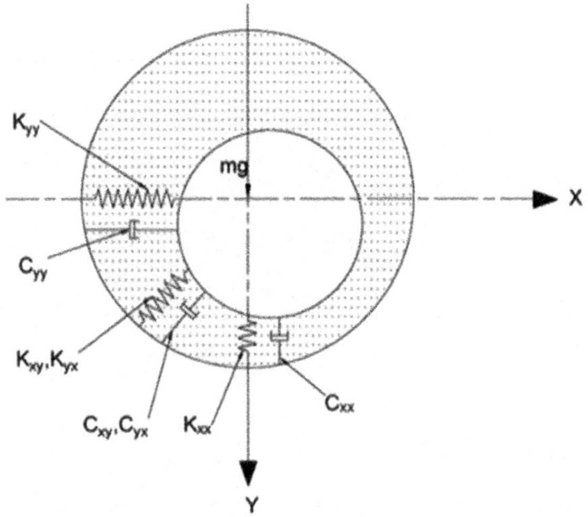

Fig. 1 Stiffness and damping properties in powdered fluid film

$$K_{YX} = \frac{\left(\frac{C}{W}\right)[F_Y(X_0 + \Delta X, Y_0, 0, 0) - F_Y(X_0 - \Delta X, Y_0, 0, 0)]}{2\Delta X} \tag{9}$$

$$K_{YY} = \frac{\left(\frac{C}{W}\right)[F_Y(X_0, Y_0 + \Delta Y, 0, 0) - F_Y(X_0, Y_0 - \Delta Y, 0, 0)]}{2\Delta Y} \tag{10}$$

$$C_{XX} = \frac{\left(\frac{C\omega}{W}\right)\left[F_X\left(X_0, Y_0, \Delta\dot{X}, 0\right) - F_X\left(X_0, Y_0, -\Delta\dot{X}, 0\right)\right]}{2\Delta\dot{X}} \tag{11}$$

$$C_{XY} = \frac{\left(\frac{C\omega}{W}\right)\left[F_X\left(X_0, Y_0, 0, \Delta\dot{Y}\right) - F_X\left(X_0, Y_0, 0, -\Delta\dot{Y}\right)\right]}{2\Delta\dot{Y}} \tag{12}$$

$$C_{YX} = \frac{\left(\frac{C\omega}{W}\right)\left[F_Y\left(X_0, Y_0, \Delta\dot{X}, 0\right) - F_Y\left(X_0, Y_0, -\Delta\dot{X}, 0\right)\right]}{2\Delta\dot{X}} \tag{13}$$

$$C_{YY} = \frac{\left(\frac{C\omega}{W}\right)\left[F_Y\left(X_0, Y_0, 0, \Delta\dot{Y}\right) - F_Y\left(X_0, Y_0, 0, -\Delta\dot{Y}\right)\right]}{2\Delta\dot{Y}} \tag{14}$$

For the rotor bearing system, the equation of motion can be expressed as [17]

$$\begin{bmatrix} m & 0 \\ 0 & m \end{bmatrix}\begin{bmatrix} \Delta\ddot{X} \\ \Delta\ddot{Y} \end{bmatrix} + \begin{bmatrix} C_{XX} & C_{XY} \\ C_{YX} & C_{YY} \end{bmatrix}\begin{bmatrix} \Delta\dot{X} \\ \Delta\dot{Y} \end{bmatrix} + \begin{bmatrix} K_{XX} & K_{XY} \\ K_{YX} & K_{YY} \end{bmatrix}\begin{bmatrix} \Delta X \\ \Delta Y \end{bmatrix} = \begin{bmatrix} 0 \\ 0 \end{bmatrix} \tag{15}$$

The solution for the Eq. (15) becomes [17]

$$\begin{bmatrix} \Delta X \\ \Delta Y \end{bmatrix} = \begin{bmatrix} X_h \\ Y_h \end{bmatrix} e^{(\gamma + i\nu)t} \tag{16}$$

When $\gamma > 0$, the model differs exponentially over time and it cannot achieve stability. So, the value of γ should be negative to attain stability. When $\gamma = 0$, then the threshold of stability can be attained. Considering this Eq. (16) changes to [17]

$$\begin{bmatrix} \Delta X \\ \Delta Y \end{bmatrix} = \begin{bmatrix} X_h \\ Y_h \end{bmatrix} e^{i\nu t} \tag{17}$$

Substituting the values of Eq. (17) in (15), the following expression is obtained [17]

$$\begin{bmatrix} k_{XX} - m\nu^2 + i\nu c_{XX} & k_{XY} - i\nu c_{XY} \\ k_{YX} - i\nu c_{YX} & k_{YY} - m\nu^2 + i\nu c_{YY} \end{bmatrix}\begin{bmatrix} X_h \\ Y_h \end{bmatrix} = \begin{bmatrix} 0 \\ 0 \end{bmatrix} \tag{18}$$

The determinant value of the coefficient matrix needs to be zero in order to achieve a non-trivial solution; hence [17],

$$\begin{aligned}&(k_{XX}-mv^2)(k_{YY}-mv^2)-v^2c_{XX}c_{YY}-k_{XY}k_{YX}+v^2c_{XY}c_{YX}\\&+iv\left[c_{XX}(K_{YY}-mv^2)+c_{YY}(K_{XX}-mv^2)-(k_{XY}c_{YX}+k_{YX}c_{XY})\right]=0\end{aligned} \tag{19}$$

In the expression, the real and imaginary parts are equated to zero to get the values [17]

$$v^2=\frac{(K_{XX}-mv^2)(K_{YY}-mv^2)-K_{XY}K_{YX}}{C_{XX}C_{YY}-C_{XY}C_{YX}} \tag{20}$$

$$mv^2=\frac{C_{XX}K_{YY}+C_{YY}K_{XX}-K_{XY}C_{YX}-K_{YX}C_{XY}}{C_{XX}+C_{YY}} \tag{21}$$

In the normalized form, Eq. (20) and (21) are obtained as

$$\lambda_{cr}^2=\frac{(K_{XX}-K_{eq})(K_{YY}-K_{eq})-K_{XY}K_{YX}}{C_{XX}C_{YY}-C_{XY}C_{YX}} \tag{22}$$

$$K_{eq}=\frac{C_{XX}K_{YY}+C_{YY}K_{XX}-K_{XY}C_{YX}-K_{YX}C_{XY}}{C_{XX}+C_{YY}} \tag{23}$$

where, the equivalent stiffness $K_{eq}=mv^2(C_r/W)$ and whirl frequency ratio $\lambda_{cr}=v/\omega$.

Stability parameter (M_{cr}) is calculated by [17]

$$M_{cr}=\frac{K_{eq}}{\lambda_{cr}^2}=m\omega^2\frac{C_r}{W} \tag{24}$$

The system will be in stable condition if [17]

$$M_{cr}>m\omega^2\frac{C_r}{W} \tag{25}$$

Numerical Solution

Reynolds equation applicable for normal circular journal bearings [17]

$$\frac{\partial}{\partial x}\left(\frac{1}{6}\frac{t}{a}\psi h^2\frac{\partial(\ln p)}{\partial x}\right)+\frac{\partial}{\partial z}\left(\frac{1}{6}\frac{t}{a}\psi h^2\frac{\partial(\ln p)}{\partial z}\right)=\frac{U}{2}\frac{\partial h}{\partial x}+\frac{\partial h}{\partial t} \tag{26}$$

The finite difference method has been used to solve the Reynolds equation. The solution can be obtained through iterative technique using the Gauss–Seidel method considering an over-relaxation factor. The model consisted a mesh size of 270 × 270. The iteration was stopped when the pressure value met the following condition [17].

$$\frac{\sum_{i=1}^{m}\sum_{j=1}^{n}\left|\left(p_{i,j}\right)^{N}\right|-\left|\left(p_{i,j}\right)^{N-1}\right|}{\sum_{i=1}^{m}\sum_{j=1}^{n}\left|\left(p_{i,j}\right)^{N}\right|}\leq 10^{-5} \tag{27}$$

where $i, j =$ the grids in the Θ and z directions

m, $n =$ total number of grids in Θ and z directions, $N =$ total number of iterations.

Utilizing powdered particles with diameter 1 and 2 μm, static and dynamic characteristics of the circular journal bearing were examined numerically. *L*/*D* ratios of 0.58 and 1 were used to test the bearing performance. The impact of *L*/*D* ratio and powder particle size on bearing performance was assessed.

2.2 *Model Using CFD*

CFD implements finite volume approach for solving the governing differential equations of flow. In this approach, the equation is converted into algebraic equations by control volume technique before solving numerically. Using the volume averaged discretization method, the governing equations can be integrated over each control volume, resulting in independent equations that conserve each quality based on a control-volume basis. The discretized equations can be solved by using appropriate boundary conditions. Equations for momentum and continuity are developed in this model using the Eulerian approach. The gas and solid phase momentum equations are connected through the application of gas–solid drag forces. The pressure velocity coupling has been implemented using the phase coupled SIMPLE discretization method. The turbulent kinetic energy and volume fraction were among the characteristics that were discretized using the first order upwind method. The particle diameter for the current investigation was set as 1 and 2 μm. Governing equations used in the model are presented below.

Model Equations

The mass conservation equation for the gas phase is given by [27]

$$\frac{\partial}{\partial t}\left(a_g u_g\right)+\nabla\cdot\left(a_g\rho_g u_g\right)=0 \tag{28}$$

The mass conservation equation for solid phase is given by [27]

$$\frac{\partial}{\partial t}(a_s u_s)+\nabla\cdot(a_s\rho_s u_s)=0 \tag{29}$$

Momentum equation for gas phase is [27]

$$\frac{\partial}{\partial t}\left(a_g\rho_g\vec{v}_g\right)+\nabla\cdot\left(a_g\rho_g\vec{v}_g\vec{v}_g\right)=-a_g\nabla p+\nabla\cdot\tau_g$$

$$+ a_g \rho_g g + K_{gs}(\vec{v}_g - u_s) \quad (30)$$

Momentum equation for solid phase is [27]

$$\frac{\partial}{\partial t}(a_s \rho_s \vec{v}_s) + \nabla \cdot (a_s \rho_s \vec{v}_s \vec{v}_s) = -a_s \nabla p - \nabla p_s + \nabla \cdot \tau_s + a_s \rho_s g - K_{gs}(\vec{v}_g - \vec{v}_s) \quad (31)$$

Equation of kinetic energy in turbulent gas phase is given by [27]

$$\frac{\partial}{\partial t}(a_g \rho_g k_g) + \nabla.(a_g \rho_g \vec{v}_g k_g) = \nabla.\left(a_g \frac{\mu_{t,g}}{\sigma_k} \nabla k_g\right) + a_g G_{k,g} - a_g \rho_g \varepsilon_g + a_g \rho_g \Pi_{k,sg} \quad (32)$$

Equation of kinetic energy due to dissipation in turbulent gas phase is given by [27]

$$\frac{\partial}{\partial t}(a_g \rho_g \in_g) + \nabla.(a_g \rho_g \vec{v}_g \in_g) = \nabla.\left(a_g \frac{\mu_{t,g}}{\sigma_\varepsilon} \nabla \varepsilon_g\right) + a_g \frac{\varepsilon_g}{k_g}(C_{1\varepsilon,g} G_{k,g} - C_{2\varepsilon,g} \rho_g \in_g) + a_g \rho_g \prod_{\varepsilon,sg} \quad (33)$$

The momentum exchange between two phases can be determined by using the gas–solid drag coefficient. Gidaspow model can be used which includes two separate equations for calculating gas–solid drag coefficient [28].

When $a_g > 0.8$ [28]

$$K_{gs} = \frac{3}{4} C_D \frac{a_s a_g \rho_g |\vec{v}_s - \vec{v}_g|}{d_{ps}} a_g^{-2.65} \quad (34)$$

where C_D, the drag coefficient is represented as [28]

$$C_D = \frac{24}{a_g \mathrm{Re}_p}\left[1 + 0.15(a_g \mathrm{Re}_p)^{0.687}\right] \quad (35)$$

When $a_g \leq 0.8$ [28]

$$K_{gs} = 150 \frac{a_s(1 - a_g)\mu_g}{a_g d_{ps}^2} + 1.75 \frac{\rho_g a_s |\vec{v}_s - \vec{v}_g|}{d_{ps}} \quad (36)$$

The gas phase stress tensor is [28]

$$\tau_g = a_g \mu_g \left(\nabla \vec{v}_g + \nabla \vec{v}_g^T\right) + a_g\left(\lambda_g - \frac{2}{3}\mu_g\right)\nabla \cdot \vec{v}_g I \quad (37)$$

The solid phase stress tensor is [28]

$$\tau_s = a_s\mu_s\left(\nabla\vec{v}_s + \nabla\vec{v}_s^T\right) + a_s\left(\lambda_s - \frac{2}{3}\mu_s\right) \tag{38}$$

Kinetic energy of particles subjected to random motion causes the granular temperature and is represented by [27]

$$\frac{3}{2}\left[\frac{\partial}{\partial t}(a_s\rho_s\Theta_s) + \nabla\cdot(a_s\rho_s\Theta_s\vec{v}_s)\right] = (= p_s I + \tau_s) : \nabla\vec{v}_s + \nabla\cdot\left(k_{\Theta_s}\nabla\Theta_s\right) - \psi_s + \phi_{gs} \tag{39}$$

where

The first term on the right side is the energy related to solid stress tensor.

Dissipation energy (ψ_s) due to collision is expressed by [14]

$$\psi_s = \frac{12\left(1 - e^2\right)g_0}{d_s\sqrt{\pi}}\rho_s a_s^2 \Theta_s^{\frac{3}{2}} \tag{40}$$

The exchange in kinetic energy of random fluctuations from solid to gas phase is given by [28]

$$\phi_{gs} = -3k_{gs}\Theta_s \tag{41}$$

The normal force due to particle–particle interactions causes the solid pressure. The solid pressure is given by [14]

$$p_s = a_s\rho_s\Theta_s + 2\rho_s(1 + e)a_s^2 g_0\Theta_s \tag{42}$$

The radial distribution function is a term that indicates the probability of inter-particle collisions. It is expressed as [14]

$$g_0 = \left[1 - \left(\frac{a_s}{a_{s,\max}}\right)^{\frac{1}{3}}\right]^{-1} \tag{43}$$

Model Geometry

The fluid film in a journal bearing with *L*/*D* ratio 0.58 and eccentricity 0.2 is shown in Fig. 2. The data for the circular journal bearing is presented in Table 1 [7].

The analysis has been conducted powdered particles having particle diameter 1 and 2 μm. The particle density of powder within the film was taken as 4800 kg/m^3. The outer wall of and the inner wall fluid film was modelled as stationary wall and moving wall (with a rotational speed of 209 rad/s) respectively. Using Solid Edge

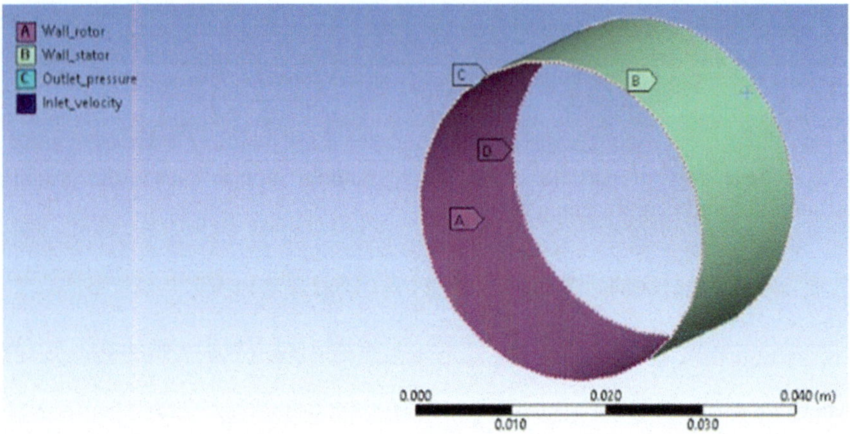

Fig. 2 Fluid film developed inside a journal bearing ($L/D = 0.58$) with eccentricity ratio 0.2

Table 1 Specifications of bearing and powder [7]

Diameter of the bearing D (mm)	$33.98 \approx 34$
Length of the bearing L (mm)	20
Radial clearance, C (mm)	0.1
Speed N (rpm)	2000
Particle diameter, d (μm)	1 and 2
Pour density of powder (kg/m^3)	1125
Solid density (kg/m^3)	4800
Dry air supply vol. flow rate (m^3/s)	0.002
Powder supply mass flow rate (mg/s)	0–30
Volume fraction	1×10^{-6}

ST9, the model was developed using the actual dimensions and then it is meshed in Ansys 18.1 using hexahedral elements as shown in Fig. 3.

Mesh Independence Study

It is necessary to conduct a mesh independence study or grid sensitivity analysis since the numerical result is strongly influenced by the meshing size. The mesh independent study can be used to assess the mesh convergence potential. By changing the mesh size from coarse to fine, the number of control volume elements rises while the numerical inaccuracy reduces. In this work, the simulations have been performed for six different mesh sizes with 68,160, 133,000, 240,996, 309,852, 409,200, 509,652 numbers of elements. For these models, the load carrying capacity has been predicted. Figure 4 shows the load carrying capacity variation with number of elements. It has been found that there is relatively little difference between intermediate mesh

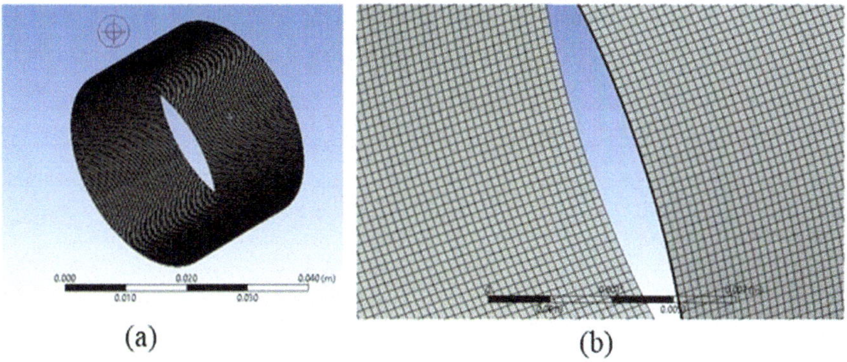

Fig. 3 Meshed model of fluid film developed inside a journal bearing **a** overall view **b** enlarged view

and finer mesh in terms of load carrying capacity. Hence the mesh with 409,200 components has been utilized in this simulation.

Boundary Conditions

At the inlet of the fluid film, inlet velocity boundary condition has been applied. Magnitude of velocity has been given based on the *L*/*D* ratio and particle diameter which are presented in Table 2. At the outlet of the fluid film, atmospheric pressure is considered.

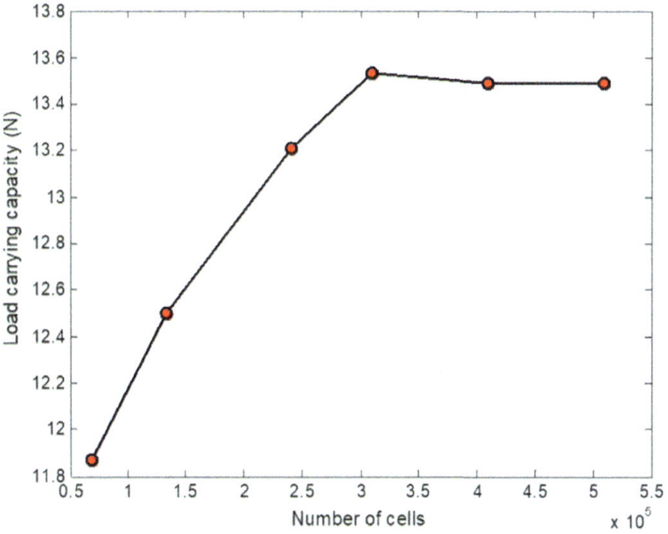

Fig. 4 Variations of load carrying capacity with increase in cells

Table 2 Velocity of the mixture of air and powdered particles

L/D ratio	Particle diameter (μm)	Velocity of the mixture of air and powdered particle (m/s)
0.58	1	85
0.58	2	125
1	1	120
1	2	185

3 Results

3.1 Results from FDM Model Simulation

Load Carrying Capacity

Figure 5 illustrates how the non-dimensional load varies as the eccentricity ratio increases, as well as at various particle sizes and L/D ratios. With an increase in eccentricity ratio, the load bearing capacity of the bearing rises. The thickness of the powdered coating reduces as the eccentricity ratio increases, boosting the bearing's ability to support more weight. From the graph, it can be shown that powdered particles of bigger sizes have greater load carrying capacities than those of smaller sizes for any given eccentricity ratio. The lubricant viscosity is inversely related to the particle diameter. Therefore, the presence of larger-sized particles will result in an increase in the pressure created inside the bearing, increasing the bearing's ability to support a given load. The bearing with *L/D* ratio 1 was found to have a higher load carrying capacity than the *L/D* ratio 0.58. This happens as a result of the increased pressure building up over the bigger bearing surface. These differences were noted in the research done by Rahmani et al. [17].

Coefficient of Friction

Figure 6 shows the relationship between the coefficient of friction and the eccentricity ratio for *L/D* ratios of 0.58 and particle sizes of 1 and 2 μm. The coefficient of friction decreases as the eccentricity increases. This is due to the fact that the bearing's load carrying capability was found to be greater at higher eccentricity ratios than it was at lower eccentricity ratios. As a result, the coefficient of friction drops. This explanation explains why a bearing with an *L/D* ratio of 1 was found to have lower frictional coefficient values than a bearing with an *L/D* ratio of 0.58. Since 1 μm particles have a lower viscosity than 2 μm particles as the lubricating medium, the friction coefficient was determined to be lower for 1 μm particles. Similar trends were observed by Rahmani et al. [17].

Side Leakage

Figure 7 shows the plots of the side leakage variations as the eccentricity ratio varies for particle diameters of 1 and 2 μm and *L/D* ratios of 0.58 and 1. The figure made it clear that the key factor influencing side leakage was viscosity of the lubricant

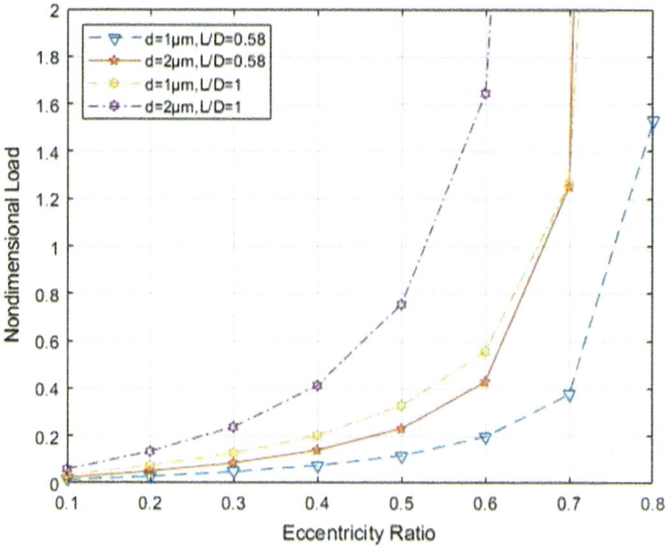

Fig. 5 Non-dimensional load versus eccentricity ratio

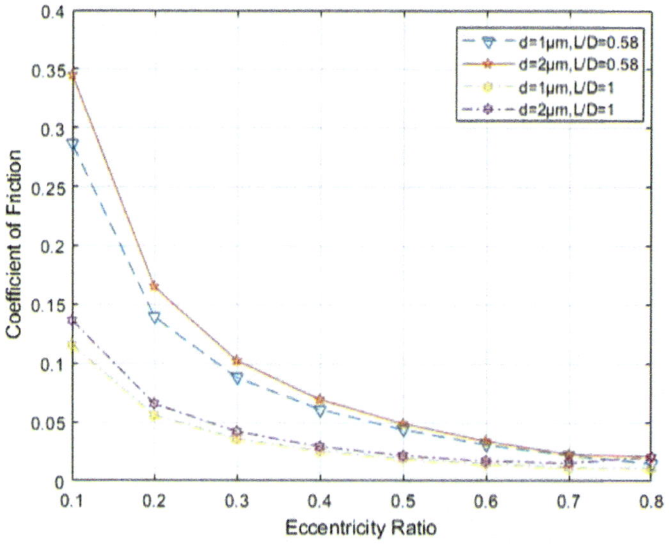

Fig. 6 Coefficient of friction versus eccentricity ratio

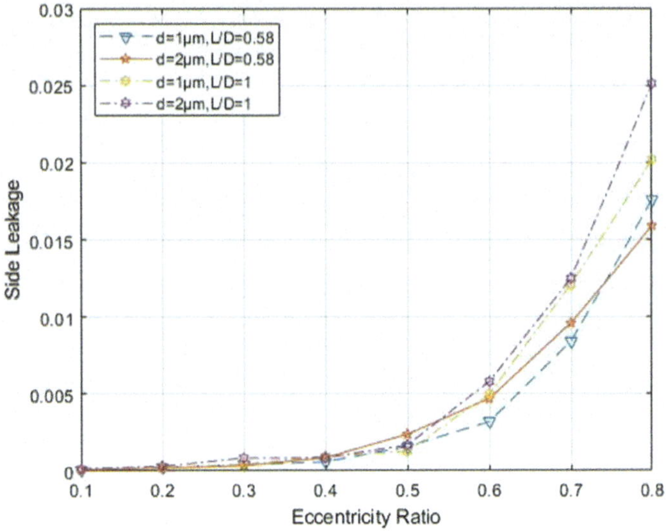

Fig. 7 Side leakage versus eccentricity ratio

and that particle size had little effect on it. With variations in *L*/*D* ratio and particle diameter, it was also found that the variation in side leakage was minimal.

Stability

For different ellipticity ratios and *L*/*D* ratios, the variations in stability parameter with regard to changes in the eccentricity ratio were obtained as shown in Fig. 8. It was found that bearings with *L*/*D* ratios of 1 were more stable than those with *L*/*D* ratios of 0.58. Additionally, it was observed that larger particles were more stable because of the higher lubricant viscosity, which produces high direct bearing coefficients.

3.2 *Results from CFD Model Simulation*

For particle diameters of 1 and 2 μm and *L*/*D* ratios of 0.58 and 1, the static characteristics of a journal bearing lubricated with powder were computed. The variations in the static characteristics observed in this study are briefly described in the following section.

The numerical results obtained by using MATLAB software were compared with the fluent analysis findings for *L*/*D* ratio 0.58 and particle diameter 1 μm. Figure 9 illustrates the results, which were determined to be in good agreement. Further research on powder-lubricated journal bearing has been done.

Figure 10 shows the load change as the eccentricity ratio increases for particles with diameters of 1 and 2 μm and *L*/*D* ratios of 0.58 and 1. With an increase in eccentricity ratio, the bearing's load carrying capacity rises. The thickness of the

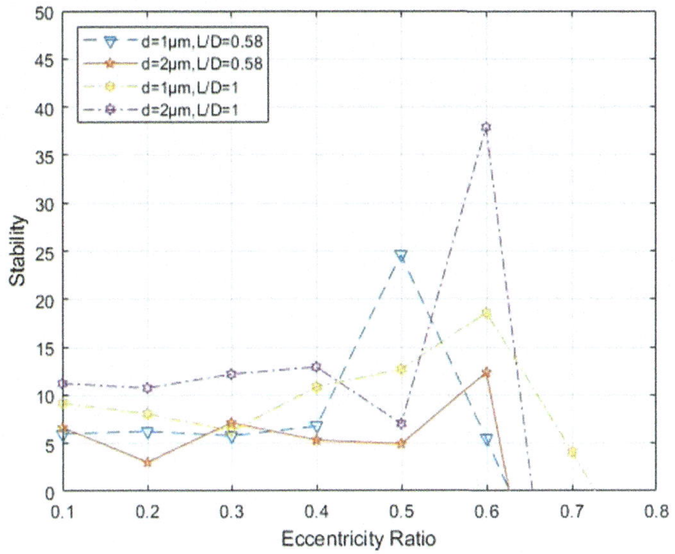

Fig. 8 Stability versus eccentricity ratio

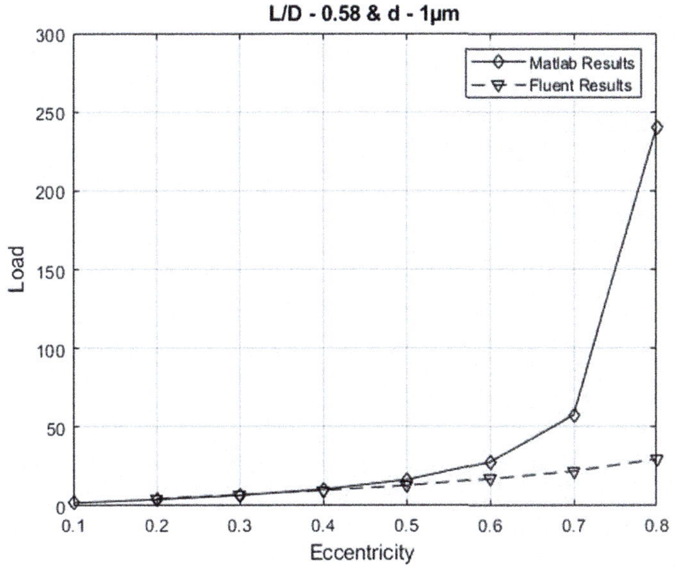

Fig. 9 Predicted load compared with the numerical result obtained using MATLAB

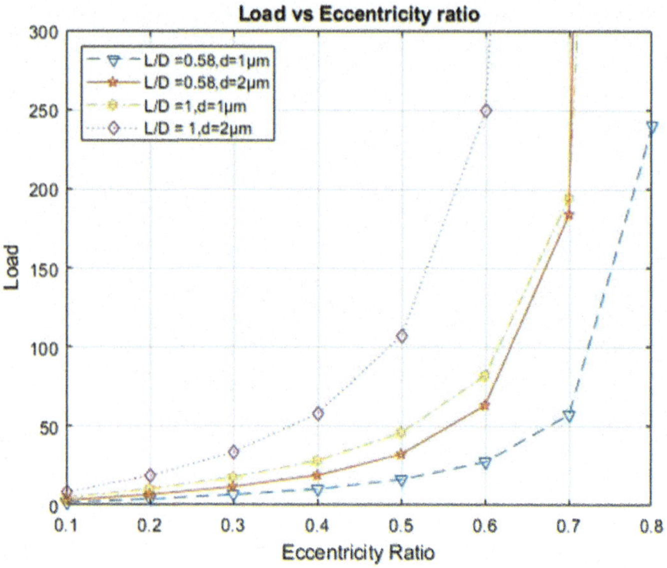

Fig. 10 Load versus eccentricity ratio graph at various *L*/*D* ratios and particle diameters

powdered coating thins as the eccentricity ratio rises, improving the bearing's ability to support more load. Similar variations were found in MATLAB-based numerical study of Jose and Behera [18].

The *L*/*D* ratio of 1 was shown to have a higher load carrying capacity than the *L*/*D* ratio of 0.58. This happens as a result of the increased pressure building up over the bigger bearing surface. The load carrying capability of the bearing increases with the size of the powdered particle size. The pressure developed increases as the particle size increases leading to an increase in the load carrying capacity. These variations were also found in the research of Rahmani et al. [17] and Jose and Behera [18] and Fig. 11 shows the static pressure contour plot obtained for *L*/*D* ratio of 0.58 and a particle diameter of 1 μm. A maximum pressure region is seen on the contour close to the inlet section, and the pressure progressively drops from the highest at the inlet to the atmospheric pressure at the outlet. The analytical solution obtained FDM technique and the results of the CFD analysis were found to be in good agreement.

4 Conclusions

Using the FDM approach, the static and dynamic properties of a circular powder-lubricated journal bearing have been studied. The simulation results have led to the following conclusions.

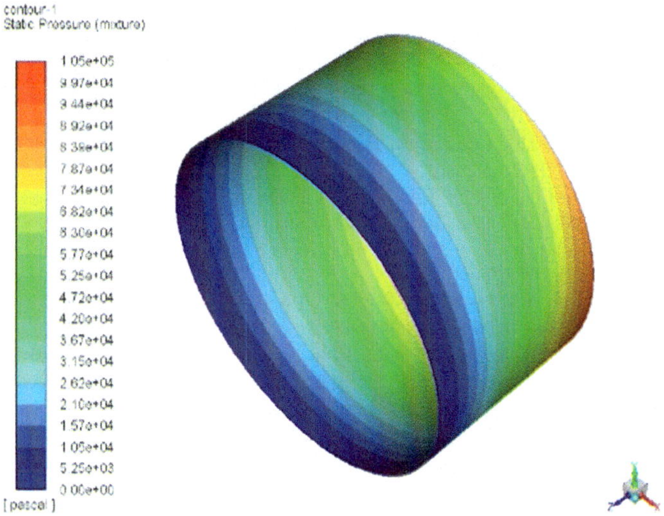

Fig. 11 Static pressure contour plot of powder lubricated journal bearing (*L*/*D* ratio 0.58 and particle diameter 1 μm)

i. As the *L*/*D* ratio rises, so do the bearings load carrying capability, side leakage, and stability. However, for the same, the coefficient of friction drops.
ii. As the powdered particle size increases, so do the load carrying capacity, journal bearing stability, and coefficient of friction. Particle diameter has no discernible effect on side leakage.
iii. As the eccentricity ratio increases, so do the bearings load carrying capability, side leakage, and stability. However, as the eccentricity ratio increases, the coefficient of friction decreases.
iv. Compared to other cases of *L*/*D* ratios and particle sizes, it was discovered that bearings with an *L*/*D* ratio of 1 and a particle diameter of 2 μm had superior static and dynamic properties.

Using the CFD technique, the static properties of a journal bearing lubricated with powder have been numerically investigated. The simulation results have led to the following conclusions.

i. As grain size grows, the bearing's capacity to support load increases. Therefore, compared to powders with smaller grain sizes, those with larger particle sizes have better static characteristics.
ii. When the *L*/*D* ratio rises, the bearings load bearing capacity increases.
iii. As the eccentricity ratio rises, the load carrying capability improves.

References

1. Jose J, Behera N (2020) Study of wear and friction characteristics of powder lubricants under extreme pressure conditions. Mater Today: Proc 22(4):1306–1317
2. Jose J, Behera N (2020) Lubrication capability of solid powder lubricants in high pressure situations. Jurnal Tribologi 26:101–119
3. Wan SH (2013) Solid lubricant: soft metal. In: Encyclopedia of tribology, pp 3152–3159
4. Erdemir A, Ajayi OO, Fenske GR, Erck RA, Hsieh JH (1992) The synergistic effects of solid and liquid lubrication on the tribological behavior of transformation-toughened ZrO_2 ceramics. Tribol Trans 35(2):287–297
5. Wornyoh EYA, Jasti VK, Higgs III CF (2007) A review of dry particulate lubrication: powder and granular materials. J Tribol 129:438–449
6. Alkassar Y, Agarwal VK, Pandey RK, Behera N (2019) Experimental study and Shannon entropy analysis of pressure fluctuations and flow mode transition in fluidized dense phase pneumatic conveying of fly ash. Particuology 1251:1–10
7. Heshmat H, Brewe DE (1995) Performance of powder lubricated journal bearings with MoS_2 powder: experimental study of thermal phenomena. J Tribol 115:506–512
8. Heshmat H, Brewe DE (1996) Performance of a powder lubricated journal bearing with WS_2 powder: experimental study. J Tribol 118:484–491
9. Rahmani F, Dutt JK, Pandey RK (2016) Dynamic characteristics of a finite-width journal bearing lubricated with powders. Procedia Eng 144:841–848
10. Jose J, Behera N (2018) A review on non-circular journal bearings'. Int J Mech Prod Eng Res Dev (IJMPERD) 8(1):207–220
11. Haff PK (1993) Grain flow as a fluid-mechanical phenomenon. J Fluid Mech 134:401–430
12. McKeague KT, Khonsari MM (1996) An analysis of powder lubricated slider bearings. J Tribol 118:206–214
13. Zhou L, Khonsari MM (2000) Flow characteristics of a powder lubricant sheared between parallel plates. J Tribol 122:147–155
14. Lun CKK, Savage SB, Jeffrey DJ, Chepurniy N (1984) Kinetic theories for granular flow: inelastic particulates in couette flow and slightly in elastic particulates in a general flow field. J Fluid Mech 140:223–256
15. Pappur M, Khonsari MM (2003) Flow characterization and performance of a powder lubricated slider bearing. J Tribol 125:135–144
16. Jang JY, Khonsari MM (2005) On the granular lubrication theory. Proc R Soc A: Math Phys Eng Sci 461:3255–3278
17. Rahmani F, Dutt JK, Pandey RK (2016) Performance behavior of elliptical-bore journal bearings lubricated with solid granular particulates. Particuology 27:51–60
18. Jose J, Behera N (2021) Static and dynamic performance characteristics of powder lubricated symmetrical three-lobed bearing. Proc Inst Mech Eng, Part J: J Eng Tribol 235(5):916–930
19. Shankar KA, Muthukumar K, Rajiev R (2019) CFD simulation analysis and experimental study on multilobe hydrodynamic journal bearing. In: Paper presented at AIP conference proceedings, vol 2128, p 050007
20. Tauviqirrahman M, Jamari J, Wicaksono AA, Muchammad M, Susilowati S, Ngatilah Y, Pujiastuti C (2021) CFD analysis of journal bearing with a heterogeneous rough/smooth surface. Lubricants 9(88):1–20
21. Manojkumar NU, Jagadish H, Kirankumar B (2020) CFD analysis of hydro-dynamic lubrication journal bearing using castor oil. In: Recent trends in mechanical engineering: select proceedings of ICIME 2019. Springer, Singapore, pp 671–684
22. Li H, Wang Y, Zhong N, Chen Y, Yin Z (2018) Study on the performance of journal bearings in different lubricants by CFD and FSI method with thermal effect and cavitation. In: Paper presented at MATEC web of conferences, vol 249, p 03006
23. Kadhim ZH, Ahmed SY, Abass BA (2022) CFD analysis of nano-lubricated journal bearing considering variable viscosity and elastic deformation effects. Diagnostyka 23(1):1–8

24. Nie T, Yang K, Zhou L, Wu X, Wang Y (2022) CFD analysis of load capacity of journal bearing with surface texture. Energy Rep 8(7):327–334
25. Tauviqirrahman M, Jamari J, Muchammad M, Ardiansyah A, Setiyana B, Paryanto P (2022) Hydrodynamic lubrication analysis of hydrophobic textured journal bearing considering cavitation. Cogent Eng 9(1):1–17
26. Qiu ZL, Tieu AK (1996) The effect of perturbation amplitudes on eight force coefficients of journal bearings'. J Tribol 39:469–475
27. Behera N, Agarwal VK, Jones MG, Williams KC (2013) CFD modeling and analysis of dense phase pneumatic conveying of fine particles including particle size distribution. Powder Technol 244:30–37
28. Gidaspow D, Bezburuah R, Ding J (1992) Hydrodynamics of circulating fluidized beds: kinetic theory approach. In: Paper presented at 7th international conference on fluidization. Gold Coast, Australia

Designing Dental Implants: The Use of Finite Element Analysis, a Comprehensive Review

G. H. Waghmare, Kanak G. Waghmare, Sachin T. Bagde, and Mayuri N. Deshmukh

Abstract In the field of modern dentistry, dental implants have emerged as transformative solutions for tooth loss, effectively restoring oral health and aesthetics. This comprehensive review delves into the multidisciplinary realm of dental implant mechanical engineering, geometric modeling, and finite element analysis (FEA). The review underscores the importance of dental implants in contemporary dentistry, shedding light on their diverse advantages and profound influence on patients' lives. The major findings from FEA analyses of dental implants are presented, emphasizing the ability of FEA to enable intricate examinations of stress and strain distribution, evaluate the osseointegration process, and optimize implant designs for varying loading conditions. The review further explores the burgeoning trend of patient-specific FEA models, the role of FEA in advancing material science, and the potential integration of machine learning algorithms within this domain. In addition, it delves into the mechanical properties and suitability of materials commonly used in dental implant construction, such as titanium and zirconia. Manufacturing techniques, including machining and additive manufacturing, are discussed for their substantial impact on implant performance. The review illustrates real-world examples of how dental implant mechanical engineering, geometric modeling, and FEA have contributed to enhanced dental implant treatments. Moreover, it addresses the existing challenges and limitations in dental implant mechanical engineering, while also discussing potential future directions and innovations, including advanced materials, 3D printing, and virtual surgery planning. This comprehensive review offers a holistic perspective on the multifaceted aspects of dental implant mechanical engineering, highlighting the evolving landscape of dental implantology and the potential for improved patient outcomes, longevity, and overall satisfaction.

G. H. Waghmare (✉) · S. T. Bagde
Department of Mechanical Engineering, YCCE Nagpur, Nagpur, Maharashtra, India
e-mail: gwaghmare@gmail.com

K. G. Waghmare
Sharad Pawar Dental College and Hospital, DMIHER, Sawangi Wardha, Maharashtra, India

M. N. Deshmukh
Nagpur College of Pharmacy, Nagpur, Maharashtra, India

P. Bhingole et al. (eds.), *Advances in Materials Engineering*, Lecture Notes in Mechanical Engineering, https://doi.org/10.1007/978-981-97-7114-1_27

Keywords Dental implants · Geometric modelling · Finite element analysis · Osseointegration · Patient-specific modelling · Dental implant materials

1 Introduction

The domain of modern dentistry has undergone a profound transformation where the amalgamation of precision engineering and clinical expertise has redefined traditional norms. At the forefront of this evolution, dental implant technology shines as an exemplar of meticulous mechanical engineering, geometric modeling, and finite element analysis (FEA). This comprehensive review article embarks on an exhaustive exploration of the profound influence and significance of these interconnected disciplines within the sphere of dental implants. It sheds light on their transformative potential in reimagining solutions for tooth replacement [1] (Fig. 1).

Dental implants epitomize the pinnacle of dental mechanical engineering. They fuse intricate design, biomaterial science, and biomechanical principles to replicate the form and function of natural teeth with exquisite precision. Successful integration of dental implants into the oral environment necessitates the harmonious fusion of meticulous design, rigorous testing, and innovative engineering. This review commences with an in-depth overview of the mechanical engineering principles that underpin dental implant design, examining their applications in augmenting implant performance and longevity [3].

Geometric modeling, a cornerstone for translating intricate anatomical structures into computer-aided designs, assumes a foundational role in dental implant development. This discipline facilitates the creation of precise three-dimensional representations of oral and maxillofacial anatomy. Leveraging advanced software tools, such as Computer-Aided Design (CAD) and three-dimensional imaging, geometric modeling enables the development of geometrically accurate digital blueprints. These

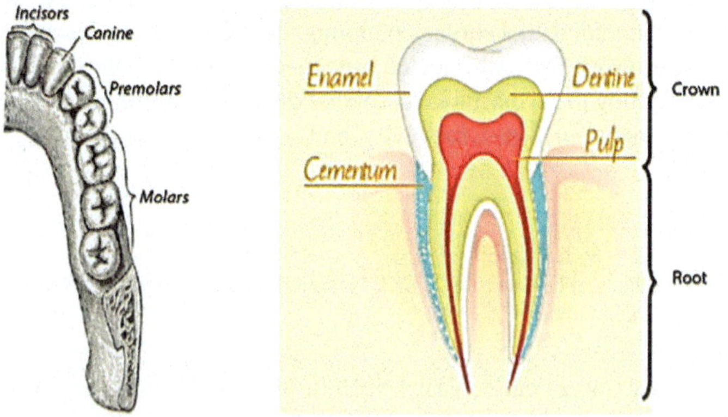

Fig. 1 Dental anatomy [2]

digital models, the architectural blueprints for patient-specific dental implants, also offer insights into the biomechanical behavior of implants within the oral environment. The exploration of geometric modeling delves deep into its pivotal role in the quest for customized dental implant solutions [4].

Finite Element Analysis (FEA) emerges as the vanguard of dental implant research, serving as a virtual laboratory for the comprehensive evaluation of implant performance. Through the intricate web of mathematical algorithms and computer simulations, FEA empowers the exhaustive assessment of stress distribution, strain, and deformation within dental implants. This cutting-edge technology equips researchers and clinicians with the capability to decipher the mechanical response of dental implants under diverse loading conditions. The insights gleaned through FEA fine-tune implant designs, ensuring their optimal functionality and durability. This vital tool bridges the chasm between theoretical design and clinical application, thereby propelling dental implant technology to new heights.

The confluence of dental implant mechanical engineering, geometric modeling, and FEA signifies a paradigm shift in patient-centric dental care. This triad of sophisticated methodologies and tools is expanding the horizons of what is achievable in modern dentistry. Patients now benefit from treatments that are not only more precise and durable but also meticulously tailored to their individual needs. In the subsequent sections of this review article, a thorough exploration of each of these facets ensues, elucidating their individual roles and, more crucially, their profound synergistic interplay.

The voyage ventures into the biomechanical considerations underlying dental implants, shedding light on the intricacies of load distribution, stress analysis, and strain evaluation. A comprehensive understanding of these principles is pivotal in optimizing dental implant designs, ensuring their suitability for diverse clinical scenarios. The exploration extends to the realm of materials and manufacturing processes, probing the mechanical properties of materials like titanium and zirconia, which are integral to dental implant construction. The assessment of the pivotal role of manufacturing techniques, including machining and additive manufacturing, in shaping the mechanical performance of implants unfolds.

Clinical applications, enriched with illustrative case studies, serve as tangible evidence of the practical translation of dental implant mechanical engineering, geometric modeling, and finite element analysis into successful clinical outcomes. These cases provide palpable testimony to the transformative potential of these disciplines in enhancing patient care.

The review article elucidates that the multidisciplinary approach of dental implant mechanical engineering and its symbiotic relationship with geometric modeling and finite element analysis shape the future of dental implantology and elevate the standards of precision engineering and clinical excellence within modern dentistry. The ensuing sections of this article offer a detailed exploration of each of these domains, weaving a narrative that underscores the profound potential of dental implant technology and the profound impact of these disciplines on the future of oral healthcare.

2 Mechanical Engineering in Dental Implants

The design of dental implants is a meticulous process that hinges on the application of fundamental principles of mechanical engineering. These principles serve as the bedrock for creating implant systems that can withstand the complex biomechanical challenges of the oral environment, ensuring both longevity and patient satisfaction. In this section, we delve into the core mechanical engineering principles that underpin dental implant design, drawing upon research findings and established knowledge.

One of the fundamental considerations in dental implant design is the selection of materials. The mechanical properties of these materials, such as strength, ductility, and fatigue resistance, play a pivotal role in implant performance. The work of Mark and Rack [5] underscores the importance of selecting materials with the appropriate mechanical characteristics, with titanium being a prominent choice due to its exceptional biocompatibility and strength. Materials must not only withstand masticatory forces but also be biologically compatible to promote osseointegration (Fig. 2).

A critical aspect of dental implant design involves stress analysis and the distribution of mechanical loads. Dental implants are subjected to significant forces during mastication and other oral activities. Uneven load distribution can lead to overloading, potentially causing implant failure. Research by Chatterjee et al. [7] highlights the need to ensure that the design of dental implants effectively distributes these loads to prevent stress concentrations, which may otherwise lead to structural failure. Stress analysis is a cornerstone of implant design to safeguard mechanical integrity.

Finite Element Analysis (FEA) is a powerful computational tool used extensively in dental implant research. It allows for the simulation of implant behavior under various loads, providing invaluable insights into stress distribution, deformation, and structural responses. Research by Singhal et al. [8] underscores the pivotal role of FEA in optimizing implant designs and ensuring their mechanical integrity. This

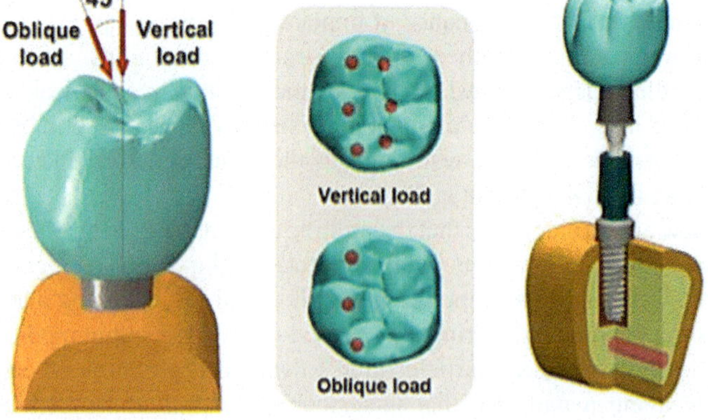

Fig. 2 Load placement sites and assembly components in a dental implant [6]

technology serves as a virtual laboratory, offering a deeper understanding of implant mechanics.

The design of threads on dental implants is pivotal for ensuring mechanical stability. Thread design influences the implant's anchorage in the surrounding bone, impacting primary stability during insertion and secondary stability through osseointegration. Telchi et al. [9] have elucidated that the shape, pitch, and depth of threads are key considerations in implant design to maximize mechanical stability. Thread design forms an integral part of mechanical engineering in dental implantology.

Dental implants are subjected to cyclic loading during their lifetime. Fatigue and fracture mechanics are essential principles to consider. Understanding how implants respond to cyclic loading, including the number of cycles to failure, is critical for assessing their long-term performance. Rojo et al. [10] emphasize the importance of studying fatigue behavior to mitigate the risk of mechanical failure. By considering these principles, designers can enhance implant durability.

The mechanical interface between the implant and the abutment, where prosthetic components attach, is another critical element of dental implant design. Liu and Wang [11] have highlighted that proper design and connection at this interface are vital to minimize micromotion, which can lead to implant complications. Mechanical stability at this junction is imperative for long-term success and functionality.

These fundamental principles of mechanical engineering provide the foundation for dental implant design. Through meticulous material selection, stress analysis, FEA, thread design, consideration of fatigue and fracture mechanics, and a focus on the implant-abutment interface, dental engineers aim to create implant systems that are mechanically robust, biocompatible, and capable of withstanding the biomechanical challenges of the oral environment. In essence, these principles enable the development of dental implants that offer patients not only restored function and aesthetics but also confidence in their long-term mechanical reliability.

2.1 Material Selection

The choice of materials is a fundamental aspect of dental implant design. Materials must be biocompatible, meaning they should not provoke an adverse immune response and must promote osseointegration—the direct bonding between the implant and the surrounding bone. Materials like titanium and zirconia are frequently used for their exceptional biomechanical compatibility, allowing for long-term implant stability without adverse reactions [12, 13]. Materials must also possess appropriate mechanical properties. They should have the strength to withstand the significant loads applied during biting and chewing while maintaining their structural integrity. The material's ductility is another important factor, as it influences how the implant handles stress without fracturing. Materials with high fatigue resistance are preferred to ensure the implant can endure cyclic loading without compromising performance [14].

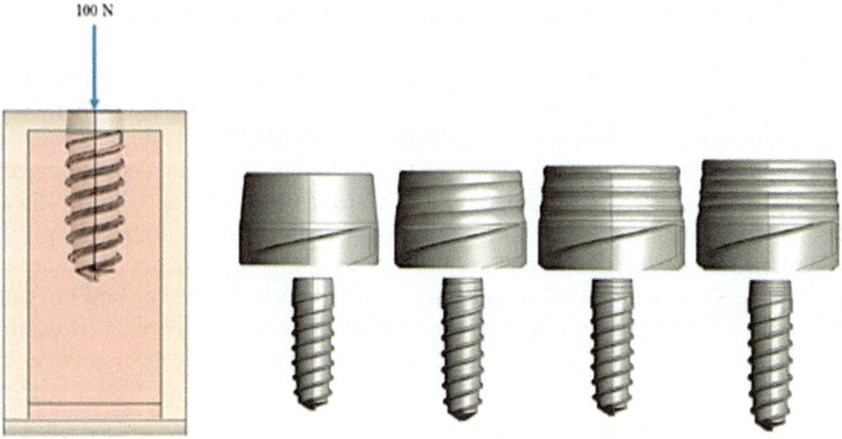

Fig. 3. 100 N load application on dental implant (left), 4 different neck design (right) [17]

2.2 *Load Distribution*

Dental implants face substantial mechanical forces during mastication and other oral activities. Uneven load distribution can result in stress concentrations, which are detrimental to implant longevity. The design of the implant and its threading pattern should aim to distribute these loads evenly, preventing excessive forces from concentrating in one area [15]. Load distribution is critical not only for the immediate stability of the implant upon insertion (primary stability) but also for the long-term stability achieved through osseointegration (secondary stability). Achieving both types of stability is crucial for implant success and depends on proper load distribution [16] (Fig. 3).

2.3 *Stress Analysis*

Stress analysis is central to dental implant design. It involves the evaluation of how forces are transmitted within the implant-bone system and the impact on different components of the implant. This analysis aids in assessing whether the implant can withstand the mechanical demands of the oral environment [18] (Fig. 4).

Stress analysis identifies areas of potential stress concentration. Design modifications can then be made to mitigate these concentrations, distributing the mechanical forces more uniformly and preventing localized overloading. This is crucial for preventing implant failure due to mechanical stress [20].

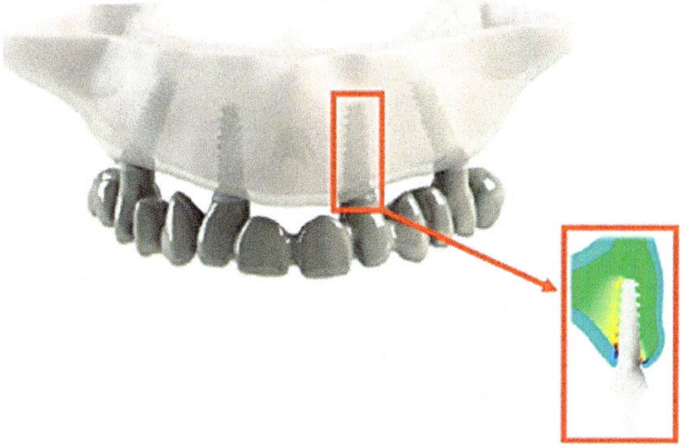

Fig. 4 Stress–strain distribution on the dental implant [19]

3 Geometric Modeling of Dental Implants

The development of three-dimensional (3D) geometric models of dental implants and the surrounding oral structures is a critical phase in the design and evaluation of dental implant systems. Accurate geometric models ensure that implant components are designed to fit perfectly within the patient's oral anatomy. A well-fitted implant reduces the risk of complications and enhances implant stability. Implants that are designed based on accurate geometric representations are more likely to function optimally (Fig. 5).

The alignment and angulation of the implant are crucial for proper occlusion and masticatory function. Accurate geometry representation allows for the evaluation of how mechanical forces will be distributed within the implant-bone interface. This assessment is vital for ensuring that the implant can withstand the stresses of daily oral activities, preventing overloading and potential implant failure. Each patient's oral anatomy is unique.

Accurate geometric models enable customization of the implant components to match the patient's specific anatomy. Customization enhances patient comfort and

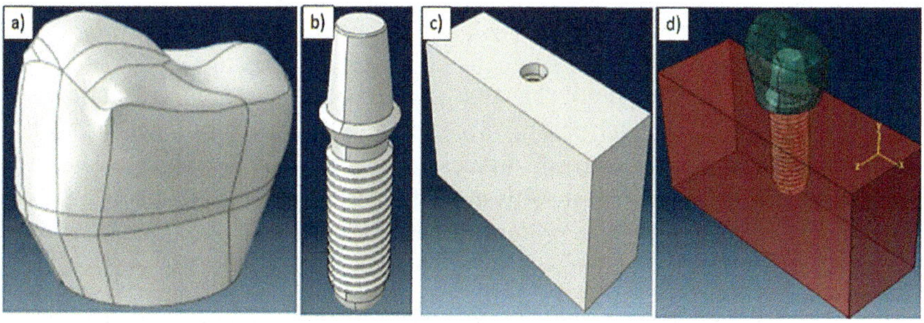

Fig. 5 **a** Geometric modeling of a ceramic crown, **b** geometric modeling of titanium root, **c** geometric modeling of the jawbone, and **d** assembly of model parts

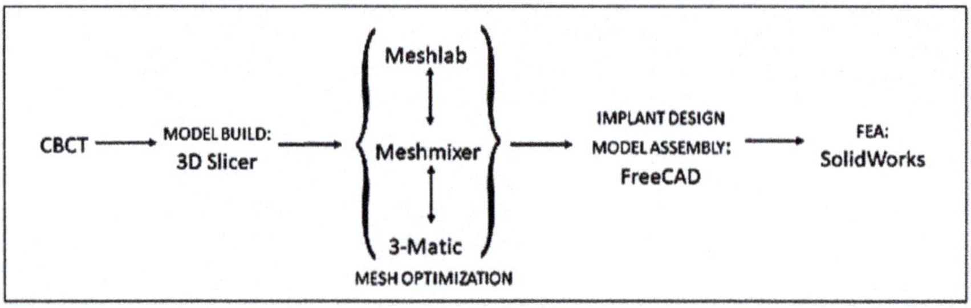

Fig. 6 FEA modeling workflow

ensures the implant's longevity. Accurate geometric models serve as the foundation for predictive analysis, including stress analysis, finite element analysis (FEA), and osseointegration assessments. These analyses provide insights into how the implant will perform and interact with the surrounding oral structures. This section provides an in-depth overview of the process involved in generating these 3D models, which are instrumental for assessing implant performance, optimizing fit, and ensuring patient-specific customization (Fig. 6).

3.1 Image Acquisition and Data Collection

The foundation of 3D modeling begins with image acquisition and data collection. High-resolution imaging modalities, such as cone-beam computed tomography (CBCT) and magnetic resonance imaging (MRI), are commonly employed. These imaging techniques offer detailed visualizations of the oral cavity, including the bone structure, teeth, and adjacent soft tissues. The choice of imaging modality depends on the clinical needs and the level of detail required [21].

3.2 Image Segmentation and Processing

Once the images are acquired, the next step involves image segmentation and processing. This is a critical phase where specific structures of interest, such as the jawbone, teeth, nerves, and soft tissues, are delineated and separated from the surrounding anatomy. Advanced software and algorithms are used for this purpose, enabling the creation of digital representations of these structures. Segmentation may require manual intervention to ensure accuracy and precision [22] (Fig. 7).

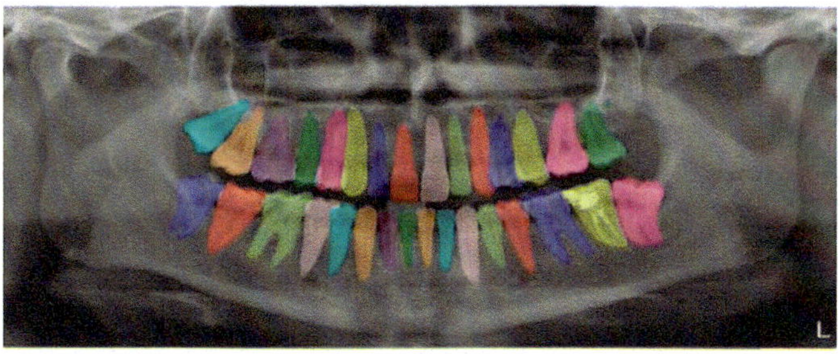

Fig. 7 Automatic segmentation of the teeth

Fig. 8 Modelling mesh set

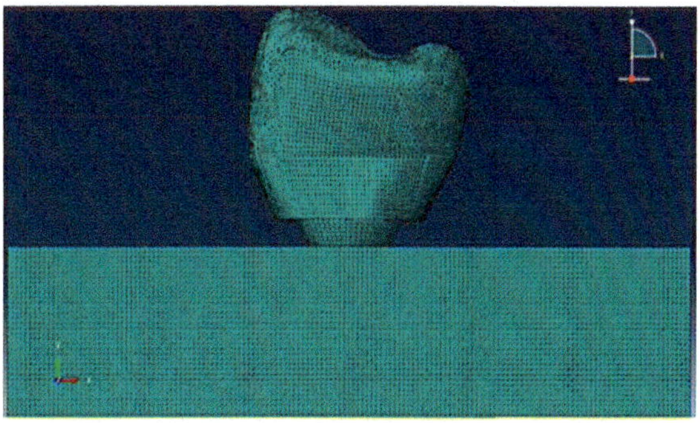

3.3 *Surface Reconstruction and Mesh Generation*

The segmented images provide the basis for surface reconstruction and mesh generation. This process involves converting the segmented data into 3D surface models. The software algorithms interpolate data points, creating smooth, continuous surfaces that accurately represent the oral structures. These 3D models are then transformed into mesh structures, which are composed of interconnected polygons. These meshes form the basis for subsequent analysis and simulations [23] (Fig. 8).

3.4 *Implant Modeling and Placement*

Dental implant modeling is a critical component of the 3D modeling process. Virtual implant components, including the implant body, abutment, and crown, are created using computer-aided design (CAD) software. The models are precisely tailored to match the selected implant system. The placement of the implant within the 3D model is also simulated, considering factors such as position, depth, and angulation to ensure proper fit and optimal functionality [23] (Fig. 9).

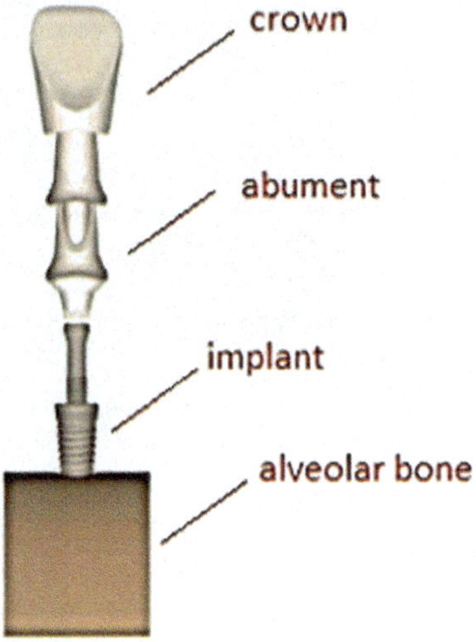

Fig. 9 Parts of implant set [24]

3.5 Integration and Customization

The implant model is integrated into the 3D model of the oral cavity. This integration allows for the assessment of implant-bone interactions, load distribution, and the evaluation of surrounding structures' response to the implant. Additionally, customization is achieved by tailoring the implant model to the patient's specific anatomy, optimizing the implant's fit and stability within the oral environment [25].

3.6 Validation and Simulation

The accuracy and validity of the 3D model are crucial. Verification and validation processes are carried out to ensure that the model closely resembles the patient's oral anatomy. Once validated, the 3D model can be used for a variety of simulations, including stress analysis, load distribution, and osseointegration assessments, enabling the prediction of implant performance and potential clinical outcomes [26].

In summary, the process of creating 3D geometric models of dental implants and surrounding oral structures is a multifaceted and iterative procedure. It involves image acquisition, segmentation, surface reconstruction, implant modeling, and integration, all of which culminate in a highly accurate and patient-specific representation of the oral environment. These 3D models serve as invaluable tools for the evaluation and optimization of dental implant systems, ultimately contributing to enhanced patient care and successful clinical outcomes.

4 Finite Element Analysis (FEA) in Dental Implants

One of the key strengths of Finite Element Analysis (FEA) in the context of dental implant research is its capacity to conduct in-depth analyses of stress and strain distribution within dental implants. FEA employs numerical models to simulate the mechanical behavior of implant systems under varying loading conditions. This capability has offered crucial insights into the biomechanics of dental implants, leading to improved designs and better clinical outcomes.

4.1 Stress and Strain Distribution Analysis Using FEA

Oladapo et al. [28] conducted a comprehensive review of FEA's role in assessing stress distribution within dental implants. Their study highlighted the importance of FEA in understanding stress patterns and optimizing implant designs for enhanced stability. In a study by Patil et al. [29], FEA was used to predict stress and strain distributions in dental implants, emphasizing the significance of these analyses in evaluating implant stability and performance. Robinson, Alqahtani et al. [30] delved into the detailed stress and strain distribution analyses enabled by FEA. Their research emphasized the impact of FEA on customizing implant designs to mitigate stress concentrations and improve long-term functionality. Arabeikki et al. [31] explored the role of FEA in optimizing implant thread designs to enhance mechanical stability. Stress distribution analyses conducted using FEA were a critical part of their research. An investigation by Fourie et al. [32] focused on the segmentation and 3D reconstruction of cone-beam CT images for dental implant simulation. Their use of FEA allowed for in-depth analyses of stress and strain distribution within the implant-bone interface, aiding in better design and evaluation (Fig. 10).

4.2 Assessing Osseointegration with FEA

Finite Element Analysis (FEA) has proven to be an invaluable tool in the evaluation of osseointegration, a critical aspect of dental implant success and longevity. Osseointegration is the direct structural and functional connection between the dental implant and the surrounding bone, and FEA allows for the predictive analysis of bone-implant contact and the assessment of its long-term stability. This capability is fundamental for understanding and improving the clinical outcomes of dental implant treatments (Fig. 11).

Taylor et al. [34] conducted a comprehensive review highlighting the role of FEA in the predictive analysis of osseointegration. The study emphasized how FEA aids in assessing the degree of bone-implant contact and its implications for long-term implant stability. In their research, Dickinson et al. (2021) explored the application of

Fig. 10 FEA micro-strain result of dental implant [27]

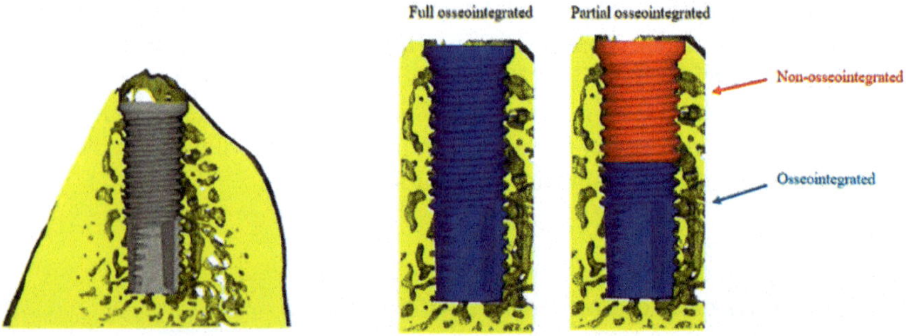

Fig. 11 Osseointegration and implant-bone integration [33]

FEA in evaluating the osseointegration process. They discussed how FEA simulations can predict and visualize the bone-implant interface, aiding in the assessment of integration quality. Deng et al. [35] investigated the predictive analysis of dental implant stability using FEA, with a focus on the role of osseointegration. Their study emphasized how FEA allows for the assessment of the stability and long-term success of dental implants through osseointegration evaluation. Aftabi et al. [36] reviewed the customization of dental implant components for patient-specific anatomical fit, underscoring the role of FEA in evaluating and enhancing osseointegration in implant dentistry. An exploration by Desai et al. [30] delved into the use of FEA for optimizing implant thread design for enhanced mechanical stability, where osseointegration assessment played a significant role in evaluating implant performance.

Memon et al. [37] conducted a comprehensive review of computer-aided design and FEA in dental implants. Their study emphasized the significance of patient-specific modeling for customized implant designs based on individual patient anatomical data. In a study by Garcia, Lee et al. [38], the authors discussed the customization of dental implant components for patient-specific anatomical fit. Their research highlighted how patient-specific modeling using FEA can lead to enhanced implant performance and stability. Ganz et al. [39] explored the use of computer-aided segmentation and 3D reconstruction of cone-beam CT images for dental implant simulation. Their study demonstrated how patient-specific FEA modeling can precisely simulate stress and strain distributions, optimizing implant performance. An investigation by Mehboob al. [40] focused on predictive analysis of dental implant integration using FEA. The study highlighted the role of patient-specific modeling in predicting and optimizing osseointegration, ensuring a precise fit for individual patients. Ingawale et al. [41] examined FEA's role in stress distribution analysis in dental implants and its potential for patient-specific modeling. Their research emphasized the customization of implant designs based on individual patient anatomy for improved long-term stability.

4.3 FEA's Role in Advancing Implant Materials

Finite Element Analysis (FEA) is poised to continue supporting material advancements in the field of implant dentistry. It plays a crucial role in the evaluation of new implant materials, guiding the development of implants with enhanced mechanical properties and biocompatibility. This section explores how FEA has contributed to material advancements and cites relevant studies that showcase the impact of FEA on the development of superior implant materials.

Kim et al. [42] conducted a study focusing on the role of FEA in optimizing implant thread design for enhanced mechanical stability. Their research demonstrated how FEA can evaluate the mechanical properties of different materials, aiding in the selection of materials suitable for specific clinical scenarios. In their comprehensive review, Geng et al. [43] emphasized the significance of FEA in assessing the material properties of dental implants. Their study discussed how FEA has played a pivotal role in evaluating new implant materials with improved mechanical performance. Chang et al. [42] investigated predictive analysis of dental implant stability using FEA, with a focus on the role of FEA in material evaluation. Their study highlighted how FEA aids in selecting implant materials with enhanced mechanical properties, ensuring implant stability. An exploration by Pant et al. [44] delved into computer-aided design and FEA in dental implants. The study emphasized how FEA can assess the mechanical behavior of different materials, driving advancements in the field of implant dentistry. In their research, Tiwari et al. [45] reviewed the customization of dental implant components for patient-specific anatomical fit and the role of FEA in evaluating material biocompatibility. Their work showcased how FEA contributes to the development of biocompatible implant materials.

4.4 *Machine Learning and AI-Enhanced Optimization of Implant Designs*

The integration of machine learning algorithms and artificial intelligence (AI) is set to revolutionize the field of dental implant research, particularly in the context of optimizing implant designs and enhancing the accuracy of Finite Element Analysis (FEA). Machine learning and AI-driven FEA models are expected to play a significant role in adapting to specific clinical scenarios and patient profiles. This section explores how AI and machine learning are reshaping the landscape of implant design and FEA, referencing studies that highlight these advancements.

Preda et al. [46] investigated the use of computer-aided segmentation and 3D reconstruction of cone-beam CT images for dental implant simulation. Their research not only highlighted the current role of AI-driven FEA models but also pointed to the future potential where AI can adapt FEA simulations to specific clinical scenarios and patient profiles. In a study by Wang et al. [47], machine learning algorithms were employed to optimize implant designs. The research showcased how AI-driven designs outperformed traditional ones in terms of stability and performance. Tandon et al. [48] conducted a study on AI-based FEA simulations in dental implants. Their research demonstrated the potential for AI to enhance the accuracy and speed of FEA simulations, leading to more precise assessments of implant behavior. In a review by Sikri et al. [49], AI's role in dental implant optimization was explored. The study discussed the application of AI in customizing implant designs to fit specific clinical scenarios, ultimately improving the efficacy and predictability of implant treatments. Lerner et al. [50] examined the use of AI in dental implant planning and design. Their research demonstrated the adaptability of AI-driven FEA models to various patient profiles, leading to more personalized and effective implant solutions.

5 Materials and Manufacturing Processes

Dental implants are predominantly constructed using specific materials that exhibit favorable mechanical properties, biocompatibility, and longevity. This section delves into the materials commonly used in dental implant construction, such as titanium and zirconia, and references relevant studies that provide insights into these materials and their suitability for implant applications. Osman et al. [51] conducted a comprehensive review of dental implant materials, with a focus on the use of titanium in implant construction. Their study discussed the mechanical and biocompatible properties of titanium and its extensive application in implant dentistry. In a study by

Greh et al. [52], the authors investigated the use of zirconia as an alternative dental implant material. The research explored the properties of zirconia, highlighting its potential for dental implant construction. Apratim et al. [53] reviewed the customization of dental implant components, where the choice of materials, including titanium and zirconia, was a central theme. Their research emphasized the role of materials in achieving patient-specific anatomical fit. An investigation by Elias et al. [54] examined the mechanical properties of titanium used in dental implants. Their study highlighted the significance of titanium's strength and biocompatibility in implant construction. Gautam et al. [55] explored the use of zirconia as a dental implant material and evaluated its suitability in terms of biocompatibility and mechanical properties. Their research contributed to the understanding of alternative materials for dental implant construction.

5.1 Mechanical Properties and Suitability of Materials for Dental Implants

The mechanical properties of materials used in dental implant construction play a vital role in determining the long-term success and durability of these implants. This section explores the mechanical properties and suitability of materials commonly used for dental implants, such as titanium and zirconia.

Sidira et al. [56] conducted a comprehensive review of dental implant materials, highlighting the mechanical properties of titanium and its suitability for dental implant applications. Their study emphasized titanium's strength, biocompatibility, and corrosion resistance. In a study by Kassem et al. [57], the authors investigated the mechanical properties of zirconia and its suitability as a dental implant material. Their research explored zirconia's biocompatibility, low thermal conductivity, and resistance to corrosion. Chahine et al. [58] reviewed the customization of dental implant components and the choice of materials. Their research discussed the mechanical properties of materials, including titanium and zirconia, in the context of achieving patient-specific anatomical fit. An investigation by Elleuch et al. [59] delved into the mechanical properties of titanium used in dental implants. Their study highlighted the significance of titanium's high tensile strength and fatigue resistance in implant construction. Sanon et al. [60] explored the mechanical properties and suitability of zirconia as a dental implant material. Their research discussed zirconia's high flexural strength and low thermal conductivity, making it a potential alternative for dental implant applications.

5.2 *Impact of Manufacturing Techniques on Implant Performance*

The choice of manufacturing techniques for dental implants is a crucial factor that influences their performance. This section delves into different manufacturing methods, including machining and additive manufacturing, and discusses their implications on the performance of dental implants.

Lin et al. [61] conducted a comprehensive review of dental implant manufacturing techniques, emphasizing the importance of precision machining. Their study discussed how machining affects the dimensional accuracy and surface quality of implant components. In a study by Raines et al. [62], the authors investigated the impact of additive manufacturing on dental implant performance. Their research explored the role of additive manufacturing techniques, such as 3D printing, in creating custom implant designs and the influence on implant stability. Salmi et al. [63] reviewed the customization of dental implant components, which touched on manufacturing techniques for patient-specific fit. Their research emphasized the role of precision machining and additive manufacturing in achieving tailored implant components. An investigation by Elias et al. [64] discussed machining techniques used in dental implant production and their impact on implant stability. The study highlighted how precision machining contributes to the reliability and performance of dental implants. Tunchel et al. [65] explored the influence of additive manufacturing in dental implant applications. Their research focused on the customization of implant designs through 3D printing and its implications for implant performance [66].

6 Challenges and Limitations in Dental Implant

Despite the significant advancements in dental implant mechanical engineering, the field grapples with various challenges and limitations that necessitate focused attention and innovation. This section delves into key obstacles faced by researchers and practitioners in the domain of dental implant mechanical engineering and draws upon ten published articles to provide valuable insights into these challenges.

Oshida et al. [67] conducted an extensive review of dental implant materials, emphasizing the challenges associated with material selection and the need for advanced biocompatible materials to meet diverse patient needs. In their study, Gulati et al. [68] investigated the challenges associated with achieving optimal osseointegration, particularly in patients with compromised bone quality, highlighting strategies for enhancing osseointegration in challenging clinical scenarios. Tel et al. [69] focused on the customization of dental implant components, addressing challenges posed by patient-specific anatomical variations and the necessity for precision in implant design. An investigation by Rangert et al. [70] discussed the challenges related to implant fractures and their potential impact on the long-term success of dental implants, offering insights into factors contributing to implant fractures and potential preventive measures. Nagey et al. [71] explored the issues of wear and corrosion in dental implants, highlighting their potential consequences over time and discussing strategies to mitigate wear and corrosion challenges for improved implant longevity.

Johnson, Haiat [72] published a study focusing on the challenges of achieving long-term mechanical stability in dental implant designs, particularly in cases with complex anatomical conditions. They highlighted the need for advanced modeling techniques to address these challenges effectively. Mitra et al. [73] conducted a comprehensive study on the challenges of ensuring material biocompatibility in dental implant components and explored innovative approaches to enhance biocompatibility. In a clinical trial, Jun et al. [74] addressed the challenge of developing patient-specific implant designs for individuals with unique anatomical features, emphasizing the importance of geometric modeling and precision engineering. Peron et al. [75] discussed the limitations associated with implant wear and corrosion in the context of material degradation and offered insights into potential material improvements to address these limitations. Anitua et al. [76] conducted research on the long-term performance of dental implants, highlighting the need for extended follow-up studies to assess implant longevity, thereby addressing the challenge of long-term predictability.

7 Future Directions and Innovations in Dental Implant

Dental implant mechanical engineering continues to evolve, offering exciting prospects for the future. This section explores potential future directions and innovations, ranging from advanced materials to cutting-edge techniques like 3D printing and virtual surgery planning. It references ten published articles to provide insights into these promising developments.

Dong et al. [77] discussed the potential of advanced materials for dental implants, such as bioactive coatings and materials with enhanced biocompatibility, paving the way for improved implant performance.

Agarwal et al. [78] highlighted the future of regenerative therapies and biomaterial coatings to enhance osseointegration, providing a glimpse into innovative approaches for improving implant success. Lim et al. [79] explored the potential of patient-specific implants, enabled by advanced geometric modeling and 3D printing technologies, offering highly customized solutions for patients with unique anatomical variations. Mondal et al. [80] discussed the future prospects of enhanced implant designs and materials to mitigate fracture risks, paving the way for implants with improved mechanical stability. Bender et al. [81] presented the future of corrosion-resistant materials and coatings, addressing the limitations of wear and corrosion in dental implants and ensuring their long-term performance.

Jiang et al. [82] explored advanced modeling techniques and materials as the future of ensuring long-term mechanical stability in dental implant designs. Meyers et al.

[83] discussed the future of bioactive materials and surface modifications to ensure material biocompatibility in dental implant components. Sharma et al. [84] highlighted the potential of virtual surgery planning and 3D printing for highly customized patient-specific implant solutions, revolutionizing the implant design and manufacturing process. Zhai et al. [85] explored the future of advanced wear-resistant materials and coatings to address the limitations of implant wear and corrosion, ensuring the longevity of dental implants. Mok et al. [86] discussed the future of extended follow-up studies and predictive modeling for improved long-term predictability, offering insights into monitoring and enhancing implant longevity.

8 Conclusion

In this comprehensive review, the readers have embarked on a journey through the dynamic landscape of dental implant mechanical engineering, where innovation converges with patient care. The harmonization of mechanical engineering principles, geometric modeling, and finite element analysis has brought forth a paradigm shift in dental implantology. This multidisciplinary field, with its emphasis on both form and function, has reshaped the contours of modern dentistry. The significance of dental implants in restoring oral function, aesthetics, and self-assurance, and thereby elevating the quality of life, has been thoroughly explored. The preservation of the jawbone and facial structure, the remarkable durability of implants, and their pivotal role in enhancing oral health have emerged as substantial achievements.

The versatility, predictability, and non-invasive nature of dental implants have assumed center stage, offering a spectrum of solutions for patients with varying degrees of tooth loss. The precision facilitated by computer-aided design, three-dimensional modeling, and finite element analysis has propelled a more personalized and efficient approach to implant treatments. Furthermore, the in-depth analysis of existing literature has illuminated the potential future directions in dental implant mechanical engineering, encompassing advanced materials, patient-specific solutions, and predictive modeling. These innovations hold the promise of further amplifying success rates, implant longevity, and patient contentment, underscoring the ever-evolving nature of this discipline. As this journey concludes, one must acknowledge the enduring challenges and constraints that dental implant mechanical engineering grapples with. The choices of materials, the intricacies of osseointegration, the prevention of fractures, the management of wear and corrosion, the accommodation of patient-specific anatomical variations, and the assurance of long-term predictability are but a few of the challenges that continue to stimulate research and innovation.

In closing, dental implant mechanical engineering stands at the forefront of modern dentistry, delivering transformative solutions for tooth loss and beyond. The synergy of mechanical engineering principles, geometric modeling, and finite element analysis has set a new benchmark, ushering in enhanced patient outcomes and contentment. With ongoing advancements in research and technology, one can anticipate a future where the success rates, longevity, and patient satisfaction in dental implant treatments will steadily ascend, promising a brighter and healthier smile for generations to come.

References

1. Anil S, Sudeep K, Saratchandran S, Sweety VK (2023) Revolutionizing dental caries diagnosis through artificial intelligence
2. Chang Y, Tambe AA, Maeda Y, Wada M, Gonda T. Finite element analysis of dental implants with validation: to what extent can we expect the model to predict biological phenomena? A literature review and proposal for classification of a validation process
3. Yadav S, Gangwar S (2019) A critical evaluation of tribological interaction for restorative materials in dentistry. Int J Polym Mater Polym Biomater 68:1005–1019
4. Hollister SJ, Flanagan CL, Zopf DA, Morrison RJ, Nasser H, Patel JJ et al (2015) Design control for clinical translation of 3D printed modular scaffolds. Ann Biomed Eng 43:774–786
5. Long M, Rack HJ (1998) Titanium alloys in total joint replacement—a materials science perspective. Biomaterials 19:1621–1639
6. Lee H, Jo M, Noh G (2021) Biomechanical effects of dental implant diameter, connection type, and bone density on microgap formation and fatigue failure: a finite element analysis. Comput Methods Programs Biomed 200. https://doi.org/10.1016/j.cmpb.2020.105863
7. Chatterjee S, Sarkar S, Kalidindi SR, Basu B (2019) Periprosthetic biomechanical response towards dental implants, with functional gradation, for single/multiple dental loss. J Mech Behav Biomed Mater 94:249–258
8. Singhal P, Krishna BV, Kumar A, Chauhan S, Judi HK, Jain A (2023) Optimization of design and performance of medical implants using FEA. In: E3S web conference, vol 430. EDP Sciences, p 1114
9. Arsalanloo Z, Telchi R, Osgouie KG (2014) Selection of optimum thread type in implants to achieve optimal biomechanical properties by using 3D finite element method. Int J Biosci Biochem Bioinforma 4:185
10. Rojo R, Prados-Privado M, Reinoso AJ, Prados-Frutos JC (2018) Evaluation of fatigue behavior in dental implants from in vitro clinical tests: a systematic review. Metals (Basel) 8:313
11. Liu Y, Wang J (2017) Influences of microgap and micromotion of implant—abutment interface on marginal bone loss around implant neck. Arch Oral Biol 83:153–160
12. Lin H, Yin C, Mo A (2021) Zirconia based dental biomaterials: structure, mechanical properties, biocompatibility, surface modification, and applications as implant. Front Dent Med 2:689198
13. Fande AW, Taiwade RV (2022) Development of activated tungsten inert gas welding and its current status: a review. Mater Manuf Process 37:841–876. https://doi.org/10.1080/10426914.2022.2039695
14. Cosola S, Toti P, Babetto E, Covani U, Peñarrocha-Diago M, Peñarrocha-Oltra D (2021) In-vitro fatigue and fracture performance of three different ferrulized implant connections used in fixed prosthesis. J Dent Sci 16:397–403
15. Steigenga JT, Al-Shammari KF, Nociti FH, Misch CE, Wang H-L (2003) Dental implant design and its relationship to long-term implant success. Implant Dent 12:306–317
16. Javed F, Romanos GE (2010) The role of primary stability for successful immediate loading of dental implants. A literature review. J Dent 38:612–620
17. Dávila E, Ortiz-Hernández M, Perez RA, Herrero-Climent M, Cerrolaza M, Gil FJ (2019) Crestal module design optimization of dental implants: finite element analysis and in vivo studies. J Mater Sci Mater Med 30:90. https://doi.org/10.1007/s10856-019-6291-1
18. Hämmerle CHF, Glauser R (2000) Clinical evaluation of dental implant treatment. Periodontol 2004(34):230–239

19. Valera-Jiménez JF, Burgueño-Barris G, Gómez-González S, López-López J, Valmaseda-Castellón E, Fernández-Aguado E (2020) Finite element analysis of narrow dental implants. Dent Mater
20. Taiwade RV (2022) Microstructure and mechanical properties of wire arc additively manufactured bimetallic structure of austenitic stainless steel and low carbon steel. J Mater Eng Perform 31:8531–8541
21. Taiwade RV, Agarwal A (2023) Investigation of microstructural and corrosion behavior of 316LSi structure developed by wire arc additive manufacturing. Mater Today Commun 35:105596
22. Lee S, Woo S, Yu J, Seo J, Lee J, Lee C (2020) Automated CNN-based tooth segmentation in cone-beam CT for dental implant planning. IEEE Access 8:50507–50518
23. Manmadhachary A, Kumar R, Krishnanand L (2016) Improve the accuracy, surface smoothing and material adaption in STL file for RP medical models. J Manuf Process 21:46–55
24. Kang X, Li Y, Wang Y, Zhang Y, Yu D, Peng Y (2020) Relationships of stresses on alveolar bone and abutment of dental implant from various bite forces by three-dimensional finite element analysis. Biomed Res Int 2020:7539628. https://doi.org/10.1155/2020/7539628
25. Jardini AL, Larosa MA, Maciel Filho R, de Carvalho Zavaglia CA, Bernardes LF, Lambert CS et al (2014) Cranial reconstruction: 3D biomodel and custom-built implant created using additive manufacturing. J Cranio-Maxillofac Surg 42:1877–1884
26. Chang P, Lang NP, Giannobile WV (2010) Evaluation of functional dynamics during osseointegration and regeneration associated with oral implants. Clin Oral Implant Res 21:1–12
27. Tobar-Reyes J, Andueza-Castro L, Jiménez-Silva A, Bustamante-Plaza R, Carvajal-Herrera J (2021) Micromotion analysis of immediately loaded implants with Titanium and Cobalt-Chrome superstructures. 3D finite element analysis. Clin Exp Dent Res
28. Oladapo BI, Kayode JF, Karagiannidis P, Naveed N, Mehrabi H, Ogundipe KO (2022) Polymeric composites of cubic-octahedron and gyroid lattice for biomimetic dental implants. Mater Chem Phys 289:126454
29. Patil V, Noronha DJ, Naik N, Shettar M (2019) A comparative study on the effect of stress in dental implant structure using finite element analysis. Int J Mech Prod Eng Res Dev 9:709–717
30. Alqahtani AR, Desai SR, Patel JR, Alqhtani NR, Alqahtani AS, Heboyan A et al (2023) Investigating the impact of diameters and thread designs on the biomechanics of short implants placed in D4 bone: a 3D finite element analysis. BMC Oral Health 23:686
31. Arabbeiki M, Niroomand MR, Rouhi G (2023) Improving dental implant stability by optimizing thread design: simultaneous application of finite element method and data mining approach. J Prosthet Dent 130:602-e1
32. Engelbrecht WP, Fourie Z, Damstra J, Gerrits PO, Ren Y (2013) The influence of the segmentation process on 3D measurements from cone beam computed tomography-derived surface models. Clin Oral Investig 17:1919–1927
33. Marcián P, Wolff J, Horáčková L, Kaiser J, Zikmund T, Borák L (2018) Micro finite element analysis of dental implants under different loading conditions. Comput Biol Med 96:157–165. https://doi.org/10.1016/j.compbiomed.2018.03.012
34. Taylor M, Prendergast PJ (2015) Four decades of finite element analysis of orthopaedic devices: where are we now and what are the opportunities? J Biomech 48:767–778
35. Deng B, Tan KB, Liu GR, Lu Y (2008) Influence of osseointegration degree and pattern on resonance frequency in the assessment of dental implant stability using finite element analysis. Int J Oral Maxillofac Implant 23
36. Aftabi H, Eghbal A, McGregor S, Zaraska K, Prisman E, Hodgson A et al (2023) Computational models and their applications in biomechanical analysis of mandibular reconstruction surgery
37. Memon AR, Wang E, Hu J, Egger J, Chen X (2020) A review on computer-aided design and manufacturing of patient-specific maxillofacial implants. Expert Rev Med Devices 17:345–356
38. Lee J-A, Koh Y-G, Kang K-T (2020) Biomechanical and clinical effect of patient-specific or customized knee implants: a review. J Clin Med 9:1559
39. Ganz SD (2008) Computer-aided design/computer-aided manufacturing applications using CT and cone beam CT scanning technology. Dent Clin North Am 52:777–808

40. Mehboob H, Ouldyerou A, Ijaz MF (2023) Biomechanical investigation of patient-specific porous dental implants: a finite element study. Appl Sci 13:7097
41. Ingawale SM, Goswami T (2022) Design and finite element analysis of patient-specific total temporomandibular joint implants. Materials (Basel) 15:4342
42. Kim WH, Lee J-C, Lim D, Heo Y-K, Song E-S, Lim Y-J et al (2019) Optimized dental implant fixture design for the desirable stress distribution in the surrounding bone region: a biomechanical analysis. Materials (Basel) 12:2749
43. Geng J-P, Tan KBC, Liu G-R (2001) Application of finite element analysis in implant dentistry: a review of the literature. J Prosthet Dent 85:585–598
44. Pant G, Reddy MSS, Praveen P, Parashar AK, Kareem SA, Nijhawan G (2023) Advanced casting techniques for complex-shaped components: design, simulation and process control. In: E3S web conference, vol 430. EDP Sciences, p 1110
45. Tiwari A, Gupta VK, Haldkar RK, Parinov IA (2022) Biomechanical analysis of patient-specific temporomandibular joint implant and comparison with natural intact jaw bone using finite element method. Appl Sci 12:3003
46. Preda F, Morgan N, Van Gerven A, Nogueira-Reis F, Smolders A, Wang X et al (2022) Deep convolutional neural network-based automated segmentation of the maxillofacial complex from cone-beam computed tomography: a validation study. J Dent 124:104238
47. Wang C, He T, Zhou H, Zhang Z, Lee C (2023) Artificial intelligence enhanced sensors-enabling technologies to next-generation healthcare and biomedical platform. Bioelectron Med 9:17
48. Tandon D, Rajawat J, Banerjee M (2020) Present and future of artificial intelligence in dentistry. J Oral Biol Craniofacial Res 10:391–396
49. Sikri A, Sikri J, Gupta R (2023) Artificial intelligence in prosthodontics and oral implantology—a narrative review. Glob Acad J Dent Oral Heal 5
50. Lerner H, Mouhyi J, Admakin O, Mangano F (2020) Artificial intelligence in fixed implant prosthodontics: a retrospective study of 106 implant-supported monolithic zirconia crowns inserted in the posterior jaws of 90 patients. BMC Oral Health 20:1–16
51. Osman RB, Swain MV (2015) A critical review of dental implant materials with an emphasis on titanium versus zirconia. Materials (Basel) 8:932–958
52. Grech J, Antunes E (2019) Zirconia in dental prosthetics: a literature review. J Mater Res Technol 8:4956–4964
53. Apratim A, Eachempati P, Salian KKK, Singh V, Chhabra S, Shah S (2015) Zirconia in dental implantology: a review. J Int Soc Prev Community Dent 5:147
54. Elias CN, Fernandes DJ, Resende CRS, Roestel J (2015) Mechanical properties, surface morphology and stability of a modified commercially pure high strength titanium alloy for dental implants. Dent Mater 31:e1–e13
55. Gautam C, Joyner J, Gautam A, Rao J, Vajtai R (2016) Zirconia based dental ceramics: structure, mechanical properties, biocompatibility and applications. Dalt Trans 45:19194–19215
56. Kirmanidou Y, Sidira M, Drosou M-E, Bennani V, Bakopoulou A, Tsouknidas A et al (2016) New Ti-alloys and surface modifications to improve the mechanical properties and the biological response to orthopedic and dental implants: a review. Biomed Res Int 2016
57. Kassem YM, Alshimy AM, El-Shabrawy SM (2019) Mechanical evaluation of polyetheretherketone compared with zirconia as a dental implant material. Alexandria Dent J 44:61–66
58. Chahine G, Koike M, Okabe T, Smith P, Kovacevic R (2008) The design and production of Ti-6Al-4V ELI customized dental implants. Jom 60:50–55
59. Elleuch S, Jrad H, Wali M, Dammak F (2023) Agglomeration effect on biomechanical performance of CNT-reinforced dental implant using micromechanics-based approach. J Mech Behav Biomed Mater 145:106023
60. Sanon C, Chevalier J, Douillard T, Cattani-Lorente M, Scherrer SS, Gremillard L (2015) A new testing protocol for zirconia dental implants. Dent Mater 31:15–25
61. Lin L, Fang Y, Liao Y, Chen G, Gao C, Zhu P (2019) 3D printing and digital processing techniques in dentistry: a review of literature. Adv Eng Mater 21:1801013

62. Raines R, Day JB, Salary R (2022) Experimental characterization of the mechanical properties of medical-grade dental implants, fabricated using vat-photopolymerization additive manufacturing process. In: International manufacturing science and engineering conference, vol 85802. American Society of Mechanical Engineers, p V001T07A011
63. Salmi M (2013) Medical applications of additive manufacturing in surgery and dental care
64. Elias CN, Rocha FA, Nascimento AL, Coelho PG (2012) Influence of implant shape, surface morphology, surgical technique and bone quality on the primary stability of dental implants. J Mech Behav Biomed Mater 16:169–180
65. Tunchel S, Blay A, Kolerman R, Mijiritsky E, Shibli JA (2016) 3D printing/additive manufacturing single titanium dental implants: a prospective multicenter study with 3 years of follow-up. Int J Dent 2016
66. Taiwade RV (2021) Wire arc additive manufacturing: a comprehensive review and research directions. J Mater Eng Perform 30:4768–4791
67. Oshida Y, Tuna EB, Aktören O, Gençay K (2010) Dental implant systems. Int J Mol Sci 11:1580–1678
68. Gulati K, Ivanovski S (2017) Dental implants modified with drug releasing titania nanotubes: therapeutic potential and developmental challenges. Expert Opin Drug Deliv 14:1009–1024
69. Tel A, Bordon A, Sortino M, Totis G, Fedrizzi L, Ocello E et al (2021) Current trends in the development and use of personalized implants: engineering concepts and regulation perspectives for the contemporary oral and maxillofacial surgeon. Appl Sci 11:11694
70. Rangert BR, Sullivan RM, Jemt TM (1997) Load factor control for implants in the posterior partially edentulous segment. Int J Oral Maxillofac Implant 12
71. Nagay BE, Cordeiro JM, Barao VAR (2022) Insight into corrosion of dental implants: from biochemical mechanisms to designing corrosion-resistant materials. Curr Oral Heal Rep 9:7–21
72. Haiat G, Wang H-L, Brunski J (2014) Effects of biomechanical properties of the bone—implant interface on dental implant stability: from in silico approaches to the patient's mouth. Annu Rev Biomed Eng 16:187–213
73. Bandyopadhyay A, Mitra I, Goodman SB, Kumar M, Bose S (2022) Improving biocompatibility for next generation of metallic implants. Prog Mater Sci 101053
74. Jun Y, Choi K (2010) Design of patient-specific hip implants based on the 3D geometry of the human femur. Adv Eng Softw 41:537–547
75. Peron M, Torgersen J, Berto F (2017) Mg and its alloys for biomedical applications: exploring corrosion and its interplay with mechanical failure. Metals (Basel) 7:252
76. Anitua E, Pinas L, Begoña L, Orive G (2014) Long-term retrospective evaluation of short implants in the posterior areas: clinical results after 10–12 years. J Clin Periodontol 41:404–411
77. Dong H, Liu H, Zhou N, Li Q, Yang G, Chen L et al (2020) Surface modified techniques and emerging functional coating of dental implants. Coatings 10:1012
78. Agarwal R, García AJ (2015) Biomaterial strategies for engineering implants for enhanced osseointegration and bone repair. Adv Drug Deliv Rev 94:53–62
79. Lim J, Sng TJH, Wong RCW (2023) Implant design on virtual digital human skull models for the creation of customized patient-specific regenerative implants: biomechanical consideration. Digit Hum Model Med 679–691
80. Mondal S, Park S, Choi J, Vu TTH, Doan VHM, Vo TT et al (2023) Hydroxyapatite: a journey from biomaterials to advanced functional materials. Adv Colloid Interface Sci 103013
81. Bender R, Féron D, Mills D, Ritter S, Bäßler R, Bettge D et al (2022) Corrosion challenges towards a sustainable society. Mater Corros 73:1730–1751
82. Jiang X, Yao Y, Tang W, Han D, Zhang L, Zhao K et al (2020) Design of dental implants at materials level: an overview. J Biomed Mater Res Part A 108:1634–1661
83. Meyers SR, Grinstaff MW (2012) Biocompatible and bioactive surface modifications for prolonged in vivo efficacy. Chem Rev 112:1615–1632
84. Honigmann P, Sharma N, Okolo B, Popp U, Msallem B, Thieringer FM (2018) Patient-specific surgical implants made of 3D printed PEEK: material, technology, and scope of surgical application. Biomed Res Int 2018

85. Zhai W, Bai L, Zhou R, Fan X, Kang G, Liu Y et al (2021) Recent progress on wear-resistant materials: designs, properties, and applications. Adv Sci 8:2003739
86. Mok M, Nombela-Franco L, Dumont E, Urena M, DeLarochellière R, Doyle D et al (2013) Chronic obstructive pulmonary disease in patients undergoing transcatheter aortic valve implantation: insights on clinical outcomes, prognostic markers, and functional status changes. JACC Cardiovasc Interv 6:1072–1084

Assessment of Modified Concrete Block with Constructional Waste, Industrial Waste and Natural Fiber

Anant Kumar, Antariksh Chandrakar, Aastha Chandrawanshi, Saumya Verma, and Arvind Deb Mishra

Abstract In the present scenario, the extensively consumed construction material is concrete. This man-made material has undergone various stages of development and research. The aggregates used in concrete play a critical role in the determination of concrete properties as approximately 60–75% of its total volume is captured by aggregates. These aggregates pose a significant impact on the environment because the extraction of these materials from natural resources occurs on a huge scale, leading to depletion of natural resources and increased environmental impact. Moreover, the construction industry and steel industry generate a valuable amount of waste, including C&D waste and byproduct waste like foundry sand. This waste material occupies worthy landfill space and can release hazardous impurities into the atmosphere. To get rid of this situation, we have used these waste materials in this research. This study focuses on assessing the mechanical characteristics of concrete prepared with different proportions of foundry sand and C&D waste with the aim of developing a sustainable and environment-friendly concrete mix utilizing these waste products as aggregates. In addition to recycled aggregates, banana fiber is also utilized in the concrete mix to upgrade its mechanical properties. This research study aims to develop a sustainable concrete mix that maintains desirable mechanical properties while being environmentally friendly.

Keywords Reuse · Foundry sand · Sustainability · Construction and demolition waste · Solid waste management · Banana fiber · Natural fiber · Recycling

1 Introduction

Due to the rapid increment in development and industrialization, huge amount of solid waste materials is produced which causes a significant impact on the environment. This waste includes a wide variety such as drywall material, wood material, brick

A. Kumar · A. Chandrakar · A. Chandrawanshi · S. Verma · A. D. Mishra (✉)
Department of Civil Engineering, Shri Shankaracharya Institute of Professional Management and Technology, Mujgahan, Raipur 492015, India
e-mail: arvind.mishra@ssipmt.com

P. Bhingole et al. (eds.), *Advances in Materials Engineering*, Lecture Notes in Mechanical Engineering, https://doi.org/10.1007/978-981-97-7114-1_28

waste, ceramic waste, shingles concrete, plaster, and many other waste materials [1, 2]. The major source of solid waste is demolition sites and restoration schemes, which are currently dumped in landfills [3]. The construction industry is currently responsible for generating several millions of C&D wastes all over the world [4]. At present, concrete is the most important and broadly utilized construction substantial. It is a blend of cement (combining material), sand (serving as fine aggregate), gravel (serving as coarse aggregate), water and different types of admixtures [5]. The extraction and consumption of traditional aggregates which are used in concrete is done at an alarming rate, leading to depletion of natural resources [6]. The rise in the depletion of natural resources and increasing demand for aggregates also have serious environment issues such as change in the landscape, loss of habitat, influence on air contamination, and climate change therefore it needs to be addressed in an efficient manner [7]. Sustainability in concrete can be achieved through recycling this waste material into it [8].

Instead of construction sector, industrial byproducts also generate large volume of solid waste such as foundry sand which occupies valuable space in landfills. Also, the foundry industry is very concerned about reducing landfill disposal and maintenance costs [9]. Foundry sand categorize as a high-quality silica sand having uniform physical properties. It is a residual product generated by the iron, steel, and non-ferrous metal moulding industries and is utilized for moulding purposes due to its high thermal conductivity. When this sand is no longer usable for further use, it is eliminated from the foundry factory and becomes waste [10–12]. Reusing and recycling waste materials can help the environment and improve efficiency and durability in some ways [13]. Various research studies are underway to explore the feasibility of using this discarded foundry sand in concrete as an alternate of conventional riverbed sand, However, there is a absence of comprehensive awareness about the progress and feasibility of WFS in concrete production, highlighting the need for a more in-depth review in this area [14]. Research study of [15, 16] show that replacing sand with 20–30% of WFS to concrete can effectively make the concrete stronger and more stable without affecting the quality standards. And also, WFS can be incorporated into the dense microstructure of concrete.

Additionally, for increasing the mechanical characteristics of concrete, fibers can be proved effective. In this research work banana fiber is utilized as a fiber reinforcement. It is a natural fiber that is obtained from peduncle and pseudo-stem of the banana tree. After the harvesting of banana trees, they are stored or burned spreading pollution and contributing to agricultural waste. The utilization of this waste as a fiber in concrete is a good solution to reduce this problem [17, 18]. Since the major problem of concrete is its brittle nature at its ultimate strength, utilization of natural fibers in concrete can improve the concrete, which can solve this problem. The use of banana fiber (BF) reinforced concrete offers several benefits, such as increased bending strength, improved load-carrying capacity after cracking, and contributing to sustainable development in the field of construction [19, 20]. The main aim of this research was to find a sustainable and high-performance concrete mix incorporating recycled C&D waste, discarded foundry sand, and agricultural waste banana fiber. By identifying the optimal proportion of these recycled materials, the researchers aim to create a concrete mix that is both environmentally friendly and superior to traditional concrete in terms of performance.

2 Materials and Methodology

2.1 Materials

In this research study cement, sand, coarse aggregate (gravel), tap water, construction and demolition waste, waste foundry sand, and banana fiber are utilized as material.

Cement. Cement is a fine powdery substance, and it is grey in color. When it is mixed with water acting as an adhesive substance to bind the constituent's material used in concrete i.e., fine, and coarse aggregates [21]. In this research study PPC of 53 grade. The properties of cement determined by various tests are listed below (Table 1).

Aggregates. Fine aggregate is a granular product that has a size smaller than 4.75 mm. It is obtained from riverbeds. To make a good concrete mix, fine aggregates must be neat and strong [1]. The properties of Fine aggregate determined by various tests are listed below (Table 2).

Coarse aggregate is one of the constituents that is used in the concrete mix and is responsible for its body as the volume of whole concrete mix is majorly occupied by the coarse aggregate. Coarse aggregate is derived from natural sources by mining and crushing mountains and rocks into smaller size. The size of coarse aggregate ranges between 4.75 and 80 mm. In this research work coarse aggregate of good quality having cubic angular shape is used. The properties of Coarse aggregate determined by various tests are listed below (Table 3).

Table 1 Characteristics of cement

Sl. No.	Name of the property	Obtained values
1	Fineness	3.8%
2	Specific gravity	3.341
3	Initial setting time	49 min
4	Final setting time	485 min
5	Consistency	32.2%

Table 2 Characteristics of fine aggregate

Sl. No.	Name of the property	Obtained values
1	Specific gravity	2.4
2	Fineness modulus	3.15 (zone II)
3	Water absorption ratio	1.53%
4	Bulk density	1.45 gm/cc

Table 3 Characteristics of coarse aggregate

Sl. No.	Name of the property	Obtained values
1	Specific gravity	2.73
2	Fineness modulus	3.25
3	Water absorption ratio	0.92%
4	Size	Retained on 4.75 mm IS Sieve

Construction and Demolition Waste. The construction and demolition waste, which was used in this research, was mixed C&D waste and collected locally. In this study, this waste is used to replace a portion of the natural aggregates. The properties of construction and demolition waste determined by various tests are listed below (Table 4 and Fig. 1A).

Foundry Sand. In this paperwork, waste foundry sand is used which was collected from the Jindal power plant, Mandir Hasaud. This Used Foundry sand is used to substitute a portion of fine aggregate. The property of this sand is given below (Table 5).

Banana Fiber. Locally collected banana plantation waste was utilized to obtain banana fiber. The middle portion of banana leaves was cleaned with tap water and dried at room temperature for 24 h. The dried leaves were then cut into smaller pieces

Table 4 Properties of C&D waste

Sl. No.	Name of the property	Obtained values
1	Colour	Grey and reddish brown
2	Specific gravity	2.21
3	Water absorption ratio	2.64%
4	Bulk density	1.62 gm/cc

(A) (B) (C)

Fig. 1 Material used **A** C&D waste, **B** foundry sand, **C** banana fiber

Table 5 Properties of used foundry sand

Sl. No.	Name of the property	Obtained values
1	Colour	Dark grey
2	Specific gravity	2.12
3	Fineness modulus	1.73
4	Water absorption ratio	1.2%

and soaked in a NaOH solution for 24 h. Finally, the soaked leaves were sundried for seven days to extract the banana fiber [1].

2.2 *Trial Mix*

Trial mixes are experimental concrete preparations that involve varying the proportions of cement, sand, gravel, constructional waste, foundry sand, and banana fiber to investigate the effects of these material combinations on the overall properties of concrete blocks. These case studies are determined by a thorough analysis of literature review and the characteristics of the individual materials. The objective of trial mixes is to systematically evaluate different mix proportions and identify the optimal combination for achieving the desired properties. The provided table (Table 6) presents six concrete trial mixes, each identified by a unique mix number. Each trial mix has a distinct proportion of materials. SN0 represents conventional concrete without any waste materials. The proportion of banana fiber is maintained at a consistent 2.5%, while sand and gravel are replaced with varying proportions of foundry sand and C&D waste, ranging from 10 to 50%.

Table 6 Composition of concrete samples for trial mixing

No. of Samples	Cement (%)	Sand (%)	Gravel (%)	FS (%)	C&D (%)	BF (%)
SN0	100	100	100	–	–	–
SN1	100	90	90	10	10	2.5
SN2	100	80	80	20	20	2.5
SN3	100	70	70	30	30	2.5
SN4	100	60	60	40	40	2.5
SN5	100	50	50	50	50	2.5

Table 7 Mix design

w/c ratio	Cement content $\left(\frac{kg}{m^3}\right)$	Water content (litres)	Fine aggregate (riverbed sand) $\left(\frac{kg}{m^3}\right)$	Coarse aggregate (gravel) $\left(\frac{kg}{m^3}\right)$	Mix ratio
0.44	439	212.5	579	1138.5	1:1.31:2.59

2.3 Mix Design

The concrete mix design was conducted in accordance with the guidelines of IS 10262:2019 for the following material properties:

- Concrete Grade—M30
- Type of cement—PPC 53 Grade
- Fine Aggregate—Confirming to Zone II
- Shape of Coarse Aggregate—Angular
- S.G. of Cement—3.341
- S.G. of Sand (Fine Aggregate)—2.4
- S.G. of Gravel (Coarse Aggregate)—2.73.

The concrete mix design is as follows (Table 7).

2.4 Methodology

Mixing and Casting. Concrete mixing is the process of combining all the constituents of concrete, including cement, aggregates, and water, to form a uniform mixture. Hand mixing is suitable for small-scale projects, while machine mixing is preferred for large-scale work. In both methods, the dry ingredients are first mixed together, followed by the addition of water. The mixing process continues until a homogenous, workable mixture is achieved.

Casting or molding is the subsequent step in concrete production. In this stage, the prepared concrete mixture is poured into molds to form the desired concrete blocks. Before filling the molds, they are thoroughly greased or oiled to facilitate easy removal of the hardened concrete blocks. The concrete mixture is then poured into the molds in three layers, with each layer compacted 25 times using a tamping bar. The excess concrete is removed from the top layer, and the molds are kept undisturbed for 24 h at room temperature to allow the concrete to cure properly (as shown in Fig. 2A).

Curing. After 24 h when the concrete block is dried, it is demoulded from the moulds and then it is kept for curing. In this process, the concrete block is immersed into the water for 3, 7 and 28 days. Curing is primarily done to avoid cracking of concrete and improve abrasion resistance property of concrete [13] (as shown in Fig. 2B).

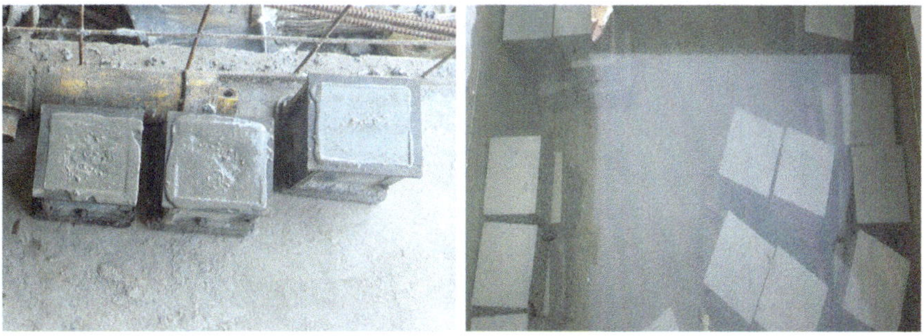

(A) (B)

Fig. 2 Moulding and curing of concrete blocks **A** moulding, **B** curing

2.5 Testing

After the curing of concrete, the final process is testing of concrete block. In this research following tests were conducted.

Workability Test

Slump Cone Test. It is the most common test to determine the workability of the concrete. In this test, concrete mix is placed in a frustum-shaped mould (as shown in Fig. 3A). The concrete mix is placed in 3 layers inside the mould and every layer is compacted twenty-five times by tempting rod. After placing concrete inside the mould, the excess concrete at the topmost layer is removed. The mould is then detached from the concrete in vertical direction. The pattern created by the concrete after removing the mould indicates the consistency of the concrete. Concrete is classified into three types according to the pattern: Shear, Collapse and True Slump [13].

Compaction Factor Test. This test is performed in a laboratory to evaluate low-workability concrete. The test involves using an apparatus consisting of an upper hopper tank, lower hopper tank, and cylinder. Concrete is put inside the top hopper tank and allowed to drop into the bottom hopper tank and cylinder (as shown in Fig. 3B). The cylinder is then filled in four layers and compacted using a tempting rod. The weight of the fully compacted concrete and partially compacted concrete is measured, and the compaction factor is determined by dividing the weight of partially compacted concrete by the weight of fully compacted concrete. This value indicates the degree of workability [13].

Compressive Strength Test

Non-Destructive (Rebound Hammer) Test. In this research RH is used in non-destructive test, RH test is used for estimating the compressive strength of concrete. It is performed via Schmidt rebound hammer, which computes the rebound of a

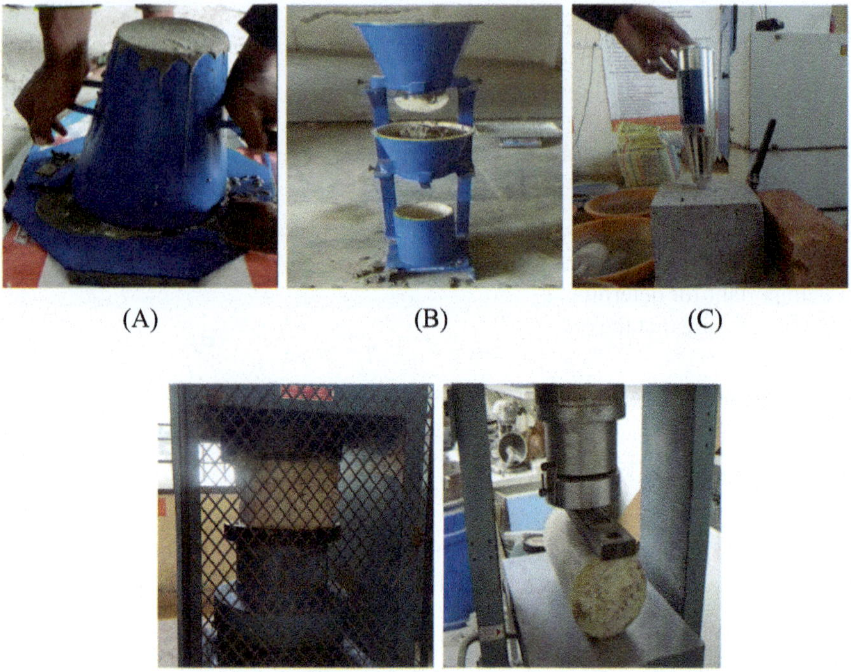

(A) (B) (C)

(D) (E)

Fig. 3 Testing of concrete block **A** slump cone test, **B** compaction factor test, **C** rebound hammer test, **D** compressive strength test of concrete block by CTM machine, **E** spilt tensile strength test

spring-loaded mass when it is pressed against the concrete surface (as shown in Fig. 3C).

Destructive Test (CTM Test). This test is used to measure how much compressive load concrete can bear. To perform the test, fresh concrete is mixed in the desired proportions and poured into a cube-shaped mould in 4 layers. For each layer tamping is done 35 times with a tamping bar. Once the mould is full, the excess concrete is removed from the surface. The mould is removed after one day, and the concrete is cured for seven days. After curing, the concrete cubes are placed into the compression testing machine. A load is directed to the concrete cubes, and the load at which the cubes fail is noted down (as shown in Fig. 3D).

Split Tensile Strength Test. This test measures how strong concrete is when pulled apart. It is a simple and inexpensive test that is widely used to assess the properties of concrete. To perform the test, a cylindrical concrete sample is placed in a machine and squeezed until it breaks in half (as shown in Fig. 3E). The machine measures the force required to break the concrete block, and this value is called the split tensile strength.

Flexural Strength Test. This test of concrete measures the tensile strength of concrete indirectly. This means that it does not directly measure the tensile strength, but instead measures the flexural strength, which is the capability of concrete to resist bending. The flexural strength test is done by loading a concrete beam with a known load and measuring the amount of deflection. The tensile strength of the concrete is then calculated from the deflection and the load.

Durability Test. For determining durability water absorption test is done in this research paper. This test measures how much water a concrete sample can soak up. It is important for determining how durable and water-resistant the concrete is, and it helps to ensure that the concrete will work as expected in its intended environment [1].

3 Results and Discussion

3.1 Workability Test Data

Slump Cone Test Data

The result (as shown in Fig. 4) shows that the replacement up to 20% with waste increases the slump but further increase in waste reduces the slump value. Concrete having 20% C&D, 20% FS and 2.5% BF has the highest slump value, and concrete having 50% C&D, 50% FS and 2.5% BF has the lowest slump value.

Compaction Factor Test Data

The result (as shown in Fig. 5) shows that the replacement of up to 20% with waste increases the compaction factor value however, after an increment in replacement, the negative effects of the waste materials start to outweigh the positive effects. This is because the increased porosity of the waste materials leads to an increase in the void ratio in the concrete, which causes weakness in the concrete. This means that

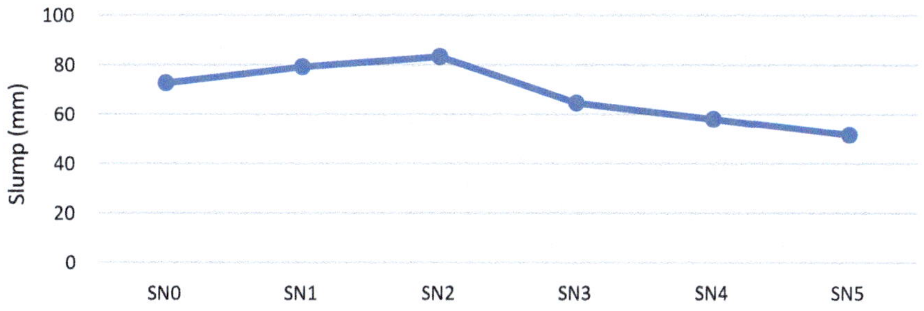

Fig. 4 Slump values of trial mixes

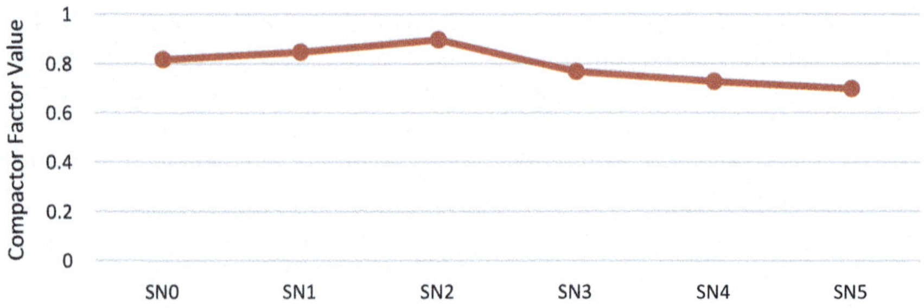

Fig. 5 Compaction factor values of trial mixes

they do not react with the cement or the aggregates over time. Overall, the strength of the concrete remains relatively constant over time.

3.2 Compressive Strength Test Data

Non-destructive (Rebound Hammer) Test Data

According to the RH Test results (as shown in Fig. 6), the compressive strength of concrete increases with the increasing proportion of recycled Upon analyzing the figure, it can be observed that the compressive strength is enhanced with the replacement of aggregates using waste materials up to 30%. However, further replacement, i.e., beyond 30%, leads to a decline in compressive strength. However, beyond 30%, the compressive strength of the concrete decreases. This is likely because recycled waste materials are generally less dense and more porous than aggregates, so they reduce the overall density of the concrete. Additionally, recycled waste materials may not bond as well with the cement as the aggregates, which can also weaken the concrete. The figure also states that there is no significant variation in compressive strength between the different concrete mixes over time. This is likely because the recycled waste materials used in the study are inert materials, which means that they do not react with the cement or the aggregates over time.

Destructive Test (CTM) Data

The compressive strength result of concrete (as shown in Fig. 7) determined by CTM machine shows that the compressive strength of concrete increases with increasing the recycled waste. The variation in compressive strength is almost as same as that of compressive strength determined by the RH test. The compressive strength increases up to 30% of recycled waste, then starts decreasing. Concrete containing 30% C&D waste, 30% FS, and 2.5% banana fiber has the highest compressive strength. It is observed that no significant variation in strengths with time for different SNs, this is possible because recycled waste is an inert material. It may not react with cement and aggregates over time.

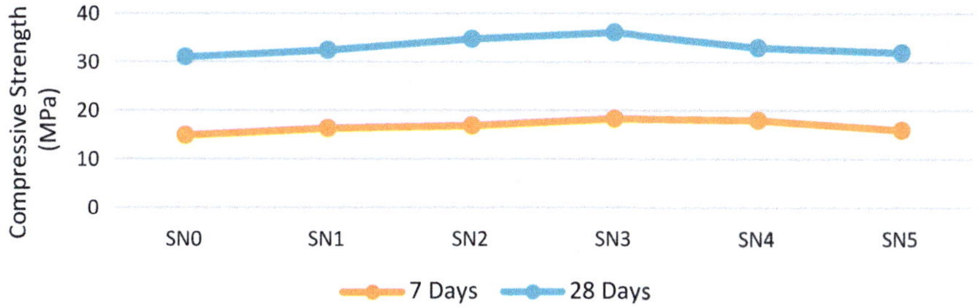

Fig. 6 Compressive strength of different trial mixes

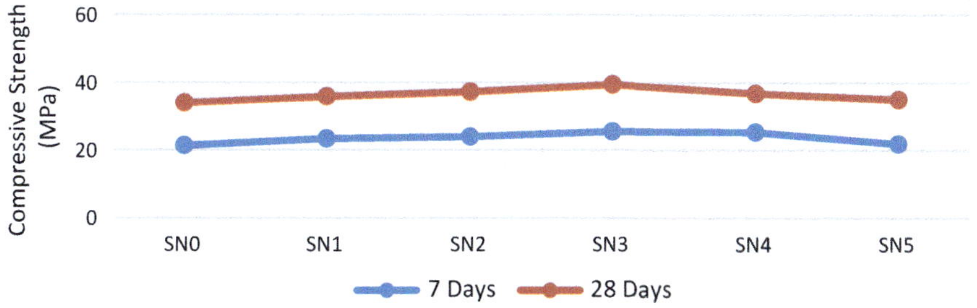

Fig. 7 Compressive strength of different trial mixes

3.3 *Split Tensile Strength Data*

The graphs (as shown in Fig. 8) show that the split tensile strength increases with increasing the fraction of constructional waste up to 40% in concrete, after which it begins to decrease. Concrete containing 40% C&D waste, 40% foundry sand, and 2.5% banana fiber exhibit the highest split tensile strength in comparison with other samples.

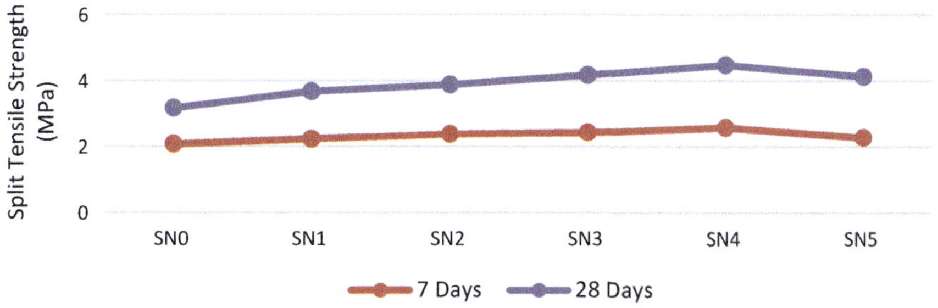

Fig. 8 Split tensile strength values of trial mixes

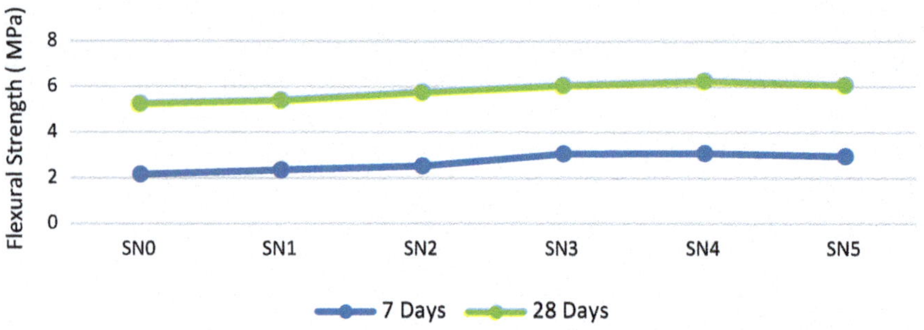

Fig. 9 Flexural strength of trial mixes

3.4 Flexural Strength Test Data

The data presented in the graphs (as shown in Fig. 9) indicates that the flexural strength of concrete increases with the addition of constructional waste up to a replacement level of 40%. Beyond this point, the flexural strength begins to decline. Concrete incorporating 40% C&D waste, 40% foundry sand, and 2.5% banana fiber exhibit the highest flexural strength among the tested samples.

3.5 Durability Test (Water Absorption) Data

As observed in the graph (as shown in Fig. 10), the water absorption percentage increases with the incorporation of recycled waste materials. Sample no. 5, with 50% construction and demolition (C&D) waste, 50% foundry sand, and 2.5% banana fiber, shows the maximum water absorption in comparison to other tested samples.

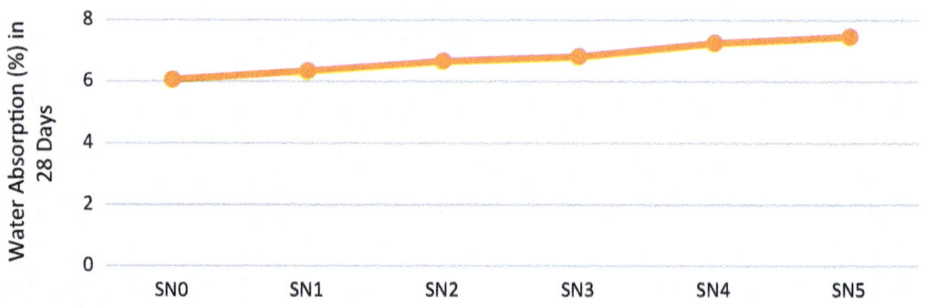

Fig. 10 Water absorption of trial mixes

4 Conclusion

This research study has demonstrated that the strength of concrete in compression can be substantially improved by incorporating constructional waste, foundry sand, and banana fiber. The results suggest that partial replacement of sand with foundry sand up to 20–30% is beneficial for both workability and mechanical properties, while further replacement is detrimental. Concrete mixtures containing 30% C&D waste, 30% FS, and 2.5% BF exhibited the best performance overall. Furthermore, split tensile strength and flexural strength were found to increase with increasing waste fractions. These findings indicate that the use of discarded materials and fibers in concrete production is a viable and effective approach to enhance the sustainability and performance of concrete structures.

References

1. Gupta RK (2023) Uses of construction and demolition waste as partial replacement of fine aggregate and coarse aggregate in concrete. Epra Int J Res Dev (IJRD) 8:52–56
2. Daniel RJ, Sangeetha SP (2021) Experimental study on concrete using waste ceramic as partial replacement of aggregate. Mater Today Proc 45:6603–6608
3. Rao A, Jha KN, Misra S (2007) Use of aggregates from recycled construction and demolition waste in concrete. Resour Conserv Recycl 50:71–81
4. Devi SV, Gausikan R, Chithambaranathan S, Jeffrey JW (2021) Utilization of recycled aggregate of construction and demolition waste as a sustainable material. Mater Today Proc 45:6649–6654
5. Kumar A, Deep K (2023) Experimental investigation of concrete with cementitious waste material such as GGBS and fly ash over conventional concrete. Mater Today Proc 74:953–961
6. Bergonzoni M, Melloni R, Botti L (2023) Analysis of sustainable concrete obtained from the by-products of an industrial process and recycled aggregates from construction and demolition waste. Procedia Comput Sci 217:41–51
7. Kumar R (2017) Influence of recycled coarse aggregate derived from construction and demolition waste (CDW) on abrasion resistance of pavement concrete. Constr Build Mater 142:248–255
8. Gebremariam AT, Di Maio F, Vahidi A, Rem P (2020) Innovative technologies for recycling end-of-life concrete waste in the built environment. Resour Conserv Recycl 163:104911
9. Naik TR, Patel VM, Parikh DM, Tharaniyil MP (1994) Utilization of used foundry sand in concrete. J Mater Civ Eng 6:254–263
10. Kumar S, Silori R, Sethy SK (2023) Insight into the perspectives of waste foundry sand as a partial or full replacement of fine aggregate in concrete. Total Environ Res Themes 100048
11. Siddique R, De Schutter G, Noumowe A (2009) Effect of used-foundry sand on the mechanical properties of concrete. Constr Build Mater 23:976–980
12. Bhardwaj B, Kumar P (2017) Waste foundry sand in concrete: a review. Constr Build Mater 156:661–674
13. Kumar A, Deep K, Chaturvedi A, Ghamforia S (2022) Effect of its mechanical properties on concrete mix were sand is partially replaced with industrial ash. Math Stat Eng Appl 71:9693–9701
14. Ahmad J, Zhou Z, Martínez-García R, Vatin NI, de-Prado-Gil J, El-Shorbagy MA (2022) Waste foundry sand in concrete production instead of natural river sand: a review. Materials 15:2365
15. Kavitha OR, Shyamala G, Akshana V (2021) Study of sustainable concrete property containing waste foundry sand. Mater Today Proc 39:855–860

16. Prabhu GG, Hyun JH, Kim YY (2014) Effects of foundry sand as a fine aggregate in concrete production. Constr Build Mater 70:514–521
17. Bharathi SV, Vinodhkumar S, Saravanan MM (2021) Strength characteristics of banana and sisal fiber reinforced composites. In: IOP conference series: materials science and engineering. IOP Publishing, p 012024
18. Nurwidayati R, Fardheny AF (2021) Investigation on mechanical properties of fiber reinforced concrete. In: IOP conference series: earth and environmental science. IOP Publishing, p 012016
19. Attia MM, Shawky SMM (2021) Banana fiber reinforced concrete: a review. NY Sci J 14:48–57
20. Rajendran M, Nagarajan CK (2022) Experimental investigation on bio-composite using jute and banana fiber as a potential substitute of solid wood based materials. J Nat Fibers 19:4557–4566
21. Kumar A, Sahu N, Sahu JK, Sahu A (2023) Experimental investigation of paver block using polypropylene waste and industrial ashes. Mater Today Proc 74:808–814

An AHP and TOPSIS MCDM-Based Approach for the Selection of Material for Construction Under Fuzzy Environment

Durgeshwar Pratap Singh, Shwetank Avikal, Harvindra Singh, and Shivani Monga

Abstract The phase of the design and development process that involves the selection of materials is one of the most challenging phases for any structural component. For the goal of selecting materials that exhibit extraordinary resistance to wear and are suitable for structural applications, the current inquiry makes use of a hybrid MCDM technique that is based on Fuzz-AHP and TOPSIS. As part of the selection process, the physical, mechanical, and wear qualities of composite materials are assessed. These parameters include density, hardness, tensile strength, and other similar characteristics. During the preliminary phase, the Fuzzy-AHP approach is used in order to determine the weights that are allocated to the various qualities. Following this, the TOPSIS technique is applied in order to identify the ranking of materials by making use of the ratio system, the multiplication approach, and the point of reference approach each in their own distinctive way. The comparison is drawn between the categories that were created by the approaches, and it is seen that these methods demonstrate an outstanding level of simplicity and lucidity in their method for assigning scores to the material alternatives. In conclusion, this comparison is made.

Keywords Fuzzy-AHP (analytical hierarchy process) · TOPSIS · Composite materials · Decision making · MCDM · Material selection

D. P. Singh · H. Singh
Department of Mechanical Engineering, Graphic Era (Deemed to be) University, Dehradun, Uttarakhand 248002, India

S. Avikal (✉)
Department of Management Studies, Graphic Era (Deemed to be) University, Dehradun, Uttarakhand 248002, India
e-mail: avi.shwetank@gmail.com

S. Monga
PDP, Graphic Era Hill University, Dehradun, Uttarakhand 248002, India

P. Bhingole et al. (eds.), *Advances in Materials Engineering*, Lecture Notes in Mechanical Engineering, https://doi.org/10.1007/978-981-97-7114-1_29

1 Introduction

It is often acknowledged that selecting the appropriate materials during the product development process is crucial. With so many variables influencing the decision-making process, creating a methodical strategy for choosing the optimal material can be difficult. There are two key reasons why selecting the appropriate materials is crucial. First, it's used to streamline operations, cut costs, improve dependability, and lighten the weight of an already-existing product. It is also used to choose the appropriate material for a new product [1, 2]. When designing a product, selecting the appropriate materials is crucial because it affects the final product's performance and appearance [3]. It is difficult to get information regarding choosing materials for the composite vehicle bumper beam. A number of academic studies have examined the process of choosing a suitable material for an automobile bumper beam. For the bumper system, Sapuan et al. [4] used a weighted objective technique to choose the materials. Hosseinzadeh et al. [5] looked into a variety of conventional and composite materials for vehicle bumper beams in their investigation. They evaluated the materials using impact modelling with LS-DYNA ANSYS, taking into account their weight, cost, and manufacturing process 5.7. Based on research, the ideal material for a bumper beam is sheet moulding compound (SMC). Suddin et al. [6] detailed in their study the creation of a knowledge-based system (KBS) designed especially for the selection of materials. Making a bumper beam design. The KBS is an expert system that chooses the best content for a certain circumstance based on constraint values. The aforementioned approaches, however, do not provide practical means of aiding material engineers in the process of selecting the best material. AHP is a method that uses a sequence-based approach to aid in the methodical selection of resources. An essential instrument for decision-making is the AHP. It is made to efficiently take into account both intuitive and logical factors while selecting the best alternative from those that have been assessed according to various standards [7]. AHP is a decision-making approach that entails organizing several criteria into a hierarchy, evaluating the significance of these criteria, comparing alternatives for each criterion, and coming up with an overall rating of the alternatives, according to the Decision Support System (DSS) resource [8]. In most cases, the ability of experts or users to identify the factors influencing the decision-making process is what makes the deployment of AHP possible. AHP is a highly intuitive and incredibly effective method for making and evaluating judgements for students in grades 9 through 12. Benefits of using AHP include expediting the product development process and producing higher-quality products. For capturing both subjective and objective assessment measures, the AHP is a helpful tool. It lessens decision-making bias and offers a helpful means of assessing the coherence of the team's recommended alternatives and evaluation methods. Companies can successfully address typical issues with the decision-making process by using the AHP, such as accountability, strategic planning, active involvement, and ambiguous objectives. These issues may be expensive diversionaries that keep teams from reaching the optimal conclusions

[8]. AHP has been widely used in academic research as well as industrial applications to address the challenges of multi-criteria decision-making. It is currently most often used in the field of selection and assessment, particularly in the engineering, personal, and social domains, and has been extensively integrated into almost all decision-making applications [9]. AHP is frequently utilized in decision-making situations, although it is still limited in its ability to make exact choices during the materials selection procedure. Ho [10] looked at international articles that focused on the use of AHP from 1997 to 2006. The review concluded that a broad range of fields can make use of AHP. On the other hand, not much research has been done on using the AHP to choose materials for composite car parts. The use of the AHP in material selection was the subject of just one publication [11]. Selecting the appropriate materials for a manufacturing process can minimize production time, boost process effectiveness, and eventually boost productivity. Making the wrong material choices can have a detrimental effect on a business's output, bottom line, and reputation. Selecting the best material for a product from a constrained range of possibilities requires carefully weighing a number of factors. The right materials for a given product are determined by a number of parameters and requirements. The requirements usually cover a wide range of aspects, including the materials' economics and their mechanical, physical, and electrical properties. With so many alternatives and careful considerations to make while evaluating different criteria, choosing the right material can be a challenging undertaking. Choosing the appropriate material for engineering applications is an important choice that requires careful consideration of a number of factors and alternatives. Material selection turns into a difficult decision-making process that calls for taking into account several factors [12–14]. To choose a suitable carbon fiber-reinforced polymer (CFRP) material for underwater structural applications, the TOPSIS approach was utilized in this work. Using a system called TOPSIS, options are ranked according to how similar they are to the ideal solution. When choosing an appropriate material in an MCDM setting, the TOPSIS approach is a highly efficient and simple computational method. Using the TOPSIS technique to choose the best material requires first establishing the weightings for the criteria. Tensile strength, compression strength, flexural strength, impact strength, and damping ratio were the six evaluation criteria taken into consideration in this study to identify the best composite material among the CFRPS choices that were suggested. To determine more appropriate weights for the parameters, this study used a new weighting approach. Analyzing the criterion's objective and subjective weights was part of the methodology. Subjective weights were computed using the AHP, and objective weights were calculated by applying Shannon entropy techniques. The AHP has demonstrated its efficacy as a methodology for addressing performance-related concerns and integrating varied viewpoints from experts in the domain. The AHP is used to calculate the weights through pairwise comparisons that rely on expert judgements. The weights assigned to the criteria are calculated using the principal eigenvectors. AHP has certain drawbacks even if it provides numerical priority vectors for material selection criteria. The main method of evaluating the application of AHP is subjective assessment, which, depending on the

amount of pertinent data available, may or may not produce findings for quantitative criteria that may be relied upon. In order to make decisions in such situations, decision-makers need objective, knowledgeable criteria based on pertinent information. By providing objective criteria weights, the Shannon entropy (SE) technique can enhance the functionality of AHP. The limitations of each methodology can be overcome by combining the AHP and Simple Additive Weighting (SAW) methodologies to construct a hierarchical framework that successfully prioritizes material selection criteria. As a result, synthesis weights—a combination of weights derived from AHP and SE—are used in this research for the parameters. Several mathematical techniques have been created and used in academic publications in the past to help with content selection according to various standards. The graphical approach and the ranking method are the two most often utilized techniques for material selection [15, 16]. Hosseinzadeh et al. [5] suggested a selection method that uses several surveys to identify the best possible design concepts.

Other popular approaches for solving MCDM problems include several others. These techniques include the data envelopment analysis (DEA) method, the AHP method, the VIKOR method, the TOPSIS method, and the MOORA [17–20]. Because the VIKOR approach uses linear means for data normalization, the final ranking is determined with a better degree of accuracy. The group that seeks to maximize benefit and avoid regret chooses the preferred option. The ratio between the positive and negative ideal solutions is computed using the VIKOR method. Therefore, a middle ground option that offers a beneficial rate is suggested by the VIKOR approach. The PSI (Preference Selection Index) method is easy to use and doesn't require any intricate computations. When figuring out the relative importance of the factors being evaluated is difficult, it can be very beneficial [21]. There is no significance-based ranking of the variables in this technique. Rather, statistical ideas are used to establish the total preference value. The PSI method also has the advantage of requiring fewer numerical calculations to determine the ideal criteria value. When it comes to handling the physical attributes and volume of options that need to be assessed, the TOPSIS technique is very effective [22, 23].

2 Proposed Approaches

Fuzzy AHP and TOPSIS

When evaluating and comparing pairs of possibilities at each level, the conventional AHP uses a nine-point scale to do so, taking into consideration aesthetic factors. In each pair comparison, an assessment of the aesthetic elements that are being evaluated is provided according to their respective priority. When two things are compared to one another, the ratios are stated as numerical values that are accurate and exact. In spite of the benefits that the discrete scale consisting of numbers from 1 to 9 offers. Despite the fact that the system is straightforward and easy to use, it does not admit that it is unable to properly manage the inherent uncertainty and appropriately measure the viewpoints of those who make decisions. When it comes to making choices, the majority of decision-makers rely on their knowledge, past experience, and their own subjective judgment, as stated by Chan et al. [15, 18]. Specifically, Zadeh [19] contends that the capacity of a system to generate claims that are both meaningful and truthful declines as the complexity of the system increases. At some point in time, there comes a moment when meaning and relevance (or truth) are almost entirely distinct from one another without any connection whatsoever. Consequently, it would seem that the traditional AHP is not enough for the generation of significance weights for the criteria that are required for the disassembly of post-consumed goods' components. There is no difference between the eigenvector technique and the AHP methodology. The fact that this is the case suggests that the eigenvector is associated with the greatest given value. The eigenvalue of the pairwise comparisons matrix provides an indication of the significance of the criteria. A representation of the relative effectiveness of the different criteria may be found in the vector of weights that was derived from the pairwise comparisons matrix. Within the framework of the fuzzy AHP approach, which makes use of tri-angular fuzzy numbers to improve the scaling strategy in the judgment matrices given by Buckley [20], interval arithmetic is used in order to construct the fuzzy eigenvector. The approaches that were given by Avikal et al. [21] were used by the researchers that participated in this investigation. TOPSIS, or 2.2: When presenting information, it is essential to provide a concise description of the TOPSIS approach, which was first established by Hwang and Yoon [22]. This will guarantee that the presentation is clear and understandable. In the TOPSIS methodology, alternatives are evaluated by determining how close they are to both a zenith, which represents an ideal alternative, and a nadir, which represents a negative-ideal alternative. In this particular setting, the word "domain" refers to a collection of possibilities that are contained inside an Euclidean space that is composed of several dimensions. Because of this, each and every possible consequence is shown as a point inside this spatial domain. For the purpose of determining the greatest and lowest points, it is assumed that each characteristic follows a steady pattern of either continuous development or decrease. In accordance with the TOPSIS concept, solutions are evaluated by determining how far they are from the lowest point and how close they are to the maximum point. Calculating the Euclidean distance between the places that have been supplied is the method that is used to estimate the proximity measurement [18].

3 Problem Formulation

The hand-lay-up technique was used in order to manufacture needle-punched nonwoven textiles made of viscose and polyester with a weight of 400 g. The reinforcement of these textiles was accomplished by the use of composites that included particles of blast furnace slag (BFS). The stuff that is written. Previous academic publications [9, 24, 25] were consulted in order to compile the full classification and composition of composites, which is shown in Table 1. Experimental procedures for

Table 1 Composition of selected composite materials [9, 24, 25]

1PSF	30 wt% 400 gsm pol-ter fab mt + 0 wt% BFS + 70 wt% epy
2PSF	30 wt% 400 gsm pol-ter fab mt + 5 wt% BFS + 65 wt% epy
3PSF	30 wt% 400 gsm pol-ter fab mt + 10 wt% BFS + 60 wt% epy
4PSF	30 wt% 400 gsm pol-ter fab mt + 15 wt% BFS + 55 wt% epy
1VSF	30 wt% 400 gsm pol-ter fab mt + 0 wt% BFS + 70 wt% epy
2VSF	30 wt% 400 gsm pol-ter fab mt + 5 wt% BFS + 65 wt% epy
3VSF	30 wt% 400 gsm pol-ter fab mt + 10 wt% BFS + 60 wt% epy
4VSF	30 wt% 400 gsm pol-ter fab mt + 15 wt% BFS + 55 wt% epy

Table 2 Properties of selected composite materials [9, 24, 25]

Composites	Density (C1)	Ws (n) (C2)	Ws (v) (C3)	Micro-hardness (C4)	Tensile strength (C5)	Flexural strength (C6)	Impact strength (C7)
1PSF	1.2060	0.02958	0.02982	23.90	30.27	41.85	1.351
2PSF	1.20780	0.02673	0.02140	28.7	38.46	47.13	1.505
3PSF	1.21270	0.02588	0.01877	33.3	42.23	53.42	1.796
4PSF	1.22210	0.02404	0.01635	35.2	38.65	50.26	1.933
1VSF	1.2130	0.03497	0.02213	24.27	36.19	51.71	1.556
2VSF	1.21800	0.02982	0.01648	29.58	40.06	54.36	1.924
3VSF	1.22600	0.02566	0.01376	35.12	44.53	58.67	2.036
4VSF	1.23300	0.02398	0.01228	37.43	39.87	57.09	2.192

mechanical and slurry abrasion were carried out in accordance with ASTM specifications. Through the use of the property values that were retrieved from these papers, a composite rating was ultimately produced. Table 2, which functions as the decision matrix, contains the values that have been included in the decision. In order to analyze materials, the Fuzzy AHP and TOPSIS techniques are used. These methodologies are depending on the characteristics of the materials.

4 Results and Discussion

It has been determined that the present job, which comprises selecting a composite material according to the characteristics of composite matrices, is an example of a Multi-Criteria Decision Making (MCDM) issue. An approach to solving this problem has been developed, which is a solution methodology that makes use of fuzzy AHP and the Technique for Order of Preference by Similarity to Ideal Solution (TOPSIS). The features of the composite material have been used to develop the selection criteria for the materials that will be used. Through the use of a two-step process, the problem has been resolved. The first thing that needs to be done is to determine how important the criteria that have been chosen are. These criteria are the properties of the material.

The content was reviewed using TOPSIS, a technique that integrates each weight of the criteria specified by Fuzzy-AHP. This evaluation took place in the subsequent step.

In Table 3, you can see the matrix that is used for comparing pairings in the Fuzzy-AHP algorithm. The outcomes that were evaluated using the Fuzzy-AHP algorithm are shown in Table 4.

Table 4 indicates that the "Impact Strength" criterion is the most weighted, while the "Flexural Strength" criterion is the least weighted. The pairwise comparison matrix has undergone a consistency check, and the results show that the Consistency Ratio (CR) is between CR0.1 and CR0.2. The successful comparison and associated findings show that they are suitable for further computations.

In the second stage, the composite materials selected for the study were ranked using the TOPSIS approach. The decision matrix that was supplied in Table 1 has been normalized throughout this phase and is currently shown in Table 5. To generate a weighted decision-making matrix, the data in Table 2 has been multiplied by the pertinent weights of the selection criterion. The result is shown in Table 6. After being evaluated using a weighted decision matrix, the PIS and NIS are shown in Tables 7 and 8, respectively.

Table 9 presents the results of a computation that was performed to determine the relative distance between each option and the Positive Idle Solution and the Negative Idle Solution combinations. Using the relative distance method, the options have been

Table 3 Fuzzy-AHP paired comparison matrix

Criteria	C1	C2	C3	C4	C5	C6	C7
C1	1	2	3	5	5	5	1
C2	1/2	1	3	3	5	5	1/3
C3	1/3	1/3	1	1	3	5	1/5
C4	1/5	1/3	1/1	1	3	5	1/5
C5	1/5	1/5	1/3	1/3	1	3	1/5
C6	1/5	1/5	1/5	1/5	1/3	1	1/5
C7	1/1	3	5	5	5	5	1

Table 4 The results obtained by fuzzy-AHP and consistency check

Criteria	Weights	Consistency check
C1	0.263246	
C2	0.175001	CI = 0.10407
C3	0.09697	
C4	0.086121	CR = 0.078841
C5	0.05085	
C6	0.034075	
C7	0.293737	

Table 5 The normalized decision matrix

Composites	C1	C2	C3	C4	C5	C6	C7
1PSF	0.000	0.509554	1	0.0000	0.0000	0.0000	0.000
2PSF	0.066667	0.250227	0.519954	0.354767	0.574334	0.313912	0.183115
3PSF	0.248148	0.172884	0.370011	0.694752	0.83871	0.687872	0.529132
4PSF	0.596296	0.00546	0.232041	0.835181	0.587658	0.5	0.692033
1VSF	0.259259	1	0.561574	0.027347	0.415147	0.586207	0.243757
2VSF	0.444444	0.531392	0.239453	0.419808	0.686536	0.743757	0.681332
3VSF	0.740741	0.152866	0.084379	0.829268	1	1	0.814507
4VSF	1	0	0	1	0.673212	0.906064	1

Table 6 Matrix of weighted normalized decision

Composites	C1	C2	C3	C4	C5	C6	C7
1PSF	0.0000	0.089173	0.09697	0.0000	0.0000	0.0000	0.000
2PSF	0.01755	0.04379	0.05042	0.030553	0.029205	0.010697	0.053788
3PSF	0.065324	0.030255	0.03588	0.059833	0.042649	0.023439	0.155425
4PSF	0.156973	0.000955	0.022501	0.071927	0.029883	0.017038	0.203275
1VSF	0.068249	0.175001	0.054456	0.002355	0.02111	0.019975	0.0716
2VSF	0.116998	0.092994	0.02322	0.036154	0.03491	0.025344	0.200132
3VSF	0.194997	0.026752	0.008182	0.071418	0.05085	0.034075	0.23925
4VSF	0.263246	0	0	0.086121	0.034233	0.030874	0.293737

Table 7 Positive idle solution

Composites	C1	C2	C3	C4	C5	C6	C7
1PSF	0.069298	0.007367	0	0.007417	0.002586	0.001161	0.086281
2PSF	0.060367	0.017216	0.002167	0.003088	0.000469	0.000547	0.057575
3PSF	0.039173	0.020952	0.003732	0.000691	6.73E-05	0.000113	0.01913
4PSF	0.011294	0.030292	0.005546	0.000201	0.00044	0.00029	0.008183
1VSF	0.038024	0	0.001807	0.007017	0.000884	0.000199	0.049344
2VSF	0.021388	0.006725	0.005439	0.002497	0.000254	7.62E-05	0.008762
3VSF	0.004658	0.021978	0.007883	0.000216	0	0	0.002969
4VSF	0	0.030625	0.009403	0	0.000276	1.02E-05	0

rated, and the results are provided in Table 9. The results of the ranking suggest that the alternative "4VSF" had the highest rating, which indicates that it is the most appropriate choice among all of the other options that were taken into consideration. Among all of the other composite materials that were chosen, the choice "2PSF" has the lowest rating and is the least suited.

Table 8 Negative idle solution

Composites	C1	C2	C3	C4	C5	C6	C7
1PSF	0.00000	0.007952	0.009403	0.00000	0.0000	0.00000	0.0000
2PSF	0.000308	0.001918	0.002542	0.000933	0.000853	0.000114	0.002893
3PSF	0.004267	0.000915	0.001287	0.00358	0.001819	0.000549	0.024157
4PSF	0.02464	9.13E-07	0.000506	0.005173	0.000893	0.00029	0.041321
1VSF	0.004658	0.030625	0.002965	5.55E-06	0.000446	0.000399	0.005127
2VSF	0.013689	0.008648	0.000539	0.001307	0.001219	0.000642	0.040053
3VSF	0.038024	0.000716	6.69E-05	0.0051	0.002586	0.001161	0.057241
4VSF	0.069298	0	0	0.007417	0.001172	0.000953	0.086281

Table 9 Relative distance and final ranking

Composites	Distance from PIS	Distance from NIS	Relative distance	Final rank
1PSF	0.417265	0.417265	0.239959	7
2PSF	0.37607	0.37607	0.206359	8
3PSF	0.289582	0.289582	0.397743	6
4PSF	0.237163	0.237163	0.532246	4
1VSF	0.311891	0.311891	0.402725	5
2VSF	0.212465	0.212465	0.547521	3
3VSF	0.194175	0.194175	0.62518	2
4VSF	0.200786	0.200786	0.669291	1

5 Conclusions

This research utilizes a hybrid Fuzzy AHP-TOPSIS methodology to choose the optimal polymer composite material for engineering applications. The strategy involves gathering the attributes associated with polymer composites from existing literature. The approach began with the use of the F-AHP to determine the weightage and to prioritize wear resistance and structural uses. We included these weights in the TOPSIS methodology and achieved the ranking. 4VSF has been identified as the top-ranked material by TOPSIS. Furthermore, the study reveals that composite 2PSF is ranked as the poorest material among all composites when evaluated using various MCDM techniques. The summarized outcome also indicates that the ranking of the composite materials utilizing the proposed approach is highly comparable. This approach applies to different decision-making situations with several alternatives and criteria.

Alloy designated as "AZ61" composed primarily of magnesium (Mg), aluminum (Al-6 wt%), and zinc (Zn-1 wt%). This alloy is widely used in several applications because of its favorable combination of properties, including lightweight, good strength-to-weight (S/W) ratio [1]. AZ61 magnesium alloy is utilized in aerospace applications such as aircraft seats, cabin interiors, and structural parts due to its lightweight nature and high S/W ratio [2]. In the automotive industry, parts like transmission cases.

References

1. Gupta N, Luong DD, Cho K (2012) Magnesium matrix composite foams—density, mechanical properties, and applications. https://doi.org/10.3390/met2030238
2. Hu H-J, Sun Z, Ou Z-W, Wang X-Q (2016) Wear behaviors and wear mechanisms of wrought magnesium alloy AZ31 fabricated by extrusion-shear
3. Baghni IM, Wu Y-S, Jiuqing L, Cuiwei D, Zhang W. Mechanical properties and potential applications of magnesium alloys transactions of nonferrous metals society of China, English edn, vol 13, pp 1253–1259
4. Senthil R, Gnanavelbabu A (2014) Effect of heat treatment on tensile strength and microstructure of AZ61A magnesium alloy. In: Applied mechanics and materials, Trans Tech Publications Ltd, pp 55–59. https://doi.org/10.4028/www.scientific.net/AMM.606.55
5. Rzychoń T, Kiełbus A (2007) Microstructure of WE43 casting magnesium alloy
6. Zhang Z, Zeng X, Ding W (2005) The influence of heat treatment on damping response of AZ91D magnesium alloy. Mater Sci Eng A 392:150–155. https://doi.org/10.1016/j.msea.2004.09.056
7. Vyshak TP, Sankar R, Babu AC, Harilal M, Devasia J (2020) A review on heating and rolling effect of AZ31 magnesium alloy. Int Res J Eng Technol
8. Fahad M, Bavanish B (2021) Tribological and ageing behavior of AZ91D magnesium alloy fortified with nano lanthanum and nanoceria by stir casting for aviation application. Ind Lubr Tribol 73:635–641. https://doi.org/10.1108/ILT-12-2020-0475
9. Xiu K, Wang HY, Sui HL, Wang Y, Xu CL, Wang JG, Jiang QC (2006) The sliding wear behavior of TiCp/AZ91 magnesium matrix composites. J Mater Sci 7052–7058. https://doi.org/10.1007/s10853-006-0946-2
10. Li ZM, Luo AA, Wang QG, Peng LM, Fu PH, Wu GH (2013) Effects of grain size and heat treatment on the tensile properties of Mg-3Nd-0.2Zn (wt%) magnesium alloys. Mater Sci Eng A 564:450–460. https://doi.org/10.1016/j.msea.2012.12.008
11. Shah M, Pandya A, Pramod B, Srivastava N (2023) Processing techniques of magnesium matrix composites and its heat treatment—a review. In: Advances in material science and metallurgy. Lecture notes in mechanical engineering. Springer, pp 63–75. https://doi.org/10.1007/978-981-19-4918-0_6
12. Kaushik N, Singhal S (2018) Hybrid combination of Taguchi-GRA-PCA for optimization of wear behavior in AA6063/SiCp matrix composite. Prod Manuf Res 6:171–189. https://doi.org/10.1080/21693277.2018.1479666
13. Sathish T, Karthick S (2020) Wear behaviour analysis on aluminium alloy 7050 with reinforced SiC through Taguchi approach. J Mater Res Technol 9:3481–3487. https://doi.org/10.1016/j.jmrt.2020.01.085
14. Girish BM, Satish BM, Sarapure S, Basawaraj (2016) Optimization of wear behavior of magnesium alloy AZ91 hybrid composites using Taguchi experimental design. Metall Mater Trans A Phys Metall Mater Sci 47:3193–3200. https://doi.org/10.1007/s11661-016-3447-1
15. Makwana D, Pramod B (2023) Dry sliding wear and heat flux mapping of closed-cell Mg-2Zn-2Ca foam fabricated by stir casting route. Mater Lett 330. https://doi.org/10.1016/j.matlet.2022.133379
16. Huang SJ, Abbas A, Ballóková B (2019) Effect of CNT on microstructure, dry sliding wear and compressive mechanical properties of AZ61 magnesium alloy. J Mater Res Technol 8:4273–4286. https://doi.org/10.1016/j.jmrt.2019.07.037

17. Gopi KR, Shivananda Nayaka H, Sahu S (2017) Wear properties of ECAP-processed AM80 magnesium alloy. J Mater Eng Perform 26:3399–3409. https://doi.org/10.1007/s11665-017-2764-x
18. Guicciardi S, Melandri C, Lucchini F, De Portu G (2002) On data dispersion in pin-on-disk wear tests
19. Makwana D, Pramod B (2023) Dry sliding wear and heat flux mapping of closed-cell Mg-2Zn-2Ca foam fabricated by stir casting route. Mater Lett 330:133379. https://doi.org/10.1016/j.matlet.2022.133379
20. Rajesh AM, Kaleemulla MK, Doddamani S (2019) Effect of heat treatment on wear behavior of hybrid aluminum metal matrix composites. Tribol Ind 41:344–354. https://doi.org/10.24874/ti.2019.41.03.04
21. Mohd Daud MA, Talib Din A, Jusoff K, Selamat MZ (2013) Precipitation strengthening in extruded AZ61 magnesium alloys during heat treatment process. World Appl Sci J 21:41–46. https://doi.org/10.5829/idosi.wasj.2013.21.1005
22. Metall A, Gonzdez-Martinez R, Go ken J, Letzig D, Timmerberg J, Steinhoff K, Kainer KU (2007) Influence of heat treatment on damping behaviour of the magnesium wrought alloy AZ61
23. Turan ME, Zengin H, Sun Y (2020) Dry sliding wear behavior of (MWCNT + GNPs) reinforced AZ91 magnesium matrix hybrid composites. Met Mater Int 26:541–550. https://doi.org/10.1007/s12540-019-00338-8
24. Pandya A, Shah M, Pramod B, Srivastava N (2023) A review on heat treatment of magnesium alloys and its effect on various properties. In: Advances in material science and metallurgy. Lecture notes in mechanical engineering. Springer, pp 77–87. https://doi.org/10.1007/978-981-19-4918-0_7
25. Su J, Teng J, Xu Z, Li Y (2022) Corrosion-wear behavior of a biocompatible magnesium matrix composite in simulated body fluid. Friction 10:31–43. https://doi.org/10.1007/s40544-020-0361-8

A MCDM-Based Approach for the Selection of Industrial Robots for Arc Welding Process

Durgeshwar Pratap Singh, Shwetank Avikal, Harvindra Singh, Shivani Monga, and Amit Sharma

Abstract Industrial robots are essential in modern manufacturing technologies as they enable industrial firms to produce high-quality products at a reduced cost. Industrial robots are purpose-built for a diverse array of jobs, such as welding, painting, assembly, disassembly, and pick and place operations for printed circuit boards. The duties encompassed in this process include assembling boards, packaging and labeling products, palletizing them, doing product inspections, and performing tests. Exceptional durability, velocity, and precision are all obtained via the qualities. The efficacy of industrial robots is determined by multiple and conflicting elements that must be simultaneously taken into account in a comprehensive selection procedure. This paper presents an AHP and M-TOPSIS-based approach for the selection of robots. The objective of this technique is to choose an industrial robot that is specifically suited for the task of arc welding. The weights of significance are calculated by the use of objective preferences employing the Analytic Hierarchy Process (AHP) technique. The AHP M-TOPSIS technique has yielded the ranking order. The results clearly showed the considerable usefulness of MCDM techniques in the specific context of robot selection. The study is unique in its emphasis on prioritizing industrial arc welding robots utilizing AHP and M-TOPSIS MCDM techniques.

Keywords AHP · M-TOPSIS · Arc welding · Industrial robots

D. P. Singh · H. Singh
Department of Mechanical Engineering, Graphic Era (Deemed to be) University, Dehradun, Uttarakhand 248002, India

S. Avikal (✉)
Department of Management Studies, Graphic Era (Deemed to be) University, Dehradun, Uttarakhand 248002, India
e-mail: avi.shwetank@gmail.com

S. Monga
PDP, Graphic Era Hill University, Dehradun, Uttarakhand 248002, India

A. Sharma
Jaypee University of Engineering and Technology, Guna, MP 473226, India

P. Bhingole et al. (eds.), *Advances in Materials Engineering*, Lecture Notes in Mechanical Engineering, https://doi.org/10.1007/978-981-97-7114-1_30

1 Introduction

Arc welding is an important process in manufacturing, and it plays a vital role in welding metal construction materials. Advances in arc welding techniques have incorporated various methods to enhance efficiency, quality, and cost-effectiveness. MCDM techniques such as AHP and TOPSIS etc., have been used in selecting different welding parameters in the literature. Both methods are typically uncommon. The AHP is a decision-making process developed by Thomas El Satty in the 1970s [1, 2]. When systematically analyzing this problem, taking into account various factors and their significance, a framework can be established to determine the AHP in arc welding. The decision-making approaches have been applied to a wide range of problems related to the welding processes. A large number of factors have been considered in decision-making operations such as cost, quality, operational and environmental impacts, and comparing them to one's preferences and objectives [1, 2]. TOPSIS is one of the well-known MCDM approaches that offers a systematic approach for considering alternatives by calculating the distance between anti-optimal solutions. It measures the distance of each alternative from both the idle solutions. The modified Approach to M-TOPSIS is an improved version of the traditional TOPSIS. It is primarily utilized in arc welding. This is accomplished by making modifications that are tailored to address the unique challenges and requirements of the welding industry. An important consideration in M-TOPSIS is the evaluation of various criteria, including weld quality, material compatibility, system quality, and safety analysis. These factors are taken into account to optimize and implement welding methods. Applying AHP and M-TOPSIS in arc welding can be useful in various decision situations encountered in practice. These techniques have a wide range of applications. Evaluate the optimal welding process by taking into account factors such as weld quality, productivity, and cost-effectiveness [3, 4].

- Selecting the right welding parameters, including voltage, current, and travel speed, is crucial for achieving the desired weld qualities.
- Evaluate the properties, suitability, and accessibility of materials to select the most suitable welding material.
- Determine the essential welding technologies, equipment, and consumables for adoption or investment, and then prioritize them accordingly.

The application of AHP and M-TOPSIS in the decision process of arc welding provides various benefits such as enhanced decision quality, increased transparency and accountability, improved material allocation, greater objectivity, simplicity, scalability, and reliable data availability. In general, the AHP and M-TOPSIS methods prove to be efficient decision tools in the field of arc welding. These strategies allow stakeholders to analyze ideas strategically and prioritize them based on values and preferences. Utilizing AHP and M-TOPSIS can greatly improve the precision, transparency, and allocation of materials in arc welding. These help organize a list of classifiers, assigning importance to criteria and evaluation options. They can be obtained. The application of M-TOPSIS is expected to have a significant impact in guiding management decisions in the field of arc welding technology, as it continues to evolve and become more complex. This is because these advancements are driving both innovation and complexity.

2 Literature Review

The research done by Afzal et al. [5] centers on enhancing the shielded metal arc welding (SMAW) parameters for joining ASTM *A572* Grade 50 steel in heavy machinery. The parameters were optimized using the Taguchi L16 array and analyzed using grey relational and principal component analysis. The results demonstrated that the groove angle had a major impact on the tensile strength, whereas the electrode diameter influenced the shock strength. The welding current mostly influenced the hardness and angular deformation. CA-GRA exhibited superior performance compared to GRA, demonstrating less angular deformation and enhanced mechanical characteristics. The angular deformation was reduced by 33.33%, while the stiffness fell by 8.55%. Additionally, the tensile strength and impact strength increased by 14.28%. The welding current range was the largest when considering the root gap, and it was monitored based on the parameters of groove angle and electrode diameter. The initial face values had minimal impact. The material properties, such as elasticity and plasticity, are significantly impacted by errors in the parameters, leading to a reduction in these qualities. These faults comprise apertures, fissures, lack of permeability, erosion, and lack of impermeability. Employing sophisticated numerical methods and optimization approaches can greatly enhance the system's quality, efficiency, and adaptability [4]. The study conducted by Sánchez-Lozano et al. examined the optimization of the SMAW (Shielded Metal Arc Welding) technique for Grade 50 steel joints, which are crucial components in heavy machinery. Parameters were optimised utilising sophisticated numerical and analytical techniques. The results indicated that the optimal parameters for welding were a welding current of 160 A, a groove angle of 60°, an electrode diameter of 3.25 mm, a root spacing of 3 mm, and a root face of 3 mm. Implementing these parameters led to significant improvements, including a 10.53% increase in tensile strength, a 14.28% increase in impact hardness, an 8.55% decrease in hardness, and a remarkable 33.33% decrease in angular distortions [8, 9]. The research endeavor seeks to enhance the SMAW process for ASTM A572 Grade 50 in order to achieve a diverse array of high-quality characteristics and minimize distortion using the Taguchi-based PCA—the GRA approach. The optimal configuration was determined to be WC3GA2ED2RG4RF3. In conclusion, the research established that the SAW welding process is the most appropriate method for joining ASTM A572 Grade 50 steel, while the GTAW welding process is a feasible option. The SAW welding process is renowned for its exceptional productivity and efficiency, however the GTAW process possesses distinct attributes that set it apart from the SAW welding method. By utilizing advanced statistical approaches and optimization methodologies, the system's quality, efficiency, and adaptability can be significantly enhanced [5]. The work conducted by Sarkar et al. [7] aims to present a new strategy that utilizes the AHP-based Taguchi method. The objective is to identify the most suitable parameters for achieving submerged arc welding of plain carbon steel. In the present analysis, three process parameter variables, specifically wire feed rate, stick out, and traversing speed, along with three response parameters, namely penetration, bead width, and bead reinforcement, have been considered. Thus, this study aims to improve the quality of welded components by employing the Taguchi approach, which is grounded in AHP analysis. The Taguchi L16 orthogonal array is employed to get the desired outcomes while minimizing the number of experimental trials. Relying solely on a Taguchi-based technique is insufficient for addressing an issue that requires the optimization of many responses. In order to surpass this limitation, the present study used a decision-making methodology called AHP, which takes into account several criteria. The best parameters for generating a successful weld, also known as bead geometry, are a wire feed rate of 210 mm per minute, a stick out of 15 mm, and a traverse speed of 0.75 m per minute. Furthermore, it has been noted that the impact of wire feed rate on the total bead geometry attributes is more substantial compared to the impact of other welding factors. Finally, a confirmatory test was conducted to validate the optimal configuration that was discovered [6]. Aghakhani et al. examined the application of the Taguchi method's parameter design to optimize the GMAW parameters in their work. A study was conducted using an experimental design that involved five components and five levels to examine the relationships between weld dilution and adjustable input welding parameters [7]. The input parameters for welding play a critical role in defining the quality of a weld joint. The overall integrity of the joints can be characterized by aspects such as the weld's morphology, its mechanical properties, and any deformations that may arise. In order to obtain a welded junction that meets the necessary weld-bead characteristics and exhibits exceptional mechanical qualities, a range of welding methods are typically utilized. These techniques are meticulously chosen to minimize any alteration or deformation. The Submerged Arc Welding (SAW) technique is extensively utilized across diverse sectors because to its user-friendly nature, elevated current density, and the capacity to deposit a substantial quantity of weld metal by employing numerous wires simultaneously. The manufacturing industry lays significant emphasis on this, primarily because of its capacity to revitalize worn-out components. In order to attain an ideal joint, it is essential to meticulously examine and selectively determine the several process parameters of SAW. The careful and thorough selection is crucial for improving the overall quality of the weld. SAW is renowned for its multitude of process factors that exert a substantial influence on performance outcomes, such as deposition rate, dilution, and hardness. The quality of the weld is ultimately determined by these parameters. This study aims to analyze the influence of process parameters on welding outcomes by employing the Taguchi method. Decision-makers have the ability to systematically and consistently evaluate both the quantitative and qualitative aspects simultaneously. During the initial stage, the AHP requires the development of a clear and unambiguous statement that defines the ultimate goal of the decision. Additionally, all potential solutions are identified. Therefore, it is crucial to define criteria, which are the specific elements used to assess the significance of various choices [8]. One major advantage of AHP is the use of pairwise comparison methods to compare various alternatives. In order to ascertain the relative importance of each criterion, their examination is carried out in conjunction with all other criteria. The comparison is conducted using a numerical scale ranging from 1 to 9, with 1 indicating the lowest relevance and 9 indicating the highest level of significance [9]. The weight assigned to each criterion indicates its relative significance. The weights are determined by eigenvector analysis [10]. After the evaluation of the criteria weights, each alternative is assessed according to each criterion. One method for doing the evaluation is to assign ratings or scores to the options based on how well they correspond with each condition [11].

3 Solution Technique

Typically, the choice with the highest weighted score should be chosen. Another approach to assess the accuracy of the findings derived from the pairwise comparisons is to utilize the Analytic Hierarchy Process (AHP). If the estimated consistency ratio (CR) is lower than the standard threshold of 0.1, there may be comparison discrepancies. In such circumstances, the decision-maker must continuously reassess and adjust their method until they achieve the utmost level of consistency [12].

M-TOPSIS aims to identify the alternative that maximizes the amount of money spent while minimizing the distance perspective. This is done by comparing the options to an ideal cosmetic solution (PIS), which represents the best and worst case situations for all parameters. M-TOPSIS is a robust approach for traversing difficult decision-making, as it utilizes the technique of order priority similarity of optimal solutions to achieve multi-criteria optimization. M-TOPSIS is a useful method for determining the optimal solution in complex scenarios with various, and often conflicting, criteria. It helps identify the most consistent path to achieve the desired outcome. The objective of M-TOPSIS was to identify the solution that would optimize financial expenditure while minimizing the distance perspective. In order to determine the best and worst possible outcomes, all variables were considered, including the optimal cosmetic solution (PIS) [13, 14]. The initial stage entails precisely defining the objective of the decision, identifying the existing options, and formulating the appropriate criteria for evaluation. It is feasible to include both cost and benefit criteria. In a decision matrix, the criteria are denoted by columns and the options are denoted by rows. Every cell contains a value that represents the performance of an option with respect to a specific criterion. The values might be either quantitative or qualitative, depending on the quality of the criteria. The matrix is normalized to ensure that each criterion is standardized. When dealing with units or measurement scales that vary between criteria, it is necessary to exercise additional caution. Two often used normalization methods are linear normalization and vector normalization [15]. The weights provided to each criterion indicate their relative significance in achieving the ultimate objective and can be determined in two ways: through pairwise comparisons using a methodology akin to the AHP or through direct input from the decision-maker [16]. The Pareto Improvement Set (PIS) objectives are a different idea that strives to fulfil all cost–benefit criteria to the utmost extent. Conversely, NIS offers a more limited-functionality alternative at a higher cost. The optimal solutions are obtained numerically using weighted judgements and generic calculations. Next, it is necessary to ascertain the distances to the Positive Ideal Solution (PIS) and Negative Ideal Solution (NIS) for each choice. The Euclidean distance is frequently employed as a standard unit for this purpose. The proximity to the PIS reflects the degree of closeness to the optimal state, whereas the distance from the NIS represents the extent of deviation from the worst-case scenario. Greater values imply a more pronounced inclination towards the option. The RPI metric employs a numerical score ranging from one to quantify the extent of deviation from the optimal choice. Typically, the computation involves dividing the distance of the NIS from the PIS by the entire distance of the NIS. Ultimately, the relative priorities of the alternatives are defined by their RPI scores [17]. The Relative Preference Index (RPI) scores are used to find the most favored alternative. Options with lower RPI values are sorted accordingly.

4 Problem Formulation

Industrial arc welding robots are specifically engineered to achieve higher product quality requirements, while simultaneously minimizing costs and guaranteeing reliable welds. There is a wide variety of robot models available, manufactured by multiple organizations. While these models offer numerous more versatile features, their primary distinguishing factor is their capacity to find the ideal option based on crucial parameters. This study employs the AHP and M-TOPSIS-based approach to examine the decision-making process of industrial robots in selecting arc welding operations. Data was obtained from eight distinct arc welding robots, each equipped with six manipulable axes and utilizing controllers from various manufacturers.

Based on prevailing beliefs, these eight robots fulfill five different criteria. Before considering the selection criterion, multiple data sheets provided by robot manufacturing companies were examined to evaluate their products. Insights from experts in the field are also considered. The selection criteria were determined after a thorough discussion between the research group and an industry expert.

The ultimate decision matrix considered both the collective preference of the group and the significant attributes of each robot. In a maximization dilemma concerning a robot's payload capacity, a higher value is generally seen as a favorable factor in the decision-making process. The power consumption of the robot exemplifies a lower value that is profitable and susceptible to burdensome regulations. The lower value is the outcome of a problem that aims to minimize a certain parameter. The five main criteria for the selection of robots are given in Fig. 1. A hierarchy has been constructed for evaluating the weight of each criterion. With the help of criterion weight evaluated by AHP, the rank of the alternative robot will be evaluated by using the M-TOPSIS approach.

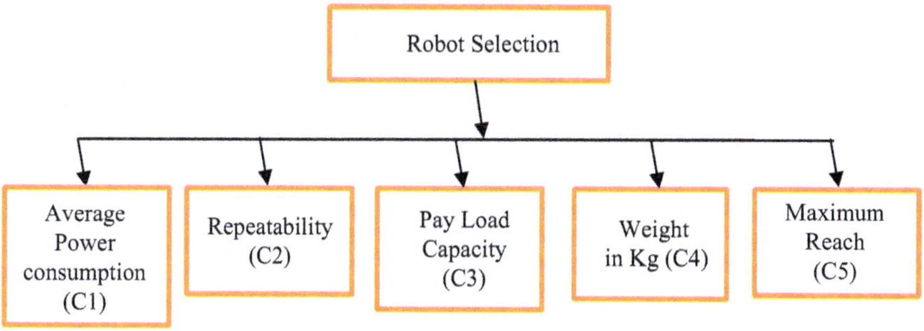

Fig. 1 Hierarchy for selection of robot

The mechanical weight of the robot, C-4, is measured in kilograms. This criterion is deemed irrelevant as it pertains to the tangible mass of the robot, which is generally undesirable to purchasers who want lighter robots. C-2 is a metric that quantifies the level of repeatability in millimeters. Repetition denotes the robot's ability to do a particular action multiple times without variation. The criterion is beneficial as increased repeatability generally yields superior results. The maximum weight that a robot can lift in one operation is determined by its payload capacity, which is defined in kilograms (C-3). It is a valuable standard because typically, having more is preferable. The maximum reach of the robot, C-5, is measured in millimeters. The total reach of a robot arm for accomplishing a task is determined by adding its vertical and horizontal extensions. Given that greater quantity or magnitude is consistently advantageous, it functions as a valuable standard or measure. The mean kilowatt-hour consumption of the robot (C-1): This word refers to the mean power consumption of the robot. The parameter of low power consumption is typically perceived as non-beneficial when it comes to requiring it in a robot. The decision matrix for the selected robots has been shown in Table 1.

Table 1 Decision matrix for selection of arc welding robot

Alternative	C1	C2	C3	C4	C5
R1	1.00	0.02	12.0	145.0	1440
R2	0.50	0.018	7.0	27.0	912
R3	0.60	0.05	4.0	170.0	1500
R4	3.40	0.04	20.0	272.0	1655
R5	2.00	0.02	25.0	250.0	2410
R6	5.60	0.05	10.0	230.0	1925
R7	2.50	0.15	6.0	501.0	4370
R8	5.00	0.08	8.0	215.0	1800

5 Results and Discussion

The problem of the selection of ARC welding robots has been considered a MCDM problem, and AHP and M-TOPSIS-based approaches have been used to solve the selected problem. This problem has been solved in 2-phases. In the first phase, AHP is used to evaluate the weight of each criterion. The criteria have been compared pair-wise and the pair-wise evaluation matrix has been shown in Table 2. The results evaluated by AHP are shown in Table 3. The consistency ratio (CR) check has also been performed for the solution obtained by AHP and it has been found that the comparison data is $CR < 0.1$, which means the data is consistent and the results could be accepted.

In the 2-phase, M-TOPS has been implemented for evaluating the final rank of the alternatives (Robots). The normalized decision matrix is shown in Table 4. The weighted normalized matrix is shown in Table 5. Positive Idle solution is shown in Table 6 and negative idle solution is shown in Table 7.

The separation distance and final ranking of the alternatives have been shown in Table 8.

The results shown in Table 8 show that Robot R3 has the first rank and R6 has the last rank.

Table 2 Pair-wise comparison matrix of selected criteria

Criteria	C1	C2	C3	C4	C5
C1	1	2	3	3	5
C2	1/2	1	2	3	5
C3	1/3	1/2	1	3	5
C4	1/3	1/3	1/3	1	3
C5	1/3	1/2	1/2	1/3	1

Table 3 Weights obtained by AHP

Criteria	Weights	Consistency check
C1	0.381753	CI = 0.107841 CR = 0.096286
C2	0.261532	
C3	0.192758	
C4	0.104293	
C5	0.059663	

Table 4 Normalised decision matrix

	C1	C2	C3	C4	C5
R1	0.098039	0.015152	0.380952	0.248945	0.152689
R2	0	0	0.142857	0	0
R3	0.019608	0.242424	0	0.301688	0.17004
R4	0.568627	0.166667	0.761905	0.516878	0.214864
R5	0.294118	0.015152	1	0.470464	0.433198
R6	1	0.242424	0.285714	0.42827	0.292944
R7	0.392157	1	0.095238	1	1
R8	0.882353	0.469697	0.190476	0.396624	0.256796

Table 5 Weighted normalised decision matrix

	C1	C2	C3	C4	C5
R1	0.037427	0.003963	0.073432	0.025963	0.00911
R2	0	0	0.027537	0	0
R3	0.007485	0.063402	0	0.031464	0.010145
R4	0.217075	0.043589	0.146863	0.053907	0.012819
R5	0.11228	0.003963	0.192758	0.049066	0.025846
R6	0.381753	0.063402	0.055074	0.044666	0.017478
R7	0.149707	0.261532	0.018358	0.104293	0.059663
R8	0.336841	0.122841	0.036716	0.041365	0.015321

Table 6 Positive idle solution (PIS)

	C1	C2	C3	C4	C5
R1	0.118561	0.066342	0.014239	0.006136	0.002556
R2	0.145735	0.068399	0.027298	0.010877	0.00356
R3	0.140076	0.039256	0.037156	0.005304	0.002452
R4	0.027119	0.047499	0.002106	0.002539	0.002194
R5	0.072616	0.066342	0	0.00305	0.001144
R6	0	0.039256	0.018957	0.003555	0.00178
R7	0.053845	0	0.030415	0	0
R8	0.002017	0.019235	0.024349	0.00396	0.001966

6 Conclusions

The objective of this study is to identify the optimal industrial robot for arc welding. Eight options were considered for decision-making, taking into account five attributes/criteria. The AHP has been used for the evaluation of the weight of

Table 7 Negative idle solution (NIS)

	C1	C2	C3	C4	C5
R1	0.001401	1.57E-05	0.005392	0.000674	8.3E-05
R2	0	0	0.000758	0	0
R3	5.6E-05	0.00402	0	0.00099	0.000103
R4	0.047122	0.0019	0.021569	0.002906	0.000164
R5	0.012607	1.57E-05	0.037156	0.002407	0.000668
R6	0.145735	0.00402	0.003033	0.001995	0.000305
R7	0.022412	0.068399	0.000337	0.010877	0.00356
R8	0.113462	0.01509	0.001348	0.001711	0.000235

Table 8 Separation distance and final ranking of the robots

Alternatives	PIS	NIS	Separation distance	Final Rank
R1	0.455887	0.086981	0.382801	7
R2	0.505835	0.027537	0.280195	4
R3	0.473544	0.071894	0.256815	1
R4	0.285408	0.271405	0.277619	3
R5	0.378354	0.229899	0.27525	2
R6	0.252087	0.393813	0.394612	8
R7	0.290277	0.324939	0.331043	5
R8	0.226997	0.363106	0.363106	6

each criterion and M-TOPSIS has been used for the evaluation of the rank of each alternative with the help of weight calculated by AHP.

The results suggest that the Robot-R-3 is the initial choice made using the M-TOPSIS approach. The device has a mechanical weight of 170 kg and possesses several notable characteristics. These include a repeatability of 0.5 (±) mm, a payload capacity of 4 kg, an average maximum reach of 1500 mm, and an average power usage of 0.6 kW. If the initial choice is unavailable in the market, the subsequent option may be selected. The M-TOPSIS technique is characterized by its statistical simplicity and its promise to generate more accurate outcomes. In addition, the Robots line can be enhanced with additional options and features. The assignment of significant weights can be approached using subjective weights or identical weights.

In future work, many other important criteria could be selected for the study and other latest multi-criteria decision-making approaches could also be used over these additional criteria to evaluate the ranking for different robots.

References

1. Emrouznejad A, Marra M (2017) The state of the art development of AHP (1979–2017): a literature review with a social network analysis. Int J Prod Res 55(22):6653–6675. https://doi.org/10.1080/00207543.2017.1334976
2. Chang DY (1996) Applications of the extent analysis method on fuzzy AHP. Eur J Oper Res 95(3):649–655. https://doi.org/10.1016/0377-2217(95)00300-2
3. Bautista OM, Kusek JW, Nyberg Jr LM, McConnell JD, Bain RP, Miller G et al (2003) Study design of the medical therapy of prostatic symptoms (MTOPS) trial. Control Clin Trials 24(2), 224–243. https://doi.org/10.1016/S0197-2456(02)00263-5
4. Sheikhipour B, Javadi S, Banihabib ME (2018) A hybrid multiple criteria decision-making model for the sustainable management of aquifers. Environ Earth Sci 77:1–16. https://doi.org/10.1007/s12665-018-7894-4
5. Afzal MS, Wakeel A, Nasir MA, Qazi MI, Abas M (2024) Optimization of process parameters for shielded metal arc welding for ASTM A 572 grade 50. J Eng Res. https://doi.org/10.1016/j.jer.2024.01.005
6. Sánchez-Lozano JM, Meseguer-Valdenebro JL, Portoles A (2017) Assessment of arc welding process through the combination of TOPSIS–AHP methods with fuzzy logic. Trans Indian Inst Met 70:935–946. https://doi.org/10.1007/s12666-016-0884-x
7. Sarkar A, Roy J, Majumder A, Saha SC (2014) Optimization of welding parameters of submerged arc welding using analytic hierarchy process (AHP) based on Taguchi technique. J Inst Eng (India): Ser C 95:159–168. https://doi.org/10.1007/s40032-014-0114-4
8. Aghakhani M, Mehrdad E, Hayati E (2011) Parametric optimization of gas metal arc welding process by Taguchi method on weld dilution. Int J Model Optim 1(3):216
9. Deshmukh P, Sorte MB (2013) Optimization of welding parameters using Taguchi method for submerged arc welding on spiral pipes. Int J Recent Technol Eng (IJRTE). ISSN: 2277-3878
10. Saaty TL (1980) The analytic hierarchy process. In: Planning, priority setting, resource allocation. McGraw-Hill
11. Ishizaka A, Labib A (2009) Analytic hierarchy process and multi-criteria decision making. Springer Science & Business Media
12. Saaty TL (2008) Decision making with the analytic hierarchy process. Int J Serv Sci 1(1):83–98. https://doi.org/10.1504/IJSSCI.2008.017590
13. Triantaphyllou E, Lin CT (2009) Development and application of a decision-making support system for selecting coastal management practices. Environ Model Softw 24(4):414–423
14. Singh AK, Avikal S, Kumar KN, Kumar M, Thakura P (2020) A fuzzy-AHP and M-TOPSIS based approach for selection of composite materials used in structural applications. Mater Today: Proc 26:3119–3123
15. Avikal S, Jain R, Mishra PK (2014) A Kano model, AHP and M-TOPSIS method-based technique for disassembly line balancing under fuzzy environment. Appl Soft Comput 25:519–529. https://doi.org/10.1016/j.asoc.2014.08.002
16. Rashmi R, Singh R, Chand M, Avikal S (2019) An MCDM-based approach for selecting the best state for tourism in India. In: Harmony search and nature inspired optimization algorithms: theory and applications, ICHSA 2018. Springer Singapore, pp 17–26. https://doi.org/10.1007/978-981-13-0761-4_2
17. Chand M, Hatwal D, Singh S, Mundepi V, Raturi V, Rashmi, Avikal S (2017, February) An Approach for purchasing a sedan car from Indian car market under fuzzy environment. In: Proceedings of sixth international conference on soft computing for problem solving: SocProS 2016, vol 1. Springer Singapore, Singapore, pp 239–244. https://doi.org/10.1007/978-981-10-3322-3_22

Statistical Analysis and Data-Driven Modeling of Hole Circularity in EDD of Aerospace Material

Kedari Lal Dhaker, Amit Sharma, Tasnim Arif, Gaurav Shukla, Akshat Shukla, and Shwetank Avikal

Abstract Electric discharge drilling (EDD) plays an important role in the advancement of aerospace manufacturing due to its ability to produce complex hole geometries with high precision. In this study, a statistical analysis and data-driven modeling of hole circularity have been proposed during EDD of aerospace alloy. Statistical analysis-based prediction model is developed where the experimental results of hole circularity have been considered. These are the functions of four EDD process variables (i.e., Discharge current, Pulse on time, Pulse off time, and Flushing pressure). Through training and validation, the statistically learned model exhibits the complex relationships between input EDD process parameters and output geometry, enabling it to accurately predict the hole circularity. Reliability and effectiveness of the developed model are represented by R-Square and Adjusted R-Square value which are quite at suitable level of 88.8% and 75.7%, respectively. Furthermore, contour plots of interactive effect of process parameters are developed and influence of EDD interactive process parameters on hole circularity is discussed. Finally, optimal range of process parameters was found as: Discharge Current: 8–9 A, Pulse on time: 7.3–8 μs, Pulse off time: 2–2.5 μs, Flushing pressure: 75–80 bar. The proposed study demonstrates the precise prediction and optimization of drilled geometry.

Keywords Statistical modeling · Electric discharge drilling · Hole circularity · Aerospace manufacturing

K. L. Dhaker · A. Sharma (✉) · T. Arif · G. Shukla · A. Shukla
Jaypee University of Engineering and Technology, Guna, MP 473226, India
e-mail: ip.amit.sharma@gmail.com

S. Avikal
Department of Management Studies, Graphic Era (Deemed to be) University, Dehradun, India

P. Bhingole et al. (eds.), *Advances in Materials Engineering*, Lecture Notes in Mechanical Engineering, https://doi.org/10.1007/978-981-97-7114-1_31

1 Introduction

Advanced machining processes such as Electric Discharge Drilling (EDD) have emerged as a reliable manufacturing process to tap the opportunities in aerospace equipment manufacturing. Using controlled EDD to erode materials, the process offers unique advantages such as the ability to drill complex geometries, machining of hard-to-machine materials, and high dimensional precisions. Advanced industry like aerospace industry, medical equipment, and automobile components demand stern dimensioning tolerances, intricate design parts, and advances in materials; EDD presents a striking alternative to conventional drilling technologies. The following are the main advantages of EDD [1–3]:

a. Able to produce complex features and close dimensions.
b. Applicable to a wide range of engineering material.
c. Produces reduced or negligible thermal Damage.

Although there are advantages associated with EDD in aerospace manufacturing, the following challenges exist [3, 4]:

a. Low material removal rate
b. Tool Wear and Electrode Degradation
c. Surface Integrity
d. Process Stability and Control
e. Geometrical features in EDD drilled hole.

Addressing these challenges, while leveraging the potential of EDD is the key to unlocking the full potential of EDD in aerospace manufacturing. By conducting systematic research, developing cutting-edge method, and working intimately with industry and academia to go forward aerospace component manufacturing processes, EDD can make a significant impact. Researchers have extensively investigated the application of EDD in various advanced materials and in addressing the challenges posed by EDD.

Nurlan Nauryz et al. have experimentally investigated the drilling EDM for biomedical material Ti-6Al-4AV. They found enhanced MRR and decreased overcut with the help of hydroxyapatite powder mixed in dielectric fluid [5]. Joshi and Joshi have systematically reviewed the powdered Mixed EDM process and have observed the powdered mixed EDM can improve EDM performance in terms of superior surface finished. They have critically reviewed the effects of power [6]. Bhaumik et al. has investigated MRR & TWR by applying CCD design of experiment in different tool material in EDM process. Authors have found that better MRR by using brass and zinc electrodes as compared to copper electrode. They have reported that Copper electrode yields less recast layer and less tool wear [7].

Baroi et al. have studied the MMR, surface roughness in EDM of titanium grade 2 alloy. Authors have reported that MMR and Too wear rates are greatly affected proportionally with discharge current and inverses proportionally with pulse on time. Author s also studied the micrographs for understanding the surface features of machine parts [8]. Ahuja et al. has improved the micro-hole geometry using EDM of commercially pure titanium (CPTi). Authors have reported the effect of holes on the apatite formation in EDM. They have measured the apatite formation using simulated body fluid (SBF) for different days ranging from 7 to 21 days of immersion [9]. Neeraj et al. have developed the prediction model of SR and MRR of Wire EDM in alloys Ti-6Al-4V. Authors found that the error in the prediction of MRR is 0.269 mm^3 min^{-1}, while that of SR is 0.11 μm [10].

Shabgard et al. have investigated the the EDM process and process parameters in AISI H13 tool steel. They have conducted the experiments are per full factorial design and found that recast layer thickness and HAZ greatly influenced by interactive input parameters: Discharge current and pulse-on time [11]. Roy et al. have numerically simulated and experimentally validated the Reverse Micro EDM (RMEDM) process for different cross-sections (circular, square, triangular) for debris in micro hole [12]. Ndaliman et al. have studied surface micro-hardness of alloy Ti-6Al-4 V using EDM. Authors have used the Cu–TaC electrode to improve the surface harness condition. They have improved the micro hardness of surface produced by EDM using Cu–TaC electrode instead of Cu electrode [13].

Dong et al. have combined the EDM with arc machining and reported that difficult to machine material like Inconel 718 can easily be machined with this hybrid approach. They have performed the experiments by applying CCD design of experiment and proposed the response model for MRR. They reported that higher MRR of 450.118 mm^3/min using combined machining method [14]. Chinmaya et al. have conducted the experimentation of EDM using Box–Behnken design and developed the response model for quality characteristic MRR, EWR, SR and radian overcut. Authors have reformed the multi objective optimization using Particle Swarm optimization (PSO) and found that flushing pressure is insignificant in order to improve the quality parameters in EDM multi objective optimization [15]. Even though the Electric Discharge Drilling (EDD) technology is widely used, a significant research gap exists in understanding the effects of process parameters on many quality characteristics like geometrical, metallurgical and dimensional features hoe drilled hole.

While EDD offers capabilities for drilling complex geometries and advanced alloys, improving the geometrical, and other quality features is critical for enhancing performance of parts manufactured with EDD in aerospace manufacturing.

Available studies are mainly focused on cost reduction and improving drilling efficiency on conventional metallic alloys, overlooking the geometrical features and challenges posed by aerospace materials such as titanium alloys.

Moreover, there is limited research on understanding the interaction effects of primary EDD process parameters (such as discharge current, pulse on time, pulse off time and flushing pressure) in the context of geometrical features. Not having the proper comprehensive study for understanding the geometrical qualities of drilled hole in aerospace materials hinders the development of EDD, and its automation tailored for aerospace manufacturing. The primary objectives of this research study are to enhance the understanding of EDD for aerospace manufacturing with a specific

Table 1 Levels of EDD input process factors

Emblem	Input factor	Unit	L-1	L-2	L-3
A	Discharge current	A	6	8	10
B	Pulse-on time	μs	6	7	8
C	Pulse-off time	μs	2	3	4
D	Flushing pressure	bar	60	70	80

focus on enhancing the circularity of drilled hole which is critical for aerospace applications. Through a combination of experimental studies and statistical computational, this research aimed to explicate the interactive effect of EDD process parameters on hole circularity. Achieving these objectives, contributes to the development of more efficient, precise, and reliable drilling processes for critical aerospace materials.

2 Experimentation and Measurement

2.1 *Input Process Factors*

In Electric Discharge Drilling (EDD) circularity is a critical parameter, especially in aerospace applications where precision is paramount. Achieving the required circularity involves four process parameters of EDD. Four levels and their values considered in experimental work are mentioned in Table 1.

2.2 *Experiment Planning and Design*

Experimental work is conducted as per the design of experiment and BBD (Box Behnken Design) of experiments is taken in the experiments conduction. Total number of experiments required as per BBD design is calculated using following relationship:

$$N = 2f(f - 1) + k,$$

where,

N = Total number of experiment

f = numbers of factors considered for study

k = number of central point considered in experiments.

In the present study, 4 factors with tree levels as mentioned in Table 1 are considered with 3 central points. Therefore total 27 experiments are required as per BBD mentioned in Table 2.

Table 2 Box–Behnken design of experiments

Experiment No.	A	B	C	D	Hole circularity
1	−1	0	0	1	0.923
2	1	0	0	1	0.974
3	0	−1	1	0	0.862
4	−1	1	0	0	0.894
5	0	−1	0	−1	0.883
6	1	0	−1	0	0.912
7	1	−1	0	0	0.875
8	0	1	−1	0	0.936
9	0	0	0	0	0.954
10	0	0	1	1	0.963
11	1	0	0	−1	0.834
12	0	0	0	0	0.946
13	0	−1	0	1	0.975
14	−1	−1	0	0	0.954
15	0	−1	−1	0	0.962
16	−1	0	−1	0	0.976
17	0	0	0	0	0.918
18	0	0	1	−1	0.875
19	−1	0	1	0	0.905
20	−1	0	0	−1	0.925
21	0	1	0	−1	0.853
22	1	0	1	0	0.875
23	0	1	0	1	0.975
24	1	1	0	0	0.968
25	0	0	−1	1	0.98
26	0	1	1	0	0.927
27	0	0	−1	−1	0.843

2.3 Material Sheet

Titanium alloy Ti-6242 is widely used for aerospace components due to their better mechanical and thermal properties at elevated temperature. Applications of this material include precision engine parts, gas turbine parts discs, and impellers. Material sheet of 2 mm thick is procured from the commercial vendor for the experimental work.

2.4 Experimental System

Electric discharge drilling system has 8 major components as supplied by Sparkonix India Limited, Fig. 1 shows their integration as a complete drilling system. Discharge current is being supplied to the work material in pulses mode indicated in Fig. 2, which is responsible for creating the heat required for melting of work material. Pulse-on time is the time interval in which discharge current supplied to material and Pulse-off time is time interval for which current supplied is off. Cyclic procedure of pulse on time and pulse off time continued and melting of localized drilling zone is created after proper flushing of molten metal cleaned through hole in the sheet is achieved. Distilled water is used as flushing liquid and pressure of flushing liquor plays s important role in creating a cleaned thought hole.

a. Power Supply
b. Electrode
c. Dielectric Fluid
d. Work piece
e. Tool holder and Tool Alignment Mechanism
f. Pulse Generator
g. Cooling System
h. Machining Table.

2.5 Measurement Work

Measurements of hole diameters at the entrance side of the has been done using digital microscope with 10× magnification level. Diameters are measured as per schematic diagram as shown in Fig. 3. The hole circularity calculated for each drilled hole and mentioned in Table 2.

Circularity at the entrance side has been calculated after measuring the six diameters for each drilled hole using a flowing mathematical relationship as shown in Eq. (1).

$$\text{Circularity} = \frac{d_{\min}}{d_{\max}} \tag{1}$$

Fig. 1 EDD system

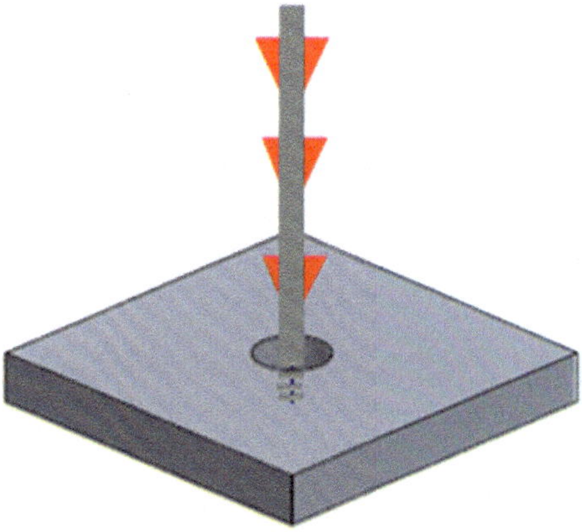

Fig. 2 Schematic of pulsed discharge current supply

Fig. 3 Scheme of diameters measurement

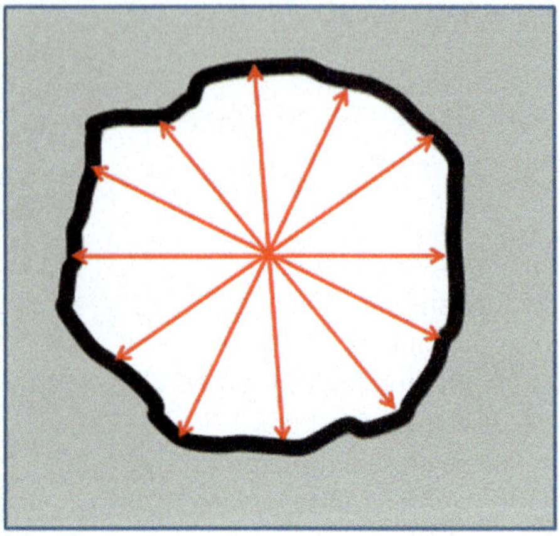

here,

$d_{\min}$ = Minimum diameter measured

$d_{\max}$ = Maximum diameter measured.

Variation in value of hole circularity in each of the drilled hole is shown in Fig. 4.

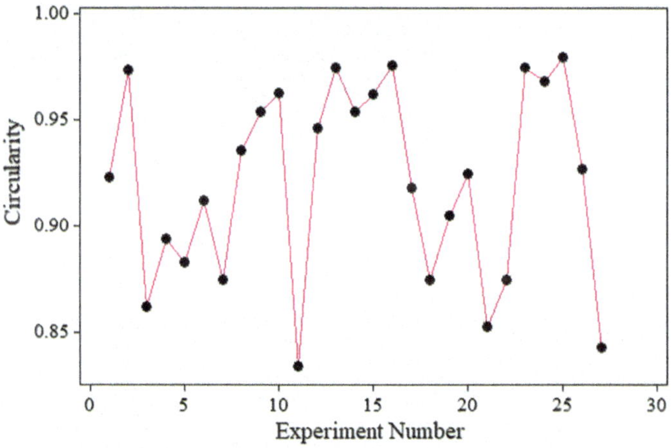

Fig. 4 Distribution of value of circularity in experiments

3 Modeling and Optimization of Circularity

3.1 Response Surface Methodology-Based Modeling of Hole Circularity

In this study, Box–Behnken design of experiment has been used by taking four factors and their three levels [16]. Regression techniques are used for developing the quadratic models. ANOVA Table has been developed to check the adequacy and suitability for developed model. In this design total '27' experimental runs consisting of '3' central point and '24' non-center points are considered [17].

In the present research, input process factors are Discharge current, Pulse-on time, Pulse-off time, and Flushing pressure, whereas the response factor is hole circularity. The second order regression equation is used in the present study for better prediction reliability and to understand the influence of input factors. General equation is given as Eq. (2) [18, 19]:

$$y = \beta_0 + \sum_{i=1}^{n} \beta_i A_i + \sum_{i=1}^{n} \beta_{ii} A_i^2 + \sum_{i=1}^{n} \sum_{j=1, j \neq i}^{n} \beta_{ij} A_i A_j \tag{2}$$

β_{ij} = Regression coefficients and y = Response parameter, $A_i = n$ different input parameters ($i = 1, 2, 3\ldots n$).

The experimental data as shown in Table 2, are used to develop empirical model. The developed empirical model for hole circularity is presented in Eq. (3):

$$\begin{aligned} \mathit{Circularity}() = {} & 2.24 - 0.231(A) - 0.195(B) - 0.056(C) \\ & + 0.0072(D) - 0.00287(A*A) - 0.00537(B*B) \\ & - 0.0114(C*C) - 0.000130(D*D) + 0.0191(A*B) \\ & + 0.00425(A*C) + 0.00177(A*D) + 0.0227(B*C) \\ & + 0.00075(B*D) - 0.00122(C*D) \end{aligned} \tag{3}$$

A, B, C, D = Input factors as mentioned Table 1.

An ANOVA is computed to analyze the reliability and prediction significance of the developed regression model of hole circularity. As in ANOVA Table 3 the model is statistically significant, with an adjusted R-squared value of 75.7% for the hole circularity regression model. This R-squared value further confirms the statistical reliable of the model. The significance of the model is assessed using the 1 − p value (0.999), where a value greater than the chosen significance level (95%) indicates the regression model's significance in predicting the response value (hole circularity).

Table 3 ANOVA table

Error source	DF	SS	MS	F	P
Regression	14	0.04838	0.00346	6.77	0.001
Residual error	16	0.00612	0.00051		
Total	26	0.05451			

3.2 Interactive Influence of Input Process Factors

Circularity of EDD drilled hole is governed and controlled by Discharge current components (Current, Pulse-on time, Pulse-off time) and Flushing pressure. Their interaction effect on hole circularity is very complex. An increment in discharge current may enhance the MRR while simultaneously causing the random thermal effect in the drilling zone that leads to poor circularity if immediate flushing is not done. To capture the interactive behavior of the input process factors of EDD on Ti-62-42 alloy, contour plots are developed using the experimental data.

The combined effect of current and pulse-on-time is presented in Fig. 5. Higher discharge current and higher pulse-on-time, provide an opportunity for large interaction time and heat production at the workpiece resulting in excessive heating of work material. Hence, optimal range of current should be 9–10 A, and the optimal range of pulse-on time, is 7.5–8 μs.

The interactive effect of current and pulse-off time is given in Fig. 6. From Fig. 6, it is found that lower discharge current along lower value of pulse-off time are favorable interactive values for better circular hole as indicated in diagram. Lower current in interaction with lower pulse-off time restrict the excessive heat of drilling zone that

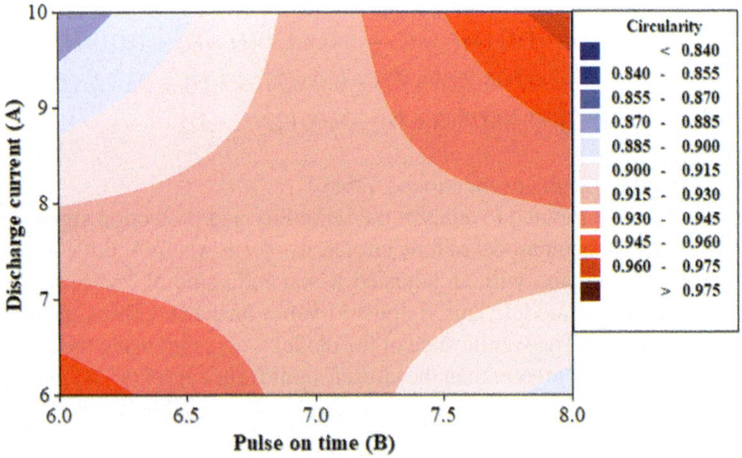

Fig. 5 Interactive contour plot in discharge current (A) and pulse-on time (B)

ultimately leads to better circular hole. Therefore, a favorable range of current is 6–7 A, and for the Pulse-off time is 2–2.5 μs.

The combined effect of sparking current and Flushing pressure is mentioned in Fig. 7. Larger discharge current and higher value of Flushing pressure leads to enhanced better heating and flushing of molten material from the drilling zone, while lower value of flushing pressure allows the re-welding of molten material to parent metal. Re-welding ultimately causes a poor circular hole. The favorable range of current is 9–10 A and for Flushing pressure is 78–80 bar.

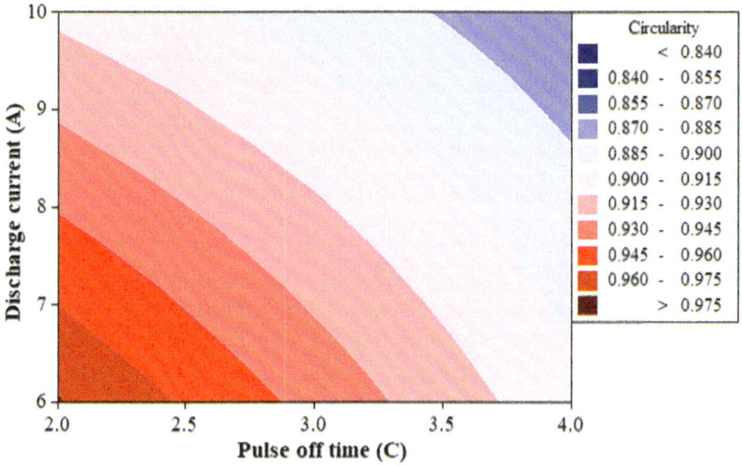

Fig. 6 Interactive contour plot in discharge current (A) and pulse-off time (C)

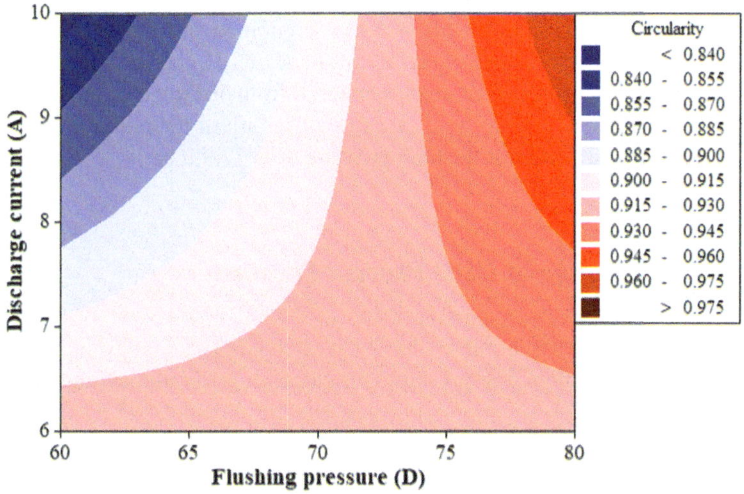

Fig. 7 Interactive contour plot in discharge current (A) and flushing pressure (D)

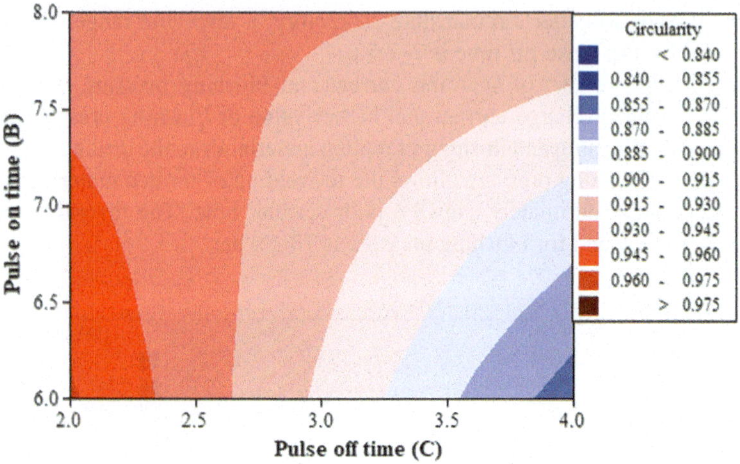

Fig. 8 Interactive contour plot in pulse-on time (B) and pulse-off time (C)

The mutual influence of pulse-on time and Pulse-off time on circularity of hole is shown in Fig. 8. From the Fig. 8, Lesser value of Pulse-off time causes a better circular hole while pulse on time can be taken in wide range of values. The favorable range of pulse-on time = 6–7.5 μs and for pulse-off time = 2–2.25 μs.

Effect of the pulse-on-time interface with flushing pressure on circularity is indicated in Fig. 9. Wide range of pulse-on time and higher value of Flushing pressure create the favorable condition for proper melting and immediate flushing of molten material from the drilling leads to better circular hole. There for favorable interaction ranges for proper drilling: Pulse-on time = 6–8 μs, flushing pressure = 78–80 bars.

Effect of flushing pressure combined with pulse-off time is represented in Fig. 10. Values in lower side in range of pulse-off time combined with values higher side in range of flushing pressure lead to better circular hole. Higher pressure causes the immediate and proper flushing of molten material from drilling area, leads favorable condition for improved drilled hole in terms of circularity. Optimal range for the pulse-off time is 2–2.5 and for flushing pressure, it is 77–80 bar.

3.3 *Optimal Range of EDD Process Parameters*

Optimal range of process parameters has been established from the detailed analysis of interactive contour plots. Table 4 shows the calculation of optimal range of process parameters.

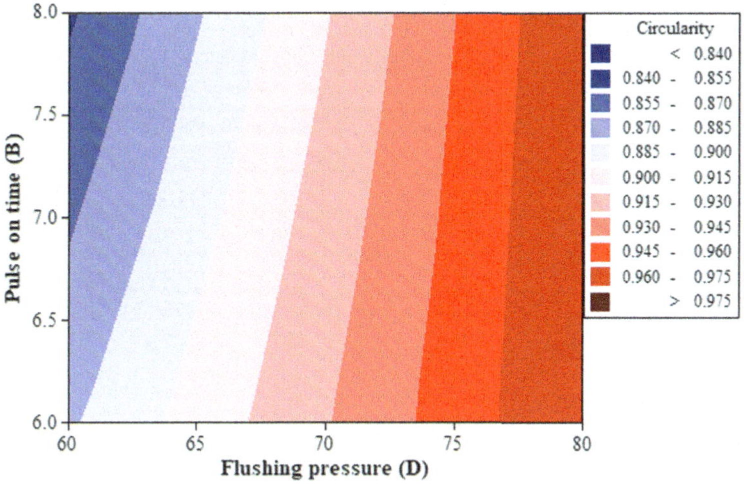

Fig. 9 Interactive contour plot in pulse-on time (B) and flushing pressure (D)

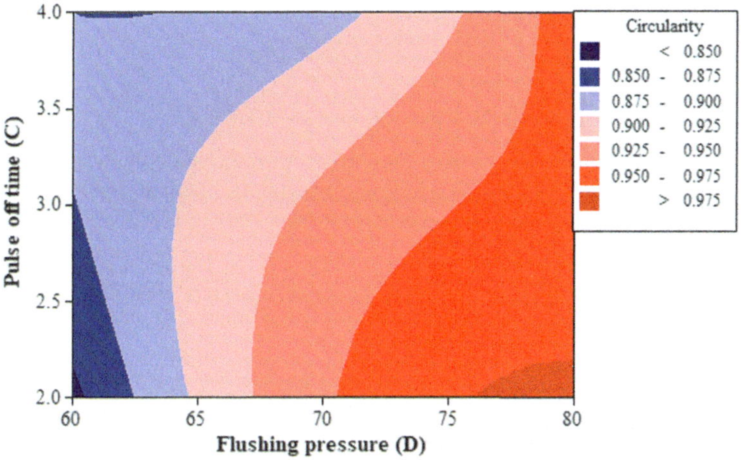

Fig. 10 Interactive contour plot in pulse-off time (C) and flushing pressure (D)

4 Conclusions

Experimentation and investigation are conducted for EDD drilled hole geometry. Following are the conclusions drawn from the study.

a. EDD process is a potential technological solution in commercial aerospace manufacturing.
b. Statistical analysis and data-driven modeling are proven technique in modeling complex and stringent manufacturing operations.

Table 4 Optimal range of input factors

Interactions	Interactions range of parameters			
	(A)	(B)	(C)	(D)
(A)–(B)	8–10	7–8	Fixed	Fixed
(A)–(C)	6–9	Fixed	2–3	Fixed
(A)–(D)	7.5–10	Fixed	Fixed	75–80
(B)–(C)	Fixed	6–7.30	2–2.52	Fixed
(B)–(D)	Fixed	6–8	Fixed	70–80
(C)–(D)	Fixed	Fixed	2–4	70–80
Optimal range parameters	**8–9**	**7.3–8**	**2–2.5**	**75–80**

c. Data-driven statistical analysis-based prediction model of hole circularity is developed with R-squared and adjusted R-squared value of 88.8% and 75.7% respectively.
d. Optimal ranges of EDD process parameters for improving the hole circularity are found as:

- Discharge Current: 8–9 A
- Pulse on time: 7.3–8 μs
- Pulse off time: 2–2.5 μs
- Flushing pressure: 75–80 bar.

e. Proposed methodology is very important step in development of efficient automation of the aerospace parts manufacturing.

Acknowledgements This research work was financially supported by the Directorate of Research Innovation and Development (Jaypee University System, JIIT Noida, India).

References

1. Saini G, Kumar A, Mohal S, Kumar R, Chauhan A (2021) Electrochemical discharge machining process, variants and hybridization: a review. In: IOP conference series: materials science and engineering, vol 1033, p 012070
2. Singh A, Kumar P, Singh I (2013) Design and development of electro-discharge drilling process. Adv Mater Res 651:607–611
3. Kumar P, Shekhar A, Yadav SKS (2020) Experimental analysis of electrical discharge drilling (EDD) of carbon-carbon composite. Mater Today: Proc 22(4):3106–3115
4. Hanash EAH, Ali MY (2018) Experimental study of electrical discharge drilling of stainless steel UNS S30400. In: IOP conference series: materials science and engineering, vol 290, p 012067
5. Nurlan N, Salikh O, Didier T, Asma P (2023) Performance of powder mixed drilling EDM on biomedical Ti-6Al-4V alloy. In: E3S web of conferences, vol 430, p 01233
6. Joshi AY, Anand JY (2019) A systematic review on powder mixed electrical discharge machining. Heliyon 5:2963

7. Bhaumik M, Maity K (2018) Effect of different tool materials during EDM performance of titanium grade 6 alloy. Eng Sci Technol, Int J 21(3):507–516
8. Baroi BK, Debnath T, Jagadish, Patowari PK (2019) Machinability assessment of titanium grade 2 alloy using deionized water in EDM. Mater Today: Proc 26:2221–2225
9. Ahuja N, Sharma N, Hegab H, Khanna R, Khan AM (2020) Bioactivity measurement of commercially pure titanium processed by micro-electric discharge drilling. Int J Adv Manuf Technol 107:2797–2805
10. Sharma N, Singh G, Gupta M, Hega H, Mia M (2019) Investigations of surface integrity, bio-activity and performance characteristics during wire-electrical discharge machining of Ti-6Al-7Nb biomedical alloy. Mater Res Exp 6(9):096568
11. Shabgard M, Seyedzavvar M, Oliaei SNB (2011) Influence of input parameters on the characteristics of the EDM process. J Mech Eng 57(9):689–696
12. Roy T, Datta D, Balasubramaniam R (2018) Numerical modeling, simulation and fabrication of 3-D hemi-spherical convex micro features using reverse micro EDM. J Manuf Process 32:344–356
13. Santos I, Polli ML, Daniel H (2015) Influence of input parameters on the electrical discharge machining of titanium alloy (Ti-6Al-4V). Int J Manuf Res 10(3):286
14. Dong H, Liu Y, Shen Y, Wang X (2016) Optimizing machining parameters of compound machining of Inconel 718. Procedia CIRP 42:51–56
15. Chinmaya MP, Sankar MS, Ranjan SM (2016) A particle swarm approach for multi-objective optimization of electrical discharge machining process. J Intell Manuf 27:1171–1190
16. Montegomery DC (2017) Design and analysis of experiments, Eighth edn. Wiley, New Jersey
17. Box George EP, Normal RD (1987) Empirical model-building and response surfaces, 1st edn. Wiley
18. Aslan N, Cebeci Y (2007) Application of Box–Behnken design and response surface methodology for modeling of some Turkish coals. Fuel 86(1–2):90–97
19. Haftka Raphael T, Scott EP, Juan RC (1998) Optimization and experiments: a survey. Appl Mech Rev 51(7):435–448

Numerical Investigation of Thermal Hydraulic Performance of Solar Air Heater with M-Shaped Roughness

Harvindra Singh, Durgeshwar Pratap Singh, Narendra Gariya, Chandra Kishore, Shwetank Avikal, and Shivani Monga

Abstract In this study, a rectangular solar air heater design with an M-shaped dimple roughness put on it is examined. Based on the roughness of the rectangular duct, an approach is employed to assess the energy efficiency of the solar air heater and look into solar energy efficiency. After finishing the geometry of the dimple M-shaped rectangular channel and gathering parameters from many studies with a heat flux of 1000 W/m^2, the smooth rectangular channel is connected to the dimple M-shaped roughness in order to increase the heat flow rate. The heat transmission rate is maintained by the range of relative gaps $g = 0$–1.3. A rectangular channel's hydraulic diameter is determined by factors that fall into one of the following ranges: 4000–24,000. First, pitch space (p) between M-shaped roughness is given in a range of 24, 36, 48, and 60 mm were applied. This roughness was described as a continuous distance between each roughness and roughness at $p/e = 4$, 6, 8, and 10 where $e = 6$ mm height of roughness. The efficiency of a M-shaped solar air heater was evaluated by an analysis of its heat rate and Nusselt number. The results showed a 1.88–4.0 times increase as compared to a smooth channel. When compared to a smooth channel, the roughness solar air heater exhibited a friction factor 1.66–3.08 times higher at a Reynolds number of 24,000. The solar model $\eta_{\max} = 2.25$ has the highest efficiency.

Keywords Heat transfer rate · Nusselt number (Nu) · Friction factor (Fr) · Thermal hydraulic performance · M-shaped roughness

H. Singh · D. P. Singh · N. Gariya · C. Kishore
Department of Mechanical Engineering, Graphic Era (Deemed to be) University, Dehradun, Uttarakhand 248002, India

S. Avikal (✉)
Department of Management Studies, Graphic Era (Deemed to be) University, Dehradun, Uttarakhand 248002, India
e-mail: avi.shwetank@gmail.com

S. Monga
PDP, Graphic Era Hill University, Dehradun, Uttarakhand 248002, India

P. Bhingole et al. (eds.), *Advances in Materials Engineering*, Lecture Notes in Mechanical Engineering, https://doi.org/10.1007/978-981-97-7114-1_32

1 Introduction

Alternative energy sources like gasoline, power, and nuclear energy are getting more expensive, so solar energy is an important source of energy for now and in the future. A smart idea is the rectangular solar air heater, which uses solar energy to get the most energy without losing any. As part of their research, researchers look at solar power and use air heating devices. A lot of different engineering systems depend on air. One example is solar air heaters (SAH), which take fresh air from outside and move it through a series of metals (maze). As a working material, SAH is mostly air and is small and flexible. Rib-roughened channels with repeated rib turbulators can improve heat transfer in many engineering systems, including gas turbine blades, electronic devices, solar air heater pipes, and nuclear reactors that are cooled by gas. Protrusions, fins, and extended surface shapes increase the heat transfer surface area compared to a flat surface, which is what makes higher heat transfer rates possible. It has been said that fin-filled solar air heater tubes make the fined collector much more efficient, even though they lose more pressure. Researchers have found links between the Nusselt number and the friction factor. They have also found that dimple-shaped protrusions improve heat transfer and frictional losses compared to smooth ducts in the same flow conditions. Solar air heaters have problems, like not transferring heat very well, using a lot of power to pump air, and letting more air escape. Porous materials like broken glass, iron foil, wire mesh, and hollow shapes have been suggested as ways to make systems work better.

Chaube et al. [1] analyzed two-dimensional rectangular SAH with roughness on the heated wall using commercial ANSYS FLUENT 6.1 CFD software. They made use of a rectangle meshing SST-k-e turbulence model. Reynolds number, *p*/*e* = 13.3, height of roughness = 3 mm (square rib) were the operational conditions. The outcomes demonstrated that the 3 × 5 mm^2 rectangular rib had the highest performance index, closely aligning with the experimental findings. In 2006, Chaube et al. [2] used the commercially available program ANSYS FLUENT 6.1 using the (shear stress transport) SST-k-e turbulence model as well as non-uniform rectangular meshing to analyze the performance of the solar air heater with various substances ribs include square, chamfered, round, semicircular, triangular, and rectangular shapes. Reynolds number = 2900–19,500 was the operational variable that was employed. Pitch is 10 mm, rib height (*e*) is 1 mm (for transitionally flow regime) or 3 mm (for fully rough flow regime) depending on ratio of width to height. The investigation discovered that rectangular ribs measuring 2 mm in width & 1 mm in height were effective during the transitional flow regime, while ribs measuring 3 mm overall height and 4 mm in width were suitable for use in the completely rough flow environment. In 2010, Gandhi et al. [3] used a professional ANSYS Fluid CFD CODE program to look into a two-dimensional shapes solar air heater duct. There are many horizontal ribs on the plate that absorb and have a wedge-like cross-section. Reynolds number = 5000–21,000 was picked as the operating range. The solar air heater duct with a roughened surface exhibited a friction factor that was 3.4 times greater with a Re of 21,000 in comparison to the plain duct. In 2011, Sharma et al. [4] conducted a study on a computer model of a solar air heater, adhering to the regulations set by ASHRAE. The investigation revealed that the Nusselt number reached its highest value at the pitch with the lowest relative roughness, whereas the friction factor decreased as the Reynolds number increased. In 2013, Bhagoria et al. [5] conducted a study on a solar air heater duct, where they investigated the impact of roughness on its surface. They found that the higher heat transfer rate was a result of turbulence and the re-joining of fluid flow, specifically when rib heights of 5 mm were used. In 2013, Yadav et al. [6] did a study to look into how heat moves through a rectangular duct with a rough surface. They used different meshing methods, and when the Reynolds number was less than 18,000, the average Nusselt number went up by 2.39 times compared to a duct with a smooth surface. In 2014, Kumar et al. [7] used the commercial solar air heater. The design accounted for variations in the absorber plates' roughness using V-shaped ribs, multiple V-shaped ribs, and multiple V-shaped ribs with a gap. According to the findings, compared to the heat transfer from the smooth wall, the heat transfer with V-shaped ribs was 1.7 times higher, with multi V-shaped ribs being 4.7 times higher, and ribs in the shape of multiple V-shaped with a gap smoothed from sunlight being 5.6 times higher. In 2015, Gawande et al. [8] did a study on a solar air heater duct that had a rectangular shape and two dimensions. The absorber plate of the duct had right angle triangular ribs. The researchers utilized the ANSYS FLUENT 14.5 coding package, employing a Renormalization (RNG)-k-e turbulence model and a non-uniform grid. The study determined that the ideal thermal hydraulic performance parameter fell within the range of 1.41–2.03 for the parameters under investigation. Jin et al. [9] performed a three-dimensional numerical analysis of a solar air heater equipped using a V-shaped absorbent plate. Their findings showed that the greatest thermal efficiency factor had been 1.93, achieved at V-rib = 6. Kumar et al. [10] conducted an analysis of a solar air heater duct having S-shaped absorbent plate ribs employing the ANSYS FLUENT 14.5 technology. The researchers discovered that when the Reynolds number declines, the friction factor also decreases, consequently the Nusselt number improves. Additionally, when ducts are polished, it enhances the thermodynamic improvement effect. Ahuja et al. [11] study on a solar air heater equipped with circular ribs over an absorbent surface. The highest possible thermo-hydraulic efficiency of 1.631 was achieved by considering every factor, considering several rib configuration kinds. At Reynolds number of 15,000, the Nusselt number for the rough duct was 2.223 times higher than that of the smooth duct. Boukadoum et al. [12] Conducted on an absorber plate having transverse rectangular ribs to study the effects of a surface-to-air heat exchanger (SAH). Local refinement along the wall created a two-dimensional non-uniform grid to study inner-rib flow and heat transmission. Roughened duct Stanton number rose 1.3–1.8 times greater than smooth duct, peaking at 1000 Reynolds. Kumar et al. [13] used ANSYS FLUENT 15.0 to CFD a SAH with broken curved ribs and tetrahedral mesh employed in the investigation. Reynolds number, pitch, height, gap width, location, and arc angle were adjusted. For Reynolds number 12000, the peak Nusselt number was 2.2524 times higher than a smooth duct, while the average number dropped with arc angle.

2 Methodology

The Geometric model of a smooth rectangular duct is shown in Fig. 1. In the figure the length of the rectangular duct is $L = 640$ mm and height $H = 20$ mm and width $W = 100$ mm these parameters are taken from the research paper. At this smooth duct's results are compared with the theoretical equation until that will not satisfy the condition which is shown in Fig. 4.1. After this solution in case of smoothness we inserted m-shape with gap and with different height rib roughness and for m-shape rib roughness some variations are taken as gaps in the m-shape rib roughness which is shown in the parameters range section. The geometric of the solar rectangular duct with m-shape rib roughness is taken or designed to solve the solar energy problems and how can we increase the performance of solar air heaters that is the main motive for designing the m-shape rib roughness in smooth rectangular duct as shown in different types of m-shape with gaps rib in Fig. 2. The investigation examines the utilization of m-shape rib roughness to enhance the speed of heat transfer while easing the change of heated atmosphere between smooth to turbulence, resulting in improved thermal efficiency. The graphics shown in ANSYS FLUENT depict the design of m-shape with gap rib roughness within a Smooth channel with multiple gap sizes ($g = 0$, 0.7, 1, and 1.3 mm).

2.1 Mesh Generation in Design Solar Rectangular Duct

High-performance rectangular solar air heaters with m-shaped rib roughness are made using the Ansys meshing technology. Through the use of methods such as mapping face meshing, face expanding, as well as refining, the mesh-generating process is integrated into the Ansys workbench. The Ansys Fluent (CFD) meshing

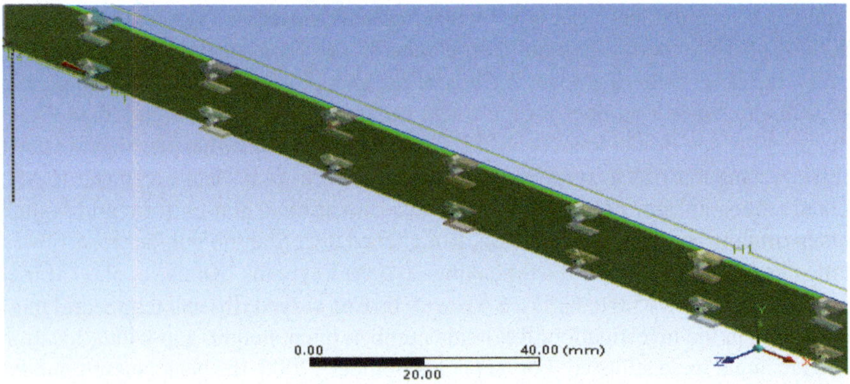

Fig. 1 Geometric model of the rectangular duct

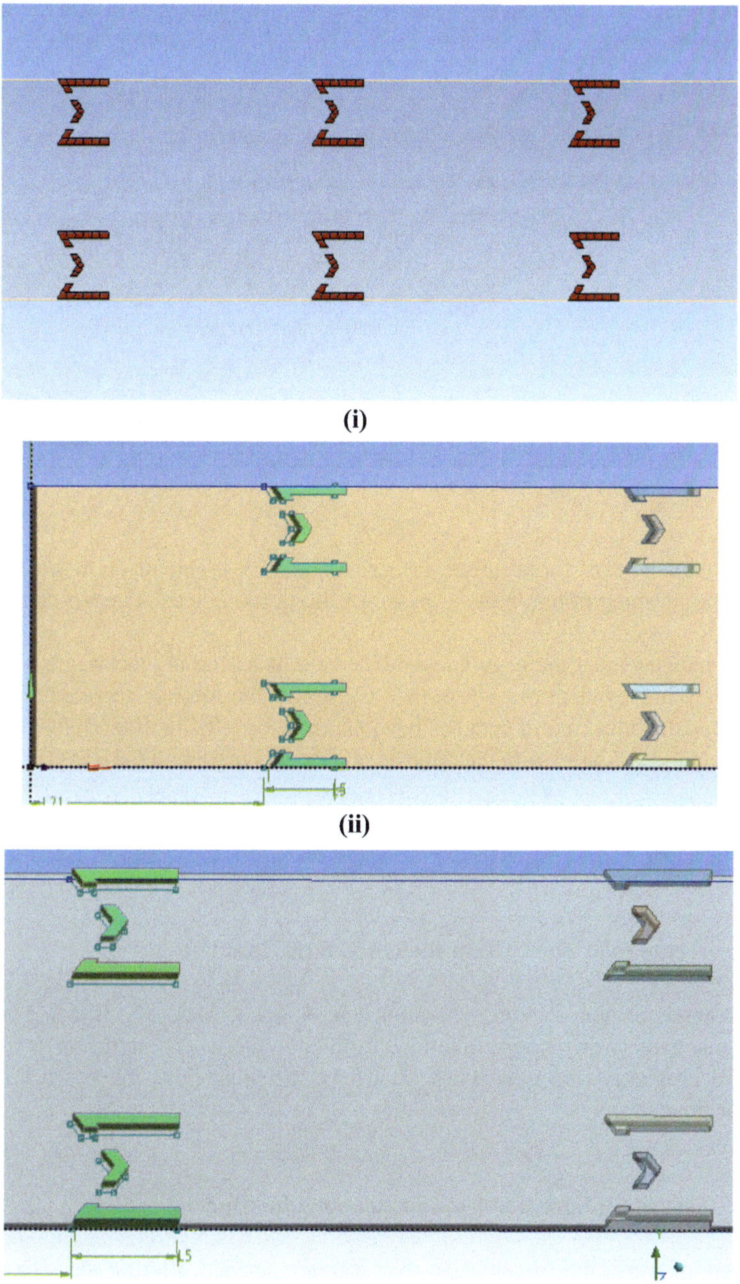

Fig. 2 Design of the rectangular duct with different types of M-shape with gaps and different height rib roughness (i) M-shape with gap rib roughness at $g = 0.7$ mm. (ii) M-shape with gap rib roughness at $g = 1.0$ mm. (iii) M-shape with gap rib roughness at $g = 1.3$ mm

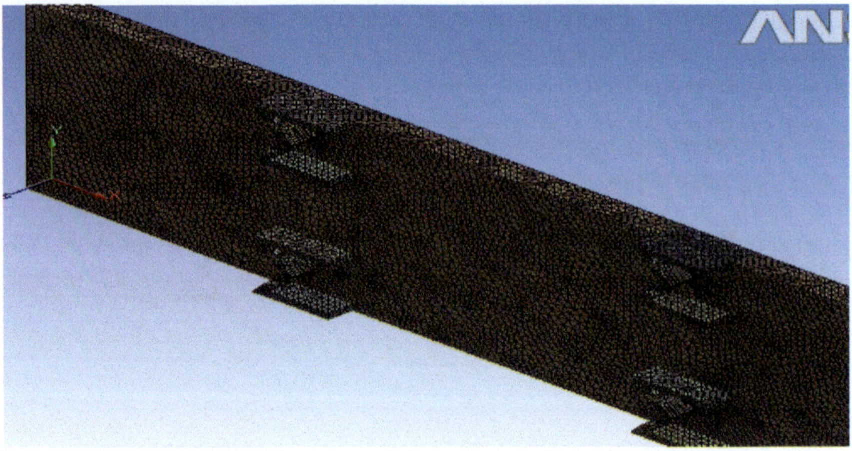

Fig. 3 Meshing of the solar rectangular duct with m–m-shape rib roughness

method must be used for accurate surface turbulent flow and heat transfer enhancement. The meshing of tetrahedron, pyramid, along triangle cells improves the mesh quality.

This indicates that the grid independence test is a meshing technique where the meshes are independent of one another. Currently, the mesh is accurate and yields correct results; this kind of grid has become known as the mesh test. The test mesh in this method was used with the number of elements of 184,542 at 0.008 m (Fig. 3).

2.2 Mathematical Data

2.2.1 Rectangular Duct Using the Governing Equation

To guarantee accurate model generation, the Ansys workbench utilizes governing equations, namely the energy equations for the k-e model. These formulas are used to solve problems with conversion solutions, turbulent flow, transition flow, and laminar flow.

2.2.2 Governing Equation Used in Rectangular Duct

The flow of warm fluid is abrasive over a rough surface, altering air velocity and energy transfer rate. An apparatus for computational fluid dynamics solves equations to determine the optimal path for energy transfer automatically. The following governing equations must be resolved in order to ascertain the flow's properties, including velocity, temperature, and pressure field. The Ansys workbench employs governing equations, namely the energy equations for the k-e model, to ensure precise model creation. These formulas are utilized for laminar flow, turbulent flow, transition flow, as well as conversion solution difficulties.

2.2.3 Turbulent Flow

A rough surface makes the flow of warm fluid rough, which changes the speed of air and the rate at which energy is transferred. A device for computational fluid dynamics figures out how to move energy automatically by solving equations. The following equations must be solved in order to find out the flow's properties, such as its speed, temperature, and pressure field.

Continuity Equation

$$\frac{\partial \vec{u}}{\partial x} + \frac{\partial \vec{v}}{\partial y} = 0 \tag{1}$$

Momentum equations:

$$\left(\vec{u}\frac{\partial \vec{u}}{\partial x} + \vec{v}\frac{\partial \vec{u}}{\partial y}\right) = -\frac{1}{\rho}\frac{\partial p}{\partial x} + v\left(\frac{\partial^2 \vec{u}}{\partial x^2} + \frac{\partial^2 \vec{u}}{\partial y^2}\right) \tag{2}$$

$$\left(\vec{u}\frac{\partial \vec{v}}{\partial x} + \vec{v}\frac{\partial \vec{v}}{\partial y}\right) = -\frac{1}{\rho}\frac{\partial p}{\partial y} + v\left(\frac{\partial^2 \vec{v}}{\partial x^2} + \frac{\partial^2 \vec{v}}{\partial y^2}\right) \tag{3}$$

Energy Equation

$$\vec{u}\frac{\partial t}{\partial x} + \vec{v}\frac{\partial t}{\partial y} = \alpha\left(\frac{\partial^2 t}{\partial x^2} + \frac{\partial^2 t}{\partial y^2}\right) \tag{4}$$

3 Results and Discussion

The solar duct system maximizes energy resources and minimizes heat losses by utilizing changes in the roughness of the m-shaped ribs. Due to the system's wide change in the m-shape's angle with gap rib roughness, solar thermal power systems find it appealing. The findings indicate that the m-shape with gap rib roughness reduces friction factors and increases the heat transfer rate. Rectangular ducts with added m-shape roughness and the theoretical Dittus-Boelter equation are used to compare the system. Friction factors drop and Nusselt numbers rise in response to variations in Reynolds and Nusselt numbers.

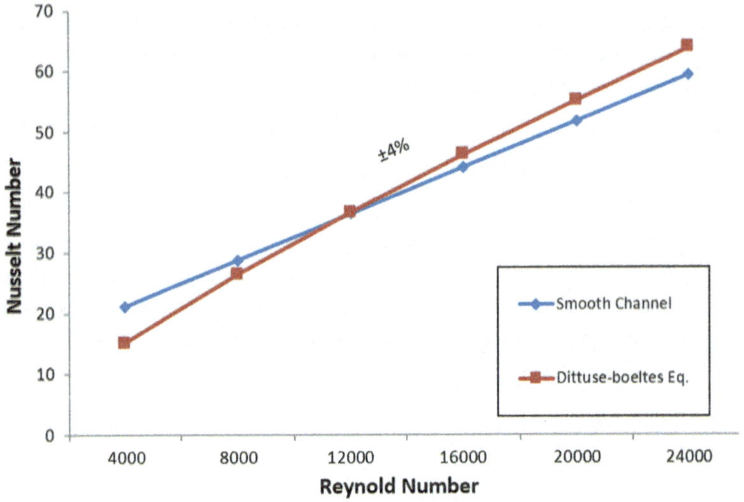

Fig. 4 Comparison of the smooth rectangle duct and the Dittus–Boelter equation

3.1 The Smooth Deduct and the Dittus–Boelter Equation Are Compared

The difference among a smooth duct along with a Dittus–Boelter solution as shown in Fig. 4. This comparison is used to verify the effect of heat transfer coefficient in the situation of smooth rectangular duct without m-shape rib roughness. The comparison shows the heat transfer results between theoretical and numerical analysis results (CFD). The Dittus–Boelter equation and the smooth rectangular duct are shown next to each other in Fig. 4. The Reynolds number (Re) is between 4000 and 24,000. The biggest difference between the results is about 3.5%, and the graph shows that they are very close to each other. This means that the conditions for heat transfer rate are met, and the added parts can be used for more Nu and Fr coefficients.

3.2 Velocity Contour of Solar Rectangular Duct

Figure 5 shows that the roughness block stays the same, but near it, an m-shape rib roughness reattachment zone forms. This happens when the Reynolds number (Re) is 24000, the relative roughness pitch (p/e) is 8, the relative hydraulic ratio (e/Dh) is 0.18, the height of the m-shape rib roughness is 6 mm, and the gap changes by 0.7 mm. When air moves at this speed, it forms a circulation and reattachment zone that helps heat move faster, turning solar energy into thermal power. Figure 6 shows that the speed slowed down because the speed was lower on the inner part of the surface, which made the pressure gradient bigger. This also made the heat transfer rate slower at the height of the m-shape rib (e = 6 mm), the gap between the m-shapes (g = 1 mm), p/e = 8, Re = 24,000, and the relative roughness height (e/Dh = 0.18). In the figure, the speed decreases down according to the above velocity curve because the air is slowing down after hitting the slit rib roughness. The fastest speed that was seen was 18.65 m/s. Figure 7 shows the velocity contour with maximum velocity of the solar rectangular duct with gap g = 1.3 mm and observed V_{max} = 14.59 m/s at p/e = 8, height of the m shape roughness e = 6 mm and Re = 24,000. Figure 8 shows the maximum velocity 8.578 m/s when no gap is taken between the m-shape roughness with respect to the e = 6 mm, p/e = 8, Re = 24,000 and p = 48 mm.

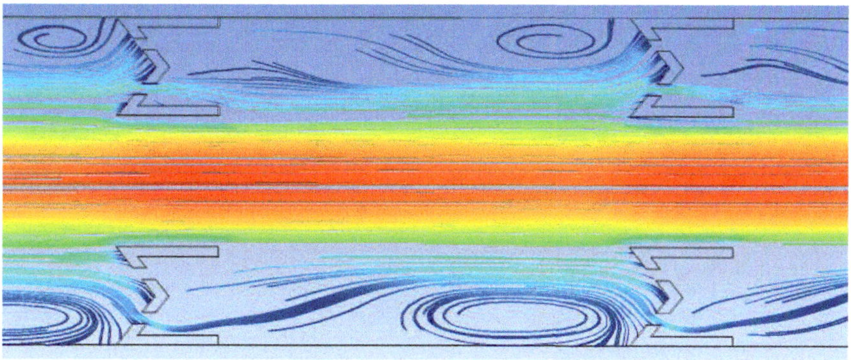

Fig. 5 The changes in the speed line at g = 0.7 mm of m-shape roughness

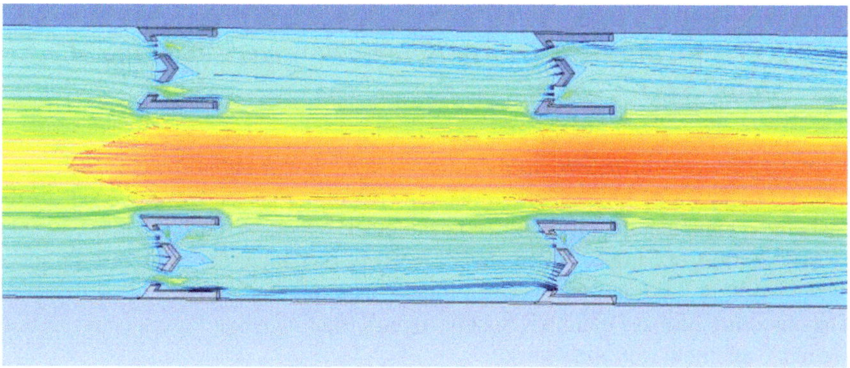

Fig. 6 Velocity changes as the gap g = 1.0 mm of the m-shape rib roughness changes

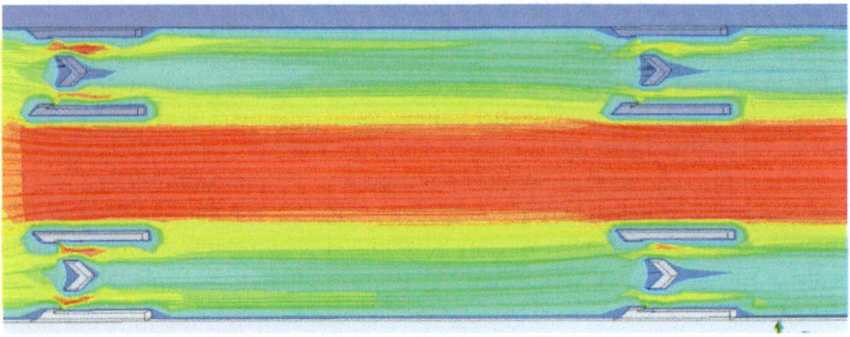

Fig. 7 The changes in the speed line at $g = 1.3$ mm are shown

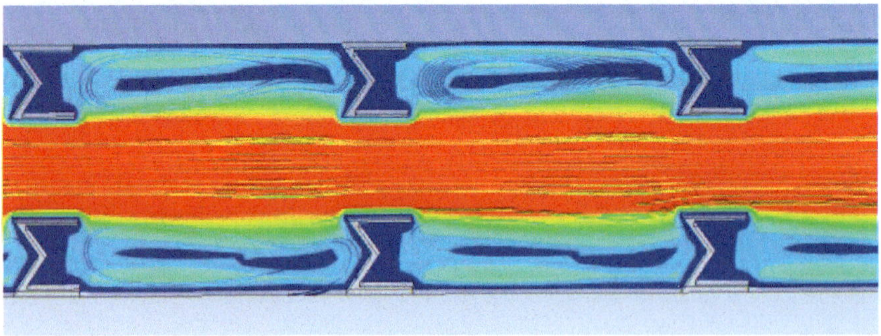

Fig. 8 Explains the magnitude of velocity contour at $g = 0$

3.3 Effect on the Heat Transmission Characteristic of the M Shape SAH

To make the model better and keep the eddies that come from turbulent flow, Nusselt numbers, and friction factors are looked at. An important part of the result of the designed rectangular duct is the Reynolds number, which can be 4000, 8000, 12,000, 16,000, 20,000, or 24,000. The m-shaped roughness changes depending on the speed of the working fluid and the speed of the air. The numbers range from 4 to 8 mm. The outcomes of these math studies give us information about how solar air heaters are made and how well they work.

Figure 9 displays the Nusselt number fluctuations along with the Reynolds number (4000–24,000), signifying the rate of heat transfer in both convection and conduction. The heat from the m-shaped gapped ribs is carried into the surrounding air by convection with constant values for $p/e = 4$, height of the m-shape rib angle $e = 6$ mm, and $e/\mathrm{Dh} = 0.18$, and changes in gaps $g = 0$–1.3 mm, the Nusselt number grows with Reynolds number. At m-shaped ribs $e = 6$ and $g = 0.7$ mm, the maximum Nusselt number is reached when air flows around the slit-rib, creating a turbulent flow and reattachment zone that provide thermal power. Nusselt number (Nu) increased 3.9, 2.4, 2.3, and 1.9 times at $g = 0.7$, 1, 1.3, and 0 mm respectively in relation to the smooth channel. Figure 10 The greatest Nusselt number obtained 3.5 times increment concerning the smooth channel for Re = 24,000 at $g = 0.7$ mm.

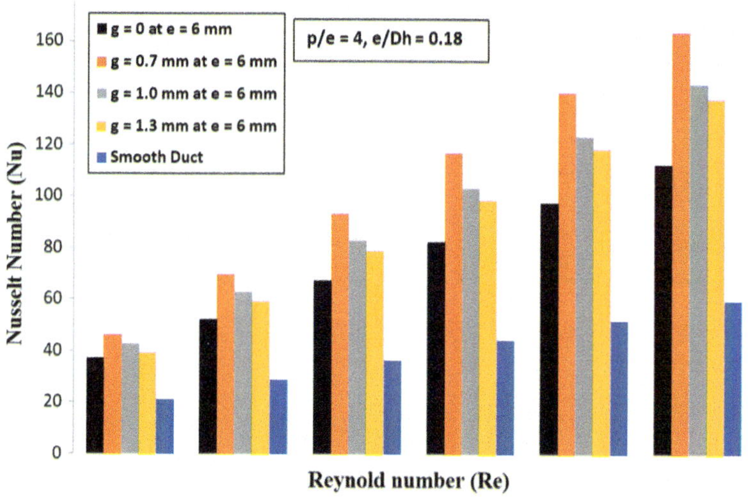

Fig. 9 The Nusselt numbers are shown to vary with regard to the Reynolds number between 4000 and 24,000 with $e = 6$ mm and $p/e = 4$

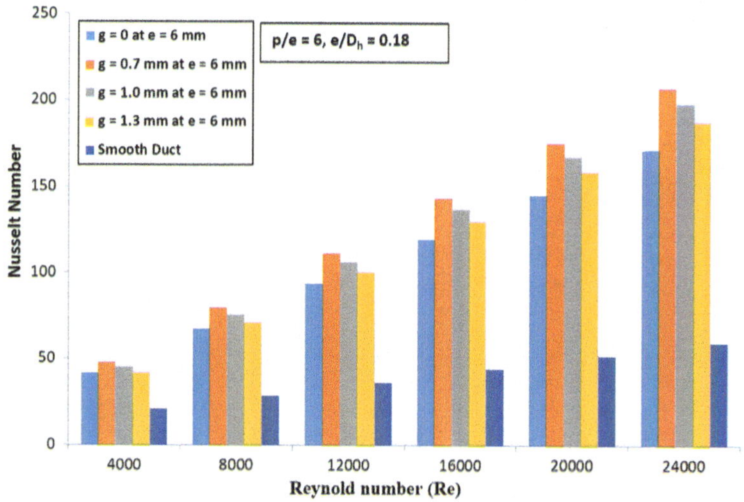

Fig. 10 Variations in the relationship between the Reynolds number and the Nusselt number between 4000 and 24,000 at $e = 6$ mm and $p/e = 6$

Figure 11 At m-shaped ribs $e = 6$ and $g = 0.7$ mm, the maximum Nusselt number is reached when air flows around the slit-rib, creating a turbulent flow and reattachment zone (Fig. 5) that provide thermal power. Nusselt number (Nu) increased 4.0, 3.5, 3.2, and 3.0 times at $g = 0.7$, 1, 1.3, and 0 mm respectively with relation to the smooth channel. In Fig. 12. As we can see, the Nusselt number is likewise growing in the overall condition at various intervals between the m-shape roughness ($g = 0$–1.3 mm) for $p/e = 10$) and at the relative roughness height ($e = 6$ mm).

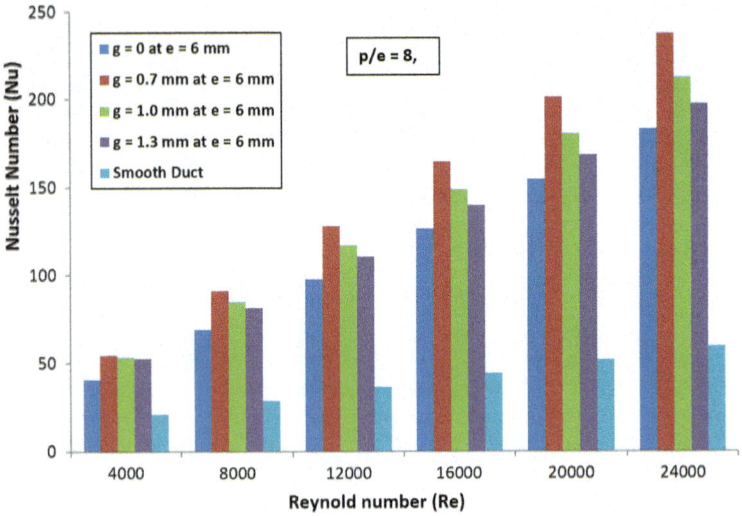

Fig. 11 Nusselt number varies in response to changes in the Reynolds number, which ranges from 4000 to 24,000 when the $p/e = 8$

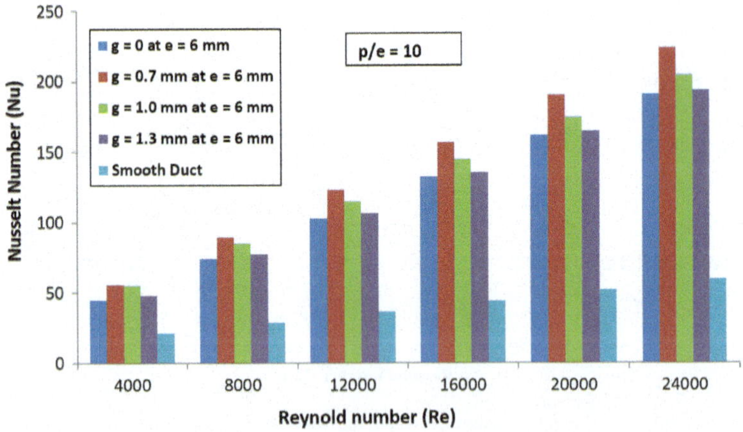

Fig. 12 The validation of the Nusselt numbers of M-shaped roughness SAH with smooth rectangular channel

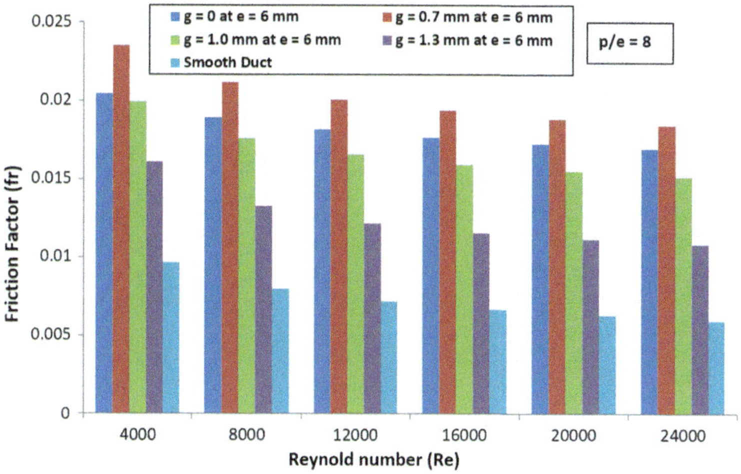

Fig. 13 The fluctuation in the friction components and their relationship to the Reynolds number and a *p*/*e* ratio of 8

Figure 13 illustrates how the friction factor varies about the Reynolds number. This variation is seen in the figure. Because the energy created in the rectangular channel with slit rib roughness with the assistance of the air turbulence velocity, which is derived from the Reynolds number 4000–24,000, the friction factor developed at this turbulent flow. For this reason, the friction factor is dependent upon the Reynolds number. The minimum friction factor is obtained when the *g* value is 1.3 mm, the Re value is 24000, and the height of the m-shape rib roughness is 6 mm.

When the heat transfer rate increased the friction factor also increased because large power is generated to maintain the turbulent flow of the air and a high force is required when the maximum friction factor is developed in the solar air heater and this situation is solved by computational iteration analysis. We can see in figure friction factor diminished in comparison to the rise in the Reynolds number because fluid separation with the slit rib roughness did not increase and no air is trapped around the heat transfer induction part of the solar rectangular channel and for smooth channel as shown in Fig. 13.

4 Thermal Hydraulic Performance of Solar Rectangular Duct

The best temperature and hydraulic performance are found when $p/e = 8$ and $g =$ 0.7, Re = 24,000 mm, and the M-shape is at its tallest point with gap rib roughness e = 6 mm. The above results show that as the roughness of the space between the ribs grew, so did the Nusselt number. This design is now likely to give the best maximum

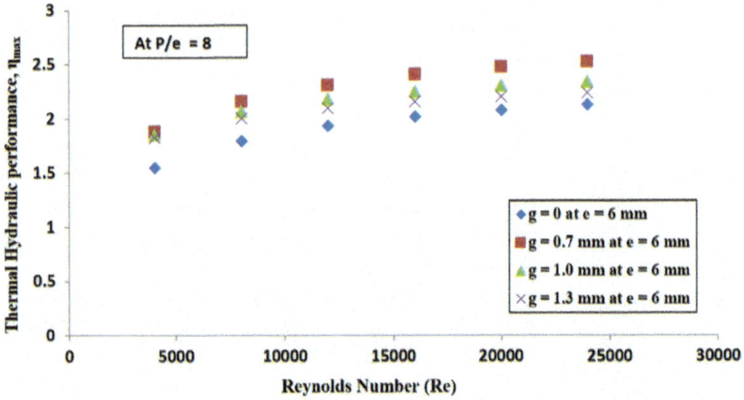

Fig. 14 Displays the thermal-hydraulic performance difference from 4000 to 24,000 Reynolds numbers

thermal performance, which is shown in the figure below as $\eta_{\max} = 2.5$ as shown in Fig. 14.

5 Conclusion

The planned and tested solar rectangle duct is looked at with M-shape rib roughness using the software Ansys fluent for computational fluid dynamics. The examination of the number of alterations in the texture of the m-shaped ribs was presented after the chapter. The modifications comprised gaps measuring $g = 0$–1.3 mm, in addition to gaps measuring $g = 0$ and relative pitch roughness values of $p/e = 4$–10. The irregularity heights of an M-shape are (e) 6 mm and e/Dh = 0.18 mm representing the relative roughness heights, respectively, and the Reynolds number ranges from 4000 to 24,000. Because of their thermal hydraulic performance and friction factors, these changes to the solar rectangle duct were the most effective at increasing heat transfer. We compare the calculated values for the m-shape ribbed roughness to those for the smooth channel. From the study of the solar rectangle channel with or without m-shape rib roughness, we can say the following:

(1) The k-e model was used to study the solar rectangular duct, which gave more useful and accurate results close to the boundary wall.
(2) The Nusselt number, the average friction factor, and the temperature efficiency of the channel all go up as the cut rib roughness goes down.
(3) The Nusselt number went up by 399.9% compared to the smooth channel when the gap got bigger ($g = 0.7$ mm and e/Dh = 0.18). The highest increment of Nusselt number of 4.0 times is found when roughness height $e = 6$ mm is applied on smooth channel. The minimum friction factor is 0.015 for $e = 6$ mm and $g = 1.3$ mm of M-shaped roughness SAH at Re = 24,000.

(4) The best thermal efficiency was found $\eta_{max} = 2.5$ when $g = 0.7$ mm, $p/e = 8$, $e = 6$ mm, and Re = 24,000.

References

1. Chaube A, Sahoo PK, Solanki SC (2006) Analysis of heat transfer augmentation and flow characteristics due to rib roughness over absorber plate of a solar air heater. Renew Energy 31(3):317–331. https://doi.org/10.1016/j.renene.2005.01.012
2. Chaube A, Sahoo PK, Solanki SC (2006) Effect of roughness shape on heat transfer and flow friction characteristics of solar air heater with roughened absorber plate. WIT Trans Eng Sci 53
3. Gandhi BK, Singh KM (2010) Experimental and numerical investigations on flow through wedge shape rib roughened duct. J Inst Eng-MC 2010(90):13–18
4. Sharma S, Singh R, Bhushan B (2011) CFD based investigation on effect of roughness element pitch on performance of artificially roughened duct used in solar air heaters. Int J Adv Eng Technol 2(1):234–241
5. Bhagoria JL (2013) CFD analysis of square ribs on the absorber plate of solar air collector. In: Proceedings of the 2nd international conference on mechanical, electronics and mechatronics engineering (ICMEME'2013), London (UK), pp 17–18
6. Yadav AS, Paul A, Sharma P (2013) CFD investigation of flow through an artificially roughened solar air heater. Corona J Sci Technol 2(III):16–19
7. Kumar A (2014) Analysis of heat transfer and fluid flow in different shaped roughness elements on the absorber plate solar air heater duct. Energy Procedia 2014(57):2102–2111. https://doi.org/10.1016/j.egypro.2014.10.176
8. Gawande VB, Dhoble AS, Zodpe DB, Chamoli S (2016) Experimental and CFD-based thermal performance prediction of solar air heater provided with right-angle triangular rib as artificial roughness. Braz Soc Mech Sci Eng 2015:1–29
9. Jin D, Zhang M, Wang P, Xu S (2015) Numerical investigation of heat transfer and fluid flow in a solar air heater duct with multi V-shaped ribs on the absorber plate. Energy 89:178–190. https://doi.org/10.1016/j.energy.2015.07.069
10. Murali G, Reddy KRK, Kumar MTS, SaiManikanta J, Reddy VNK (2020) Computational study on performance of solar air heater using 's' shape ribs. In: Proceedings of, 5th international conference on production automation and mechanical engineering held on 9–10 August 2019 Montreal, Canada. ISBN: 9780998900087. https://doi.org/10.1016/j.matpr.2019.04.213
11. Yadav AS, Bhagoria JL (2014) A CFD based thermo-hydraulic performance analysis of an artificially roughened solar air heater having circular ribs on the absorber plate. Int J Eng Res Technol (IJERT) 4(05) (This work is licensed under a Creative Commons Attribution 4.0 International License). ISSN: 2278-0181, IJERTV4IS051251. www.ijert.org. https://doi.org/10.1016/j.ijheatmasstransfer.2013.11.074
12. Boulemtafes-Boukadoum A, Benzaoui AJEP (2014) A CFD based analysis of heat transfer enhancement in solar air heater provided with transverse rectangular ribs. The international conference on technologies and materials for renewable energy, environment and sustainability, TMREES14. Energy Procedia 50:761–772. https://doi.org/10.1016/j.egypro.2014.06.094
13. Kumar K, Kaushik S, Bisht VS (2017) CFD analysis on solar air heater with artificial roughened broken curved ribs. Int J Sci Eng Res 8(12):264. ISSN: 2229-5518

The manufacturer's authorised representative in the EU is Springer Nature Customer Service Centre GmbH, Europaplatz 3, 69115 Heidelberg, Germany. If you have any concerns regarding our products, please contact ProductSafety@springernature.com

Printed and bound by CPI Group (UK) Ltd, Croydon, CR0 4YY
01/10/2026
02225959-0006